**HUMBER LIBRARIES LAKESHORE CAMPUS**
3199 Lakeshore Blvd West
**TORONTO, ON. M8V 1K8**

# Digital Product Management, Technology and Practice:
## Interdisciplinary Perspectives

Troy J. Strader
*Drake University, USA*

**BUSINESS SCIENCE REFERENCE**

Hershey · New York

HUMBER LIBRARIES LAKESHORE CAMPUS
3199 Lakeshore Blvd West
TORONTO, ON. M8V 1K8

| | |
|---|---|
| Director of Editorial Content: | Kristin Klinger |
| Director of Book Publications: | Julia Mosemann |
| Acquisitions Editor: | Lindsay Johnston |
| Development Editor: | Mike Killian |
| Publishing Assistant: | Jamie Snavely & Keith Glazewski |
| Typesetter: | Casey Conapitski & Keith Glazewski |
| Production Editor: | Jamie Snavely |
| Cover Design: | Lisa Tosheff |

Published in the United States of America by
Business Science Reference (an imprint of IGI Global)
701 E. Chocolate Avenue
Hershey PA 17033
Tel: 717-533-8845
Fax: 717-533-8661
E-mail: cust@igi-global.com
Web site: http://www.igi-global.com

Copyright © 2011 by IGI Global. All rights reserved. No part of this publication may be reproduced, stored or distributed in any form or by any means, electronic or mechanical, including photocopying, without written permission from the publisher. Product or company names used in this set are for identification purposes only. Inclusion of the names of the products or companies does not indicate a claim of ownership by IGI Global of the trademark or registered trademark.

Library of Congress Cataloging-in-Publication Data

Digital product management, technology and practice : interdisciplinary perspectives / Troy J. Strader, editor.
       p. cm.
  Includes bibliographical references and index.
  Summary: "This book covers a wide range of digital product management issues and offers some insight into real-world practice and research findings on the technical, operational, and strategic challenges that face digital product managers and researchers now and in the next several decades"--Provided by publisher.
  ISBN 978-1-61692-877-3 (hbk.) -- ISBN 978-1-61692-879-7 (ebook) 1. Electronic industries. 2. Digital electronics. 3. Digital media. 4. New products. I. Strader, Troy J., 1965- II. Title.

  HD9696.A2D54 2011
  621.38068--dc22

2010016384

British Cataloguing in Publication Data
A Cataloguing in Publication record for this book is available from the British Library.

All work contributed to this book is new, previously-unpublished material. The views expressed in this book are those of the authors, but not necessarily of the publisher.

# Digital Product Management, Technology and Practice:
## Interdisciplinary Perspectives

Troy J. Strader
*Drake University, USA*

**BUSINESS SCIENCE REFERENCE**

Hershey · New York

| | |
|---|---|
| Director of Editorial Content: | Kristin Klinger |
| Director of Book Publications: | Julia Mosemann |
| Acquisitions Editor: | Lindsay Johnston |
| Development Editor: | Mike Killian |
| Publishing Assistant: | Jamie Snavely & Keith Glazewski |
| Typesetter: | Casey Conapitski & Keith Glazewski |
| Production Editor: | Jamie Snavely |
| Cover Design: | Lisa Tosheff |

Published in the United States of America by
Business Science Reference (an imprint of IGI Global)
701 E. Chocolate Avenue
Hershey PA 17033
Tel: 717-533-8845
Fax: 717-533-8661
E-mail: cust@igi-global.com
Web site: http://www.igi-global.com

Copyright © 2011 by IGI Global. All rights reserved. No part of this publication may be reproduced, stored or distributed in any form or by any means, electronic or mechanical, including photocopying, without written permission from the publisher. Product or company names used in this set are for identification purposes only. Inclusion of the names of the products or companies does not indicate a claim of ownership by IGI Global of the trademark or registered trademark.

Library of Congress Cataloging-in-Publication Data

Digital product management, technology and practice : interdisciplinary perspectives / Troy J. Strader, editor.
  p. cm.
 Includes bibliographical references and index.
 Summary: "This book covers a wide range of digital product management issues and offers some insight into real-world practice and research findings on the technical, operational, and strategic challenges that face digital product managers and researchers now and in the next several decades"--Provided by publisher.
  ISBN 978-1-61692-877-3 (hbk.) -- ISBN 978-1-61692-879-7 (ebook) 1.  Electronic industries. 2. Digital electronics. 3. Digital media. 4.  New products.  I. Strader, Troy J., 1965- II. Title.
  HD9696.A2D54 2011
  621.38068--dc22

                                                            2010016384

British Cataloguing in Publication Data
A Cataloguing in Publication record for this book is available from the British Library.

All work contributed to this book is new, previously-unpublished material. The views expressed in this book are those of the authors, but not necessarily of the publisher.

## Editorial Advisory Board

Stephen Burgess, *Victoria University, Australia*
William Cheng-Chung Chu, *TungHai University, Taiwan*
Michele Gribbins, *University of Illinois - Springfield, USA*
Gary Hackbarth, *Northern Kentucky University, USA*
Anthony Hendrickson, *Creighton University, USA*
Philip Houle, *Drake University, USA*
Ric Jentzsch, *Compucat Research Pty Ltd, Australia*
Chip Miller, *Drake University, USA*
Matthew Nelson, *Illinois State University, USA*
Juergen Seitz, *Baden-Wuerttemberg Cooperative State University Heidenheim, Germany*
Michael Shaw, *University of Illinois, Urbana-Champaign, USA*
Reima Suomi, *University of Turku, Finland*
Yi-Minn (Minnie) Yen, *University of Alaska, USA*
Hans-Dieter Zimmermann, *FHS St. Gallen University of Applied Sciences, Switzerland*

## List of Reviewers

Stephen Burgess, *Victoria University, Australia*
Richard Carter, *Iowa State University, USA*
William Cheng-Chung Chu, *TungHai University, Taiwan*
James Dodd, *Drake University, USA*
Stephen Gara, *Drake University, USA*
Gary Graham, *Manchester University, UK*
Michele Gribbins, *University of Illinois - Springfield, USA*
Ric Jentzsch, *Compucat Research Pty Ltd, Australia*
Delaney Kirk, *University of South Florida - Sarasota-Manatee, USA*
Juergen Seitz, *University of Cooperative Education Heidenheim, Germany*
Reima Suomi, *Turku School of Economics, Finland*
Yi-Minn (Minnie) Yen, *University of Alaska - Anchorage, USA*
Hans-Dieter Zimmermann, *FHS St. Gallen University for Applied Sciences, Switzerland*

# Table of Contents

**Foreword** ........................................................................................................................... xii

**Preface** .............................................................................................................................. xiv

**Acknowledgment** ............................................................................................................. xix

### Section 1
### Technology

**Chapter 1**
Digital Technology: Capabilities and Limitations ............................................................... 1
    *Philip A. Houle, Drake University, USA*

**Chapter 2**
DRM Protection Technologies ............................................................................................ 19
    *Gary Hackbarth, Northern Kentucky University, USA*

### Section 2
### Business Functions

**Chapter 3**
Legal Issues Facing Companies with Products in a Digital Format .................................. 32
    *J. Royce Fichtner, Drake University, USA*
    *Lou Ann Simpson, Drake University, USA*

**Chapter 4**
Pricing in the Digital Age .................................................................................................. 53
    *Chip E. Miller, Drake University, USA*

**Chapter 5**
Financing Digital Product Companies .................................................................................................. 73
    *Richard B. Carter, Iowa State University, USA*
    *Frederick H. Dark, Iowa State University, USA*

**Chapter 6**
Accounting for Digital Products ............................................................................................................ 85
    *Yasemin Zengin Karaibrahimoğlu, Izmir University of Economics, Turkey*

## Section 3
## Issues and Strategies

**Chapter 7**
It's All about the Relationship: Interviews with the Experts on How Digital Product Companies
Can Use Social Media ............................................................................................................................ 96
    *Delaney J. Kirk, University of South Florida - Sarasota-Manatee, USA*

**Chapter 8**
Digital Convergence and Horizontal Integration Strategies ................................................................ 113
    *Troy J. Strader, Drake University, USA*

**Chapter 9**
The Role of the Internet in the Decline and Future of Regional Newspapers ..................................... 142
    *Gary Graham, University of Manchester, UK*

**Chapter 10**
Software as a Service and the Pricing Strategy for Vendors ............................................................... 154
    *Nizar Abdat, Utrecht University, The Netherlands*
    *Marco Spruit, Utrecht University, The Netherlands*
    *Menne Bos, Accenture, The Netherlands*

**Chapter 11**
The Private Copy Issue: Piracy, Copyright and Consumers' Rights ................................................... 193
    *Pedro Pina, Polytechnic Institute of Coimbra, Portugal*

**Chapter 12**
Service Systems as Digital Products .................................................................................................... 206
    *Hsin-Lu Chang, National Chengchi University, Taiwan*
    *Michael J. Shaw, University of Illinois at Urbana-Champaign, USA*
    *Feipei Lai, National Taiwan University, Taiwan*

# Section 4
# Visions for the Future

**Chapter 13**
Transitioning to Software as a Service: A Case Study .................................................................. 218
    *Dave Sly, Proplanner.com, USA*

**Chapter 14**
Digital Media: Future Research Directions ................................................................................... 225
    *Anthony Hendrickson, Creighton University, USA*
    *Trent Wachner, Creighton University, USA*
    *Brook Matthews, Creighton University, USA*

**Chapter 15**
Digital Technology in the 21st Century ......................................................................................... 235
    *Troy J. Strader, Drake University, USA*

**Compilation of References** ...................................................................................................... 263

**About the Contributors** ........................................................................................................... 286

**Index** ......................................................................................................................................... 291

# Detailed Table of Contents

**Foreword** ............................................................................................................................. xii

**Preface** ................................................................................................................................ xiv

**Acknowledgment** ............................................................................................................... xix

### Section 1
### Technology

*This section includes chapters that describe existing digital technology capabilities, limitations, and methods for protecting digital products. It provides background information necessary for understanding the issues discussed in the other sections.*

**Chapter 1**
Digital Technology: Capabilities and Limitations ............................................................... 1
    *Philip A. Houle, Drake University, USA*

A review of existing digital technology capabilities and limitations. Historical evolution and current standards are discussed for digital representation of text, numbers, images, audio, and video, as well as telecommunications. This chapter provides sufficient background for understanding the digital product management issues addressed by each of the chapters that follow.

**Chapter 2**
DRM Protection Technologies ............................................................................................ 19
    *Gary Hackbarth, Northern Kentucky University, USA*

An overview of issues related to protecting digital products. Digital rights management (DRM) is discussed along with the important issue of balancing user access and owner rights. Other technical and non-technical protections are also discussed.

## Section 2
## Business Functions

*In this second section the focus is on how the unique characteristics of digital products impact tactical level management (business functions) for all digital product industries. The business functions encompassed in this section include legal, marketing, finance, and accounting.*

### Chapter 3
Legal Issues Facing Companies with Products in a Digital Format ................................................... 32
*J. Royce Fichtner, Drake University, USA*
*Lou Ann Simpson, Drake University, USA*

A survey of legal issues relevant to digital products. Historical developments and the current legal environment relevant to digital product managers are discussed. Topics include four categories of intellectual property law: patents, trade secrets, trademarks, and copyrights.

### Chapter 4
Pricing in the Digital Age ................................................................................................................... 53
*Chip E. Miller, Drake University, USA*

Focus is on the unique issues faced when pricing digital products. Digital product pricing is complex because online customers can easily price compare and they know that digital products have low marginal costs, while sellers realize that digital products are easily pirated. Traditional pricing strategies are discussed along with how they may be incorporated into the appropriate pricing of digital products.

### Chapter 5
Financing Digital Product Companies ............................................................................................... 73
*Richard B. Carter, Iowa State University, USA*
*Frederick H. Dark, Iowa State University, USA*

Identifies financing issues facing digital product companies. The focus is on one financing option—initial public offerings (IPOs)—and how digital product fixed and marginal costs at different points in a company's life cycle may impact the appropriate timing for IPOs. Empirical findings and case studies are analyzed and managerial and research implications are discussed.

### Chapter 6
Accounting for Digital Products ........................................................................................................ 85
*Yasemin Zengin Karaibrahimoğlu, Izmir University of Economics, Turkey*

An overview of accounting issues facing digital product companies. Traditional accounting topics are surveyed along with how they may apply to accounting for digital products. The chapter addresses digital product accounting issues for recognition, measurement, valuation, reporting, and taxation.

# Section 3
# Issues and Strategies

*This section includes chapters focusing on other issues that are relevant to digital product managers. Some of the digital product management issues addressed are: the impact of social media, the range of strategic integration strategies available to digital product companies, changes in the regional newspaper industry, the software as a service model, a European view of copyright issues, and similarities between managing digital products and managing a healthcare service system in Taiwan.*

### Chapter 7
It's All about the Relationship: Interviews with the Experts on How Digital Product Companies Can Use Social Media........................................................................................................................ 96
    *Delaney J. Kirk, University of South Florida - Sarasota-Manatee, USA*

Provides insights from interviews with several social media experts. Suggestions for how social media may be used in support of digital products are discussed along with specific techniques that work, and do not work.

### Chapter 8
Digital Convergence and Horizontal Integration Strategies ................................................................ 113
    *Troy J. Strader, Drake University, USA*

Describes the concept of digital convergence and how it provides opportunities for digital product companies. A wide range of strategic alternatives are discussed along with their associated benefits and risks for companies in each of the digital product industries.

### Chapter 9
The Role of the Internet in the Decline and Future of Regional Newspapers .................................... 142
    *Gary Graham, University of Manchester, UK*

Discusses the role of the Internet in the decline of regional newspaper influence. The chapter suggests ways that new technology may be utilized for co-creating content with readers. Insights are provided from interviews with management from several different newspapers.

### Chapter 10
Software as a Service and the Pricing Strategy for Vendors............................................................... 154
    *Nizar Abdat, Utrecht University, The Netherlands*
    *Marco Spruit, Utrecht University, The Netherlands*
    *Menne Bos, Accenture, The Netherlands*

Provides an introduction to the concept of Software as a Service (SaaS). The chapter discusses scientific and business perspectives on this issue. An in-depth discussion of the benefits and risks associated with SaaS is provided from the perspectives of all potential participating organizations. The chapter concludes with a review of issues related to vendor pricing strategies.

**Chapter 11**
The Private Copy Issue: Piracy, Copyright and Consumers' Rights ................................................... 193
  *Pedro Pina, Polytechnic Institute of Coimbra, Portugal*

Provides a European perspective on copyright issues. The chapter describes the conflicts between the exclusive right of the owner to exploit a digital product and the user's private copy issues. The chapter provides suggestions for how to balance the rights of digital product providers and consumers.

**Chapter 12**
Service Systems as Digital Products ..................................................................................................... 206
  *Hsin-Lu Chang, National Chengchi University, Taiwan*
  *Michael Shaw, University of Illinois at Urbana-Champaign, USA*
  *Feipei Lai, National Taiwan University, Taiwan*

Presents the concept of services as digital products. The chapter discusses issues such as development of service value models, development of service metrics, and management of service systems, based on a remote healthcare platform in Taiwan.

**Section 4**
**Visions for the Future**

*The final section provides views of the future. It includes insights from a software company case study, ideas for future digital product research themes, and a survey of future digital technologies.*

**Chapter 13**
Transitioning to Software as a Service: A Case Study ......................................................................... 218
  *Dave Sly, Proplanner.com, USA*

A case study that provides an executive's perspective on benefits, challenges, and lessons learned during an engineering software company's transition from a software as product model to software as a service. Technical issues and recommendations are also discussed based on real-world experiences at Proplanner.

**Chapter 14**
Digital Media: Future Research Directions .......................................................................................... 225
  *Anthony Hendrickson, Creighton University, USA*
  *Trent Wachner, Creighton University, USA*
  *Brook Matthews, Creighton University, USA*

Identifies directions for future digital product management research. The chapter presents several themes for future research that address issues for how technological innovations will impact the global workplace. These themes include digital products as operant resources, issues for user-generated content and data quality, impact of network externalities, and challenges for digital product business models.

**Chapter 15**
Digital Technology in the 21st Century ........................................................................................... 235
   *Troy J. Strader, Drake University, USA*

Discusses a wide range of digital technologies that are currently being developed that may impact digital product management in the 21st century. The chapter describes the potential impact of these new technologies on digital product industries in addition to identifying relevant research and philosophical questions that must be addressed. Impact of new digital technologies on societal institutions such as healthcare, government services, higher education, political campaigns, cybercrime law enforcement, and life at home are also discussed.

**Compilation of References** ............................................................................................................ 263

**About the Contributors** ................................................................................................................. 286

**Index** ................................................................................................................................................. 291

# Foreword

The emergence and rapid growth in digital technologies, particularly over the past twenty years, have led to greater concentration within industries, performance differentiation within industries, and market turbulence (McAfee & Brynjolfsson, 2008). Digital Technologies (DT) have fundamentally changed the competitive landscape and will continue to do so. This is the landscape in which an organization lives and moves as it strives to survive, while staying true to its mission. The unceasing DT innovations disrupt environments, presenting both daunting challenges and unprecedented opportunities for the organizations that inhabit them. Relative to the industry in which it operates, such disruption results in an organization being more or less productive, more or less agile, more or less innovative, and more or less reputable than its competitors. Mastery of DT, including cleverly applying it, is one key to an organization's quest for not only survival in a hypercompetitive world, but also sustained excellence.

Here, we have a book that is a substantial contribution to cutting-edge mastery of DT. Digital Product Management, Technology, and Practice: Interdisciplinary Perspectives offers a state-of-the-art treatment of digital technologies. This well-organized, clearly written book is packed with information and insights for researchers, practitioners, educators, students, and vendors – not only in the DT field, but also relevant to such topic areas as competition strategies, knowledge management, process management, and organizational networking. Its coverage ranges from what digital technologies are and encompass, to how DT work, to identification and resolution of emergent DT issues that need to be resolved, to DT applications, to new directions for DT management.

There is a particular emphasis on digital products, including both goods and services. The digital goods are comprised of knowledge representations that a processor (human or computer-based) regards as being useful. The digital services are computer-based actions performed at the request of, or on behalf of, a client processor (human or computer-based). Digital products flesh out an ever more pervasive virtual world that shadows our actual world. In some dimensions, this virtual world is replacing facets of the actual world. In other ways, digital products complement the actual world. In some cases, the digital product is quite novel in the sense of having no actual world counterpart. This book helps us understand nature, scope, and prospects the world of digital products enabled by DT.

The DT issues explored include societal (e.g., environment effects on use of DT and vice versa), legal/ethical (e.g., consumer and producer rights), financial (e.g., financing and accounting for digital products), and marketing (e.g., product pricing) issues. Resolving such issues is of great practical importance for effectively harnessing the potential of digital products. The book considers a nice variety of DT applications that illustrates the richness of digital product possibilities. These range from news delivery to software delivery to service delivery systems. The book also addresses management topics such as limitation of DT, strategies for using DT, digital convergence, leveraging social media to enhance digital product success, and creating value via digital products.

In my first read of the book, I was repeatedly struck by the way in which its content aligns with principles of Knowledge Management (KM) and concepts in the fledgling Science of Competitiveness (SoC). In perusing the book's content, I found chapter after chapter provoking new and interesting questions about linkages between knowledge management and DT (Holsapple, 2005). Moreover, I found multiple chapters of high relevance to further developing the SoC, within which DT form an integral component (Holsapple & Jin, 2007). Within the SoC, DT foster the convergence of an organization's knowledge, networks, and processors for undertaking advantageous competitive moves within a turbulent environment. Thus, I recommend the book not only to those interested in DT advances in general, and digital products in particular, but also to those interested in pushing forward KM and SoC research and practice.

In all, Digital Product Management, Technology, and Practice is a welcome and unique addition to the scholarly literature – furnishing timely coverage of an important phenomenon that is an essential ingredient for business success.

## REFERENCES

Holsapple, C. W. (2005). The inseparability of modern knowledge management and computer-based technology. Journal of Knowledge Management, 9 (1), 42-52.

Holsapple, C. W., & Jin, H. (2007). Connecting some dots: Electronic commerce, supply chains, and collaborative decision making. Decision Line, 38(5), 14-21.

McAfee, A., & Brynjolfsson, E.(2008). Investing in the IT that makes the competitive difference. Harvard Business Review, 86(7/8), 98-107.

*Clyde W. Holsapple*
*University of Kentucky, USA*

**Clyde W. Holsapple** *is a Fellow of the Decision Sciences Institute and holds the Rosenthal Endowed Chair at the University of Kentucky. His research focuses on supporting knowledge work, particularly in decision-making contexts, and on systems for multiparticipant collaboration. He has authored 150 research articles in journals such as* Decision Sciences, Operations Research, Decision Support Systems, Journal of Management Information Systems, Group Decision and Negotiation, Journal of Operations Management, Organization Science, Communications of the ACM, Journal of American Society for Information Science and Technology, Knowledge and Process Management, Journal of Knowledge Management, Entrepreneurship Research and Practice, *and IEEE journals. His many books include* Foundations of Decision Support Systems, Decision Support Systems: A Knowledge-Based Approach, the Handbook on Knowledge Management, *and the* Handbook on Decision Support Systems. *He is Editor-in-Chief of the* Journal of Organizational Computing and Electronic Commerce.

# Preface

For centuries, products have been tangible and physical items. They could be viewed and touched at a market prior to purchase, and the creation of these products required some combination of raw materials, people, machines, time and money. It took a great deal of effort to move products from where they were created to where they were sold and consumed. Sellers and buyers knew that the price paid had to cover all of the fixed and marginal costs to make the product or the producer would not survive for very long. The economics and operational decision-making associated with these tangible goods have been extensively studied which resulted in a pretty good understanding of how these products should be managed. Today, traditional products are not gone; in fact they still make up most of the products sold worldwide. But a few decades ago something changed. An array of computer and information technologies created a world where some products could be digitized. Most impacted by this change were industries where the product was information content, multimedia, or software. Newspapers, magazines, book publishers, music, movies, games, and all forms of software have all been dramatically affected by digital technologies and this transformation will continue in the future.

Digital product managers are not able to rely on the same rules that would have been used to manage traditional products. Digital products can be stored on a computer or other digital device for very little cost, new units can be easily and cheaply produced by using a computer to make another copy of the product, and the digital product can be quickly and cheaply distributed to consumers worldwide. Digital products typically have high fixed development costs, but their marginal costs are near zero. These differences impact most business functions (accounting, finance, legal, marketing, manufacturing, logistics, customer service, and so forth) and strategies (operations, tactical, and strategic-level) because they introduce new opportunities and challenges. Companies found that they could get some benefits from translating their product into a digital form, but the best companies knew that this was just the beginning. Digital products can be delivered in many different ways to many different devices. They may also be extended by combining existing digital products with other digital products or services. Digital product companies can cross over into other digital product industries, and they can also create entirely new industries and business models. Further, the explosion of activity surrounding digital products will also have a profound impact on individuals and our global society today and in the future.

Digital Product Management, Technology, and Practice: Interdisciplinary Perspectives covers a wide range of digital product management issues and offers some insight into real-world practice and research findings. Experts in several disciplines from around the world offer their views on the technical, operational, and strategic challenges that face digital product managers and researchers now and in the next several decades. The following are some of the broad questions addressed in this book:

- What is digital technology? What are its capabilities and limitations?
- How can digital assets be protected?
- What are the legal issues associated with digital products?
- How does a company determine the appropriate price for a digital product?
- What are the unique financing issues facing digital product companies?
- How is accounting for digital products different from accounting for traditional products?
- How should social media technology be used by digital product managers?
- What is digital convergence? How does digital convergence enable digital product horizontal integration strategies? What are some of the other strategic alternatives that may lead to a competitive advantage?
- How has digital technology affected the role of regional newspapers?
- Should software be offered as a service? How should software as a service be priced?
- How can the rights of digital product producers and consumers be balanced?
- Can service systems be managed like digital products?
- What are some of the lessons learned from running a software company that uses a "software-as-a-service" business model?
- What are the important digital product research issues that need to be addressed in the future?
- How will new digital technologies impact businesses, people, and society, in the 21st century? What are the digital technology trends for the 21st century?

The following describes how the chapters in this book are organized and what topic is covered in each chapter. The topics and sections are shown in Figure 1.

The foundation for all digital product companies is the digital technology itself. This first book section is about the characteristics of digital technology and how digital products can be protected.

In Chapter 1, *Digital Technology: Capabilities and Limitations*, Philip Houle from Drake University provides a survey of digital technology capabilities including how they can be used to represent simple informational forms such as text and numbers, and how they are also capable of representing

*Figure 1. Book sections and chapter topics*

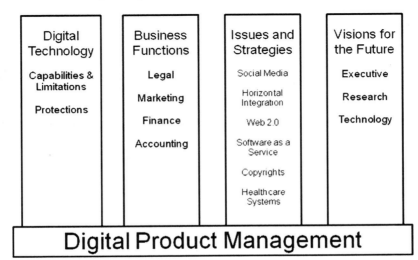

more complex multimedia such as still images, moving images, and sounds. The chapter also identifies some of the limitations for current digital technology. This chapter provides the background necessary to understand the issues and challenges discussed throughout the remainder of the book.

In Chapter 2, *DRM Protection Technologies*, Gary Hackbarth from Northern Kentucky University provides a review of issues associated with protecting digital products. Digital rights management (DRM) is concerned with ownership and access to information. Methods used for securing digital products are discussed along with the important issue of balancing access and security. The chapter discusses current protection issues and future directions for digital security.

Digital products have an impact on all business functions. The second book section includes chapters about digital product business functional issues from four perspectives: legal, marketing, finance, and accounting.

In Chapter 3 entitled *Legal Issues Facing Companies with Products in a Digital Format*, J. Royce Fichtner and Lou Ann Simpson from Drake University provide a survey of legal issues associated with digital products and technology. Digital product companies face unique legal challenges as they sell or license products that are in an informational, multimedia, or software form and can be easily copied and reproduced. This chapter surveys four categories of intellectual property law: patents, trade secrets, trademarks, and copyrights, and describes how laws in these areas apply to digital products. The chapter concludes with a discussion of how views of copyright infringement are changing and directions for future research.

Chapter 4, *Pricing in the Digital Age*, by Chip Miller from Drake University, focuses on the unique issues faced when pricing digital products. Online markets provide vast amounts of information to consumers and this has increased their ability to do price comparisons. Consumers also know that the cost of each additional digital product is very low which may impact the price they feel they should pay. From the seller perspective, pricing strategies must take into account these issues, but also they must factor in the ease with which their digital products may be pirated. The chapter begins with an overview of traditional pricing strategies. This is followed by a discussion of versioning, windowing, bundling, and unbundling, with examples of pricing strategies used by digital product companies. Novel digital product pricing strategies are also identified.

In Chapter 5, *Financing Digital Product Companies*, Richard Carter and Frederick Dark from Iowa State University discuss some of the financing issues associated with managing a digital product company. The unique cost structure of a digital product company impacts their financing decisions in both the short-term and long-term. The primary focus of this chapter is on one financing option – the initial public offering (IPO) – and the factors that impact timing for IPOs. They utilize an empirical and case study methodology to identify the results produced by IPOs that were too early in a company's life cycle, and those that were too late. These are important issues because many digital product companies require financing early in the life cycle to cover their large fixed cost expenditures. A successful IPO can provide funding through the turbulent early days and give them a chance to survive and prosper in the marketplace.

In Chapter 6, *Accounting for Digital Products*, Yasemin Zengin Karaibrahimoğlu from Izmir University of Economics in Turkey provides an overview of accounting issues faced by digital product companies. One issue addressed in this chapter is the accounting treatment for digital products including regulations for recognition, measurement, valuation, reporting and taxation. Suggestions are proposed for how to properly account for digital products given their differences from traditional physical goods. The chapter concludes with implications and directions for future research.

Moving beyond the functional issues addressed in the previous section, the third section encompasses a set of chapters for other important digital product issues and overall strategies.

In Chapter 7, *It's all About the Relationship: Interviews with the Experts on How Digital Product Companies Can Use Social Media*, Delaney Kirk from the University of South Florida at Sarasota-Manatee provides insights from several interviews with social media experts. The lessons they have learned are discussed along with ideas for how social media may best be utilized in support of a digital product. Techniques that work, and do not work, are discussed. The chapter concludes with a discussion of managerial implications and directions for future research.

Chapter 8, titled *Digital Convergence and Horizontal Integration Strategies*, describes the concept of digital convergence. The unique characteristics of digital technology enable three forms of digital convergence: technological convergence, content convergence, and industry convergence. The strategic management process is briefly reviewed in the chapter and digital convergence is shown to be both a potential external threat and an external opportunity for digital product companies as they strive to achieve some form of competitive advantage. Many companies have determined that horizontal integration is the best strategy for taking advantage of digital convergence opportunities. Real-world horizontal integration strategy examples from several digital product industries are discussed including the rationale for why these strategies were chosen. In addition, a wide range of other potential digital product strategic alternatives are discussed along with their associated benefits and risks.

In Chapter 9, *The Role of the Internet in the Decline and Future of Regional Newspapers*, Gary Graham from the Manchester Business School in the UK discusses the impact of digital technology on the newspaper industry. The chapter discusses the role of the Internet in the declining social and business influence of regional newspapers. It then provides an assessment of the impact of new technology developments, such as Web 2.0, on the future of regional newspapers. The chapter concludes with suggestions for how new technology may be utilized by news media organizations to co-create content with their readers.

Chapter 10 is titled *Software as a Service and the Pricing Strategy for Vendors*. In this chapter Nizar Abdat and Marco Spruit from Utrecht University in the Netherlands, and Menne Bos from Accenture in the Netherlands, provide an in-depth description of the phenomenon called Software as a Service (SaaS) where software is installed in data centers and delivered as a service instead of selling software like a traditional product. The chapter discusses the scientific and business perspectives on SaaS and identifies managerial and research implications. It discusses the benefits and risks associated with SaaS from the perspectives of all of the organizations that could be involved. The chapter concludes with a framework that can be used by vendors to identify appropriate prices for their SaaS services.

In Chapter 11, *Piracy, Copyright and Consumers' Rights: A European Perspective on the Private Copy Issue*, Pedro Pina from the Polytechnic Institute of Coimbra in Portugal provides a European perspective on digital copyright issues. The chapter describes the conflicts between the exclusive right to exploit a digital product and the private copy issues. The chapter concludes with suggestions for how the rights of digital product developers and consumers can be balanced.

In Chapter 12, *Service Systems as Digital Products*, Hsin-Lu Chang from the National Chengchi University in Taiwan, Michael Shaw from the University of Illinois, and Feipei Lai from the National Taiwan University present the concept of services as digital products. In the chapter they focus on a remote healthcare platform developed at National Taiwan University Hospital. The chapter identifies several important issues in this area including the development of service value models, the development of service metrics, and the management of service systems.

The fourth and final section in this book includes visions for digital product management, research, and technology in the future.

Chapter 13 by Dave Sly, President of Proplanner.com, is titled *Transitioning to Software as a Service: A Case Study*. The chapter provides an executive's perspective on the benefits, challenges and lessons learned from managing a digital product (software) company. The case describes the transformation of the company from a traditional "software as a product" engineering software company into a company that utilizes a "software as a service" business model. Technology issues and recommendations are discussed based on actual company experiences.

In Chapter 14, *Digital Media: Future Research Directions*, Anthony Hendrickson, Trent Wachner and Brook Matthews from Creighton University provide some directions for the future of digital media research. They explore how digital technologies are challenging business models and processes. They conclude by identifying several themes for future research that address important issues for how technological innovations will impact the global workplace. These themes include digital products as operant resources, issues for user-generated content and data quality, impact of network externalities, and challenges for digital product business models.

Chapter 15, titled *Digital Technology in the 21st Century*, provides a vision for the future of digital technology. The chapter begins with a review of current digital technology used for basic input, process, output, storage, and wide-scale networking (Internet and Web) tasks. A number of new digital technologies that are currently under development are identified and their potential benefits are discussed. These new technologies will create opportunities and threats for digital product managers in the coming decades and several possible impacts on digital product industries are discussed. These technologies will also impact societal issues such as healthcare, government services, higher education, political campaigns, cybercrime law enforcement, and life at home. The chapter concludes with several research issues and philosophical questions that must be addressed to determine the best uses for the relative advantages of technology and humans in the future.

Technology's future is exciting and scary. Digital product companies are operating in a hypercompetitive world where everyone has access to the same technology, differentiation is becoming more difficult, consumers have increasingly high expectations, and the lines between industry sectors is blurring because digital technology enables companies to grow by crossing traditional boundaries. I hope that this book can provide some insights and suggestions for how to manage digital products in the future. I also hope that the book has generated ideas for digital product management research so that answers can be found for important technical, business and social science questions.

*Troy J. Strader*
*Drake University, USA*

# Acknowledgment

This book is the culmination of the work and support of a number of people and organizations. I would like to thank Drake University for allowing me to take a sabbatical leave to work on this project. I would also like to acknowledge all of the chapter authors and reviewers for their contributions along with the assistance provided by the editorial staff.

*Troy J. Strader*
*Drake University, USA*

# Section 1
# Technology

## Chapter 1
# Digital Technology:
## Capabilities and Limitations

**Philip A. Houle**
*Drake University, USA*

## ABSTRACT

*Digital technologies are at the heart of all modern communication and information systems. The technologies have evolved from expensive component devices to current inexpensive systems used by everyone. This chapter examines digital technology in terms of its capabilities and limitations. It attempts to answer the following questions: What is digital technology? How does digital technology represent characters, symbols, and language? How does digital technology represent sound, images, and video? How does digital technology store and transmit these types of information? Finally, this chapter explores limitations of digital technology.*

## INTRODUCTION

The first purpose of this chapter is to introduce the fundamental ideas about how information is represented when using digital technologies. The second purpose is to identify the capabilities and limitations of these representations and to suggest avenues for future research and development as well as implications for a manager working with, and dependent on, digital technologies.

Digital technologies have become the core building block of all modern information and communication systems. Digital computer systems use databases and processing algorithms to store and deliver digital information on a vast array of applications. Networks, most notably the Internet, interconnect these computer systems across the world enabling the exchange of data and information. More importantly, perhaps, these networks allow applications to be created that are resident on multiple computers in the network facilitating distributed applications with processing power of enormous magnitude.

The importance of digital technologies suggests that an understanding of what makes a system digital is critical to assessing its capabilities and

DOI: 10.4018/978-1-61692-877-3.ch001

limitations. The natural world is not digital. Yet much of what modern consumers and businesses do to conduct the activities of daily commerce is based on digital technologies. One purpose of this chapter is to show how the digital world represents the natural world.

To explore how information is represented in digital systems, we start with basic idea how information arises in natural systems. We then look at the evolution of symbols, alphabets, and other representations of information. The advent of digital representation of information starts with binary digits and the organization of these digits to represent numeric values, alphabets, and symbols. The basic ideas of such representations are outlined and specific standards are then presented.

We next examine how images can also be represented using combinations of binary digits. The evolution of the standards for representations of images is presented ending with a discussion of modern standards.

The objective of this chapter is paint a picture of how binary digits are used to represent information, whether the information is in the form of language or in the form of images. At the same time, the goal is to show how these representations facilitate and limit what modern processing systems do.

## REPRESENTING INFORMATION

Each of us gets information through the use of our five basic senses: hearing, sight, smell, touch, and taste. First-hand information comes from what each of us personally experiences via these basic senses. Each of us interacts with humans via these same senses, which allows us to obtain information from others. We do this using images and language. Language can be spoken or it may be written alphabetic representations of the words used in the language.

Memory allows each of us to retain information obtained from experience and from others. Memory is part of the human experience. However, memory also exists when pictures, symbols and alphabetic representations are placed where they persist through time. For example, pictures can be draw in the sand or painted on a canvas. Memory allows us to retrieve information that relates to events that have occurred the past.

In this chapter we will focus primarily on information represented by language and images. This means we are primarily interested in the senses of sight and hearing. As we shall see, the use of digital representations is almost entirely limited to these two senses.

In the following sections we start with an examination of natural systems and representations of information in these systems. We then examine the nature of digital systems and representations. Then we look at how these ideas are applied to represent sound, images, and languages.

## Analog vs. Digital

The starting point for thinking about what digital means is to explore the contrasts between representations that are analog and those that are digital. Analog representations are measures that are natural and continuous. Digital representations are measures that are invented and discrete.

Analog measures vary over a span of characteristics where the values are continuous. For example, consider a glass containing some amount of water. How much water is in the glass? It may be full or it may be empty. Or it may be somewhere between. The possible states ranging between empty and full are infinite in number. The amount of water in the glass is an analog measure.

Digital measures are representations with discrete values. Some digital representations are exact. For example, the number of coins in a purse would be an exact measure. However, other digital representations are approximations. For example water in a one-cup measuring cup with a graduated scale showing popular fractions would be both analog, the amount of water, and digital, the closest fraction for the portion of the cup filled with water.

Automobiles typically have speedometers, which show the speed of the car as it moves. The speed of the car is inherently analog in that it can range continuously over a range of values. Some speedometers show the speed with a needle that moves across a scale of values, such as 30, 40, 50, etc. Other speedometers use a digital display to show the speed. If the digital display is an integer, then the actual speed must be approximated to the nearest integer value. Some digital speedometers might display fractions, making the approximation more accurate. However, in either case, the range of values displayed is discrete while that actual speed is continuous.

In order to have digital systems, all information must be represented by discrete values. This means that natural things must be approximated with representations that are discrete. These representations must be "good enough" so that the results are acceptable to the users of the digital systems. Music can be digitized. This is acceptable if the digital system can play back the music to satisfy the human ear. Photographs can be digitized. This is acceptable if the digital system can display the image to satisfy the human eye.

## Symbols and Alphabets

Probably the most fundamental types of representations used by humans involve symbols that make up alphabets that can be used to represent written language. For example, the English language contains words all of which can be spelled using a combination of letters from the alphabet A through Z. Alphabets also contain other symbols that can be used for punctuation and for other notations. Examples of punctuation symbols include the period (.), the comma (,), the semicolon (;), etc. Examples of other symbols include the dollar sign ($), the ampersand (@), etc.

In addition to letters and special symbols, alphabets include digits that can be used to represent numeric values. For example the standard decimal number system includes the digits 0, 1, through 9.

Information can be classified as numeric or non-numeric. Numeric information is the same as numbers. Non-numeric information can be considered text. A text is a sequence of symbols from an alphabet that cannot necessarily interpret as a number.

In the following sections we consider how digital technologies represent both numeric values and text.

## Numbers and Digits

Numbers are represented by using the symbols for the digits. Implicit in the representation is a base for expressing larger values. For example, the decimal number system uses the base of ten. Integer values represented in the decimal system have a units digit, a tens digit, a hundreds digit, etc. Fractional values in the decimal system have a tenth digit, a hundredth digit, etc. Algebraically, a numeric value in the decimal systems can be expressed as

$$\ldots d_n 10^n + \ldots + d_2 10^2 + d_1 10^1 + d_0 10^0 + d_{-1} 10^{-1} + d_{-2} 10^{-2} + \ldots d_{-m} 10^{-m} + \ldots$$

The values of $d_i$ above are the symbols for digits; 0, 1, etc. In the above, the implication is that the sequence of digits $d_i$ extends infinitely in both the left and right directions. However, the digits will be zero to the left at some point and convention is to not write them. The digits to the right may also be zero at some point, which means we need not write them. However, many values have infinitely extending sequences of digits to the right, which means we are approximating the actual value when we stop writing the digits. Examples of such values are pi = 3.14159... and 1/3, which in the decimal system, is 0.33333....

The above discussion has assumed decimal as the base of the number system. Other bases can be and are used. If a base different than ten is used, then the occurrences of ten in the above algebraic equation would be replaced with the alternative base. In addition, the number of different symbols used for digits would be restricted such that it matched the base. For example, if the base was three, then only three digit symbols would be used.

Common bases other than ten that are used include two, eight, and sixteen. The following table summarizes the characteristics of each of these bases and compares each to the decimal system in terms of representing the decimal value 43.

Modern digital processing systems use base 2, or binary, number systems. That is, all information is stored in terms of the digits 0 and 1. The reason for this is because it is easiest to build electronic devices that represent two states, which then can be associated with the digits 0 and 1. An electromagnetic switch can be either on or off.

As is evident from Table 1, values stored in binary are relatively inefficient in terms of notational space. Therefore it is common practice to use groupings of binary digits as values. For example, a group of eight binary digits is commonly called a byte. Eight binary digits can be divided into two parts, each with four binary digits. Each part can be considered a hexadecimal digit. Thus a value stored in a byte can be expressed as two hexadecimal digits. The decimal value in the above table can be stored in a byte as 00101011. This value can be represented as 2B in base 16.

## Codes

The idea behind codes is to represent the symbols in the alphabet of interest in terms of sequences of things that can be physically stored and transmitted. The Morse code is an early example of such coding. The physical things used in Morse code were dots and dashes. Given that modern processing systems use binary system, the next type of code is to use numeric values that can be represented in terms of ones and zeros, where the ones and zero can be embedded into a physical medium.

*Table 1. A comparison of differing base systems for numeric values*

| Base | Digits Used | Decimal 43 |
|---|---|---|
| Binary (2) | 0 and 1 | 101011 |
| Octal (8) | 0, 1, 2, ..., 6, and 7 | 53 |
| Decimal (10) | 0, 1, 2, ..., 8, and 9 | 43 |
| Hexadecimal (16) | 0, 1, ..., 9, A, B, ..., F | 2B |

In modern computers, a standard has evolved leading to something called a universal character set. Each alphabetic character is assigned a code that corresponds to its symbolic value. These symbolic values are assigned to keys on keyboards for manual input. Pressing a key generates the corresponding code value. In the reverse, a display system uses the code to look up the appropriate font value from a table, which is then placed in the display device.

Standard coding schemes have evolved. Two important schemes in the English language are ASCII and EBCDIC, which are discussed in the following section. ASCII uses seven-bit codes and EBCDIC uses eight-bit code values. ASCII limits the number of different codes to 127 and EBCDIC limits the number of different codes to 255. Because non-English languages may require more code values, the Unicode standard as also evolved. Unicode code values are sixteen bits in size, which dramatically expands the number of possible values.

## ASCII and EBCDIC

Early developments of code schemes for representing alphabets used only six bits per code. This limited the number of values to 64, which was too small to allow both upper and lower case representations of the English alphabetic letters. For example, the early code scheme used six bits and all letters were upper case only.

The early six-bit scheme was replaced with the ASCII code standard, which used seven bits

*Table 2. Values for ASCII and EBCDIC coding for selected symbols*

|      | Code Values | | | Code Values | | | Code Values | |
| --- | --- | --- | --- | --- | --- | --- | --- | --- |
| Char | ASCII | EBCDIC | Char | ASCII | EBCDIC | Char | ASCII | EBCDIC |
| (sp) | 32 | 64 | A | 65 | 193 | a | 97 | 129 |
| ( | 40 | 77 | B | 66 | 194 | b | 98 | 130 |
| ) | 41 | 93 | C | 67 | 195 | c | 99 | 131 |
| + | 43 | 78 | D | 68 | 196 | d | 100 | 132 |
| - | 45 | 96 | E | 69 | 197 | e | 101 | 133 |
| . | 46 | 75 | F | 70 | 198 | f | 102 | 134 |
| 0 | 48 | 240 | G | 71 | 199 | g | 103 | 135 |
| 1 | 49 | 241 | H | 72 | 200 | h | 104 | 136 |
| 2 | 50 | 242 | I | 73 | 201 | i | 105 | 137 |
| 3 | 51 | 243 | J | 74 | 209 | j | 106 | 145 |
| 4 | 52 | 244 | K | 75 | 210 | k | 107 | 146 |

per code value. IBM extended the six-bit code scheme to eight bits per code with the introduction of the extended binary coded decimal interchange code, or EBCDIC. Examples of the code values for a subset of the English symbol set are show in Table 2 (Haralambous, 2007).

Both ASCII and EBCDIC became important standards. Standards organizations and most vendors embraced the ASCII code schedule. However, IBM dominated its markets during the decades of the 1970's and 1980's, which meant that the EBCDIC also became a de facto standard.

## Extended Alphabets

As stated above, Unicode uses sixteen-bit codes and, therefore, expands the number of codes available to represent alphabetic symbols. However, it does more than that. Quoting from the www.unicode.org web site (Unicode Consortium, 2009),

*"Unicode provides a unique number of every character no matter what the platform no matter what the program no matter what the language"*

The Unicode standard has been used by most important software systems. Most modern operating systems and all modern web browsers know about and work with Unicode. The key idea behind Unicode is a standardization process that is embraced by a large number of users. Once the standardization is developed and adopted by most users, things work the way we expect them to do so. In addition, the standard makes things easier for the software system developers, since they can achieve systems that recognize a wide range of alphabets and languages within that single standard.

## Digital Systems and Representations

In our discussion thus far, we have looked at alphabets with letters or symbols or digits that can be used to express written language. We have also explored the case of digits, which can be combined to produce numeric values. However, modern digital systems go beyond the representation of symbols, words, sentences, and numeric values.

Modern digital systems include representations of sound, images, and video. In what follows, we explore how digital ideas can be applied to these types of media.

## Sound

Sound is what we hear. Sound occurs in nature and as a product of human action. It is speech as produced humans. It is noise and it is music. It is alarms, sirens, and the rushing wind.

Sound in nature is propagated by pressure waves in the atmosphere. The waves travel through air at a speed of approximately 1,125 feet per second. The actual speed depends on the temperature and pressure of the air.

Sound can also travel though other mediums such as water and certain solids. However, humans hear sounds via the ear, which senses the pressure waves in air. Thus our discussion will focus on sound in air.

All modern technology depends on devices known as microphones to convert the air pressure waves representing sound into electrical signals. The wave forms represented in the pressure waves are converted into matching wave forms in the electrical signals. Since the pressure waves in air vary in a continuous manner in amplitude and frequency, the resulting wave forms are analog.

Modern technology has lead to devices that can synthesize sound. That is, electrical signals can be created that correspond to sound without having the sound starting as air pressure waves. For example, such devices can create synthesized music and computer voices.

The electrical signals representing sound are converted back to air pressure waves using speakers. Modern speaker technologies create pressure waves as sound in the air in wide variety of systems.

Using representations of sound as electrical signals allows creation of systems to transmit sound across long distances not possible with air pressure waves. For example, telephone systems allow sound to be transmitted across the world using electrical connections such as physical telephone lines. Radio technologies can be used to transmit sound without the need for physical wires. However, both telephone and radio transmit sound in real time. That is, the source of the sound is connected in time with the hearer of the sound. The need to store sound as it is produced so that it can be heard at a later time will lead us to recording technologies.

## Analog Representation

As mentioned earlier, sound is, by nature, analog phenomena. This means that pressure waves and the corresponding electrical signals vary in frequency and amplitude in a continuous manner as a function of time. In order to record sound, technologies were needed to find a way to store the continuous signal characteristics in a physical medium such that the signal could be retrieved at a later time.

Two early technologies were invented that could store sound. The first was the phonograph or gramophone. The second was magnetic tape.

The history of the phonograph starts with the work of Thomas Edison and leads up to more modern high quality systems (Gelatt, 1955). The earliest devices were mechanical rather than electrical. Edison used a physical diaphragm that vibrated from sound waves that "dented" tin foil. The dents in the tin foil represented the wave forms of the sound. Later devices used wax cylinders instead of tin foil. As phonographs developed, the cylinder was replaced with disc shapes, which ultimately lead to the LP record industry.

While early phonographic devices where mechanical, later devices replaced the mechanical linkages with electronic circuitry. However, the recording of sound was ultimately accomplished by embedding the wave forms into the medium of the LP disc.

Magnetic tape is created by coating a thin strip of plastic with an iron oxide material that can be magnetized by passing it near a magnet. If the magnet is created via electrical signals, then the magnetism imparted into the passing oxide material will match the signal characteristics of the electrical signals. To retrieve magnetized signal, the oxide material can be passed near a wire resulting in electrical signal in the wire (Begun, 1949).

## Digital Representation

Digital representation of sound involves a process of converting the time dependent analog representation into a stream of binary bits. This is accomplished by using a sampling technique where the amplitude of the analog signal is measured at regular intervals and the resulting digital values are used to approximate the actual signal. Suppose we choose to sample the analog signal amplitude 8000 times a second. Also assume that each sample produces a number in the range 0 to 255, which is a byte of data or eight bits. This process replaces the analog curve with a series of rectangular boxes where the width of each box is 1/8000 of a second and the height of each box ranges from 0 to 255.

It turns out that a sample rate of 8000 per second is a standard used by the telephone system for converting voice signals to digital values. Since there are eight bits per sample, this sample rate produces a bit rate of 64,000 bits per second. This rate is the standard for the transmission of telephone traffic across the digital trunk network. However, the question must be asked, why 8000 samples per second?

The telephone system has traditionally used bypass filtering to limit the input into the system. Human speech contains frequencies from zero to approximately 20,000 cycles per second. However, the filter used by the phone company blocks frequencies lower than 300 cycles per second and higher than 3,400 cycles per second. Experience has taught us that the human ear hears this range and considers the quality acceptable for telephone use.

Nyquist and Shannon have shown that using a sample rate that is twice the highest frequency will produce a quality signal when the digital values are changed back into analog signals for the human ear (Jerri, 1977). Since the highest frequency allowed in the telephone system is 3,400 cycles per second, the sample rate of 8000 is justified.

What if we want to sample such that the quality is acceptable to the ear that can hear the full spectrum of frequencies? Most human ears can hear frequencies as high as 20 to 22 thousand cycles per second. Because of this, the standard for digitizing CD-quality music is to sample at 44.1 thousand times per second. The other factor that changes in digital music is the number of levels of amplitude that result from the sample. In the telephone system, the number of levels was 255. In CD-quality music, the number of bits per sample used is 16, which allows over 65,000 different levels to be represented.

The device or process that creates digital representation of analog sound and then returns it to analog sound is called a CODEC. Most modern CODEC processes involve digital processors that execute the algorithms to do the sampling. There are a wide variety of standards for CODECs depending on their application. These standards determine the sample rates and the number of bits produced for each sample.

An additional function of CODECs is to compress digital information that is produced by the sampling process. Engineers have noticed that there is significant redundancy in the digital values that represent the sound. For example, if the tone remains the same for some length of time, then the values produced over that interval will all be the same. An important example of one such compression standard is MP3.

The MP3 standard is used in many modern devices to store and play music. The MP3 representation of a piece of music may require only 10 percent of the bits as the original CD-quality recording. However, the quality of the music may be somewhat reduced from the CD-quality original.

## Images

Images are two-dimensional objects that can be seen and interpreted by the human eye. Images are typically represented on paper or other materials where colors can be applied. If the image is black and white only, then the only color is black since white is the absence of black.

Images have historically been captured by cameras using film. The lens of the camera focused the image on a material that had chemicals that reacted to the wavelengths of the incoming light, which gives color. However, modern cameras use digital representations to record images.

The representation of color is achieved through the use of primary colors. Primary colors are those colors that can be combined to create a range of colors. The combining process can be either subtractive or additive. For example, when mixing paint, the process is subtractive. This is because the color seen by the human eye is a reflection of the environmental light from the painted surface and the color is determined by which parts of the color spectrum are absorbed. On the other hand, when using a computer monitor, the process is additive. This is because the color is determined by the light sources within in the monitor.

The primary colors commonly used differ by application. Early photographic process used orange, green, and violet. Cyan, magenta, and yellow are used when pigments or dyes are mixed, such as with painting or printing. The traditional set of primary colors used in art and design are red, yellow, and blue. As we will see below, when digital images are used, the primary colors are red, green, and blue.

There are two major methods for representing graphic information (images) using digital ideas. These are bitmap images and vector images. Since bitmap images are used in displays, printers, and standard digital photography, this presentation will deal only with bitmap graphics.

The fundamental idea behind bitmap images is to consider the image as a rectangular grid of pixels. Each pixel in the grid has data associated with it that determines which color it displays. All of the pixels together create the image.

The number of pixels in the grid determines the quality of the image. For example, most computer monitors display 70 to 100 pixels per inch. Obviously, the size of the grid and the number of pixels in the grid determine the density of the pixels. For example, HDTV has a standard grid of pixels of 1920 by 1080. This means a larger HDTV screen will have fewer pixels per inch on the screen than a smaller HDTV screen.

The number of color levels encoded also determines quality. Each pixel is made of variable intensity of three colors: red, blue, and green. If all three colors are intense, the result is black. If all are absent, then the color is white. The full color palette is defined by the varying levels of intensity of the three colors.

A modern digital camera might produce a grid of pixels such as 3648 by 2736. This would amount to slightly under 10 million pixels, hence a 10 mega-pixel camera. Each pixel would require multiple bytes to represent the color values of that pixel. Yet images from such a camera are typically much smaller than 10 megabytes in size. This occurs because most pictures have significant redundancy, which allows compression to reduce the actual size needed to store the picture. For example, many adjacent pixels may have identical color values.

There are many different standards for representing images. Examples include: GIF, JPEG, JPG, PICT, TIFF, among others.

## Video

We use the term video to refer to images that show moving pictures. Video images had a beginning with analog representation of images both in the motion picture theater and with the development of television. In the following sections, we visit each of these ideas and show how digital representations are used.

*Digital Technology*

## Moving Pictures

Perhaps, one of the earliest and easiest forms of a moving picture was achieved by sketching a sequence of images on sheets of paper such that when the sheets were fanned across one's field of vision, the sequence of images was transformed into a single moving image. This idea is inherent all forms of moving images or pictures. It exploits a property of the human eye and brain that converts the sequence of images into apparent motion.

The idea of creating a moving image, or an animation, depends on having a way to present the sequence of images at an appropriate speed. This speed is called the frame rate. What constitutes an appropriate frame rate is complex. Determining the appropriate frame rate depends on the human eye and brain. It also depends on the rate at which elements of the image change. If there are objects in the image that are changing position at rates that resonate with the rate, undesirable effects may result. For example, wheels may appear to rotate the wrong direction or portions of the image will be blurry.

The film industry used cameras and projectors that create and display individual frames at the appropriate rate using colloid film technologies. Television used an analog scanning process to transfer the images on to a cathode ray tube. Black and white television used light and dark with a single beam. Color television used three beams, one for each or the three colors, red, blue, and green. For many years, both the film and television industries used analog technologies to create, record, and display the images used on the frames created. More recently, both have adopted digital technologies.

## Digital Representations for Video

The basic concepts for representing images or graphics have been presented in an earlier section. To obtain digital representations for video, which are moving images or graphics, these ideas have to be extended to include the concept of frames and frame rates, which creates the appearance of motion. If audio is to be part of the video presentation, then the concepts must include standards for representing both audio and images.

The ideas of a pixel and codes for presenting the color attribute of each pixel explained for images above extend into the realm of video coding standards. What is added is the idea of representing the sequence of frames that must be presented to create the moving picture. Frame rate standards have evolved in television and in the film industry. Both use a frame rate of approximately 30 frames per second. This rate creates an acceptable moving picture for the human eye.

An important issue involved in the evolution of standards for digital video is the issue of compression. This is important for two reasons. If we visualize a scene in a moving picture and consider the frame sequences that create the scene, many, if not most, of the pixels from one frame to the next remain unchanged. This means there is enormous opportunity for compression techniques that exploit this redundancy, which dramatically reduces the number of bits required to represent the video presentation. The second reason that compression is important is that digital video requires, without compression, a very large number of bits, which translates into a very demanding storage requirement and a very high bit rate to sustain the frame rates for the video presentation. For example, if each frame of a video required a megabyte of data and the video frame rate was 25 frames per second, then each second of video would require 25 megabytes of data. Each minute would require 1,500 megabytes, or 1.5 gigabytes. Each hour would require 90 gigabytes. A standard DVD only has a capacity of approximately 4 gigabytes. From this example, the need for compression is very evident.

There are many different formats for representing digital video with compression in computer applications. Examples include AVI, QuickTime, Flash, MPEG, etc. There are many other standards for representing audio/video materials. A complete treatment of these standards is beyond the scope of this work.

## STORING AND TRANSMITTING INFORMATION

Our discussion thus far in this chapter has centered on digital representations of information. The discussion has focused on the techniques used to create representations that can be seen and heard. In this section, we will move to how technology system can be used to remember or store information and how to transmit information from one location to another.

If someone hears or sees, the information content is observed at an instance of time. If the information content was created at an earlier time, then there must be some sort of memory device that recorded the information and replayed the information at the time it was heard or seen. If distance is involved, then there must also be some sort of transmission facility to move the information from one place to the next.

In the following sections, we first look at how digital systems store information. We will examine this starting with a historical view and moving to contemporary technologies. Early technologies combined the idea of storage and moving information. Finally, we look at how digital systems move information from one place to the next.

### Paper and Printing

The very oldest forms of storing information go back to men making images on objects or on surfaces. Early techniques included carving, sculpture, and painting. The media were wood, clay, stone, etc. Later the idea of ink and sheets of material upon which ink could be placed arrived. These types of techniques were very labor intensive and not very efficient. For example, duplication of materials was slow and prone to errors. However, all of this changed with the invention of the printing press by Gutenberg in 1440.

The printing press, along with its successors, became the backbone technology for recording information. In addition, printed materials could easily be transported from place to place, which enable transmission of information. Books were distributed and sold. Mail delivery systems were established to move printed materials (letters) from one place to the next. Perhaps one of the more famous examples of this in the history of the United States is the Pony Express of the Old West.

The typewriter came along in 1867. This machine enabled individual writers to create documents with characters similar to those produced by the printing press. This invention led to the mass adoption of the keyboard as the primary mode of writing, which has continued even into modern times.

There are several limitations with the use of printing technologies such as the printing press and typewriters. First, the medium is paper, or similar materials, which are subject to deterioration with time. Secondly, information on a printed medium is not readily available for machine input.

### Computer Storage

Materials created by the printing presses and typewriters are available to the human eye. The materials are stored such that they can be read at later times. However, the advent of the digital processing systems required that the materials stored be accessible to machines. This leads us to the ideas of digital storage techniques.

### Punch Cards

Perhaps the most significant invention in the world of digital processing system came from Herman Hollerith. Hollerith conceived the use of the punch card for storing and processing information. He conceived and built machines that created, sorted, and used punch cards to process information digitally. His ideas were used in processing the 1890 census, which reduced an estimated ten-year task into on that took three months (Herman Hollerith, 2009).

The use of punch cards as the medium for recording data and programs for digital machines lasted into the decade of the 1970's. Data were recorded using keypunch machines. Early digital computers read the punch cards, performed processing, and created reports and new punch cards, which became part of the input to be read into the next processing cycle.

Early programs were created by Hollerith using wiring panels. Later, even after the advent of the digital computer, programs were recorded using punch cards. Even with the advent of magnetic storage technologies, punch cards continued to be the manner in which many programs and dataset were prepared for input into digital computer systems.

## Magnetic Memory, Tape and Disc Memories

The invention of the digital computer required that there be some type of memory connected with the processor from which program instructions could be retrieved and where data could be retrieved and replaced. This memory was constructed using arrays of iron cores that could be magnetized in one of two states, each state representing either a one or a zero in the binary number system.

Memories used within digital computer systems for storage of program instructions and data used directly by the processing elements of the systems are considered primary memories. Primary memories had to be fast enough to supply instructions and data at processor speeds. The contents of primary memory are transitory in that it is not saved for long periods of time. In contrast, secondary memories are used to save content for later use, the example being the punched cards described above.

Magnetic tape was first developed in 1928 and allowed digital content to be stored on the tape and retrieved by electro-mechanical methods. The tape material was coated with an iron oxide material that could be magnetized to record bits. The bits could then be read later by moving the magnetized spots past a device where electrical signals were produced by induction. Magnetic tape allowed digital systems to process sequential data files. Algorithms were developed for sorting large data sets stored on tape. Old master files were read from an input tape, updated with new transactions read either from tape or punch cards, and then written to a new master file stored on an output tape.

Magnetic tape was used by the earliest digital computers as the secondary storage device of choice until early in the 1960's when disk storage units were introduced into the commercial marketplace. Disk storage units used the same magnetic principles, but had the advantage that the read/write mechanism could retrieve and write bits in randomly addressable locations. This was in contrast to magnetic tape where the bits were arrayed sequentially on the tape and it was not possible to write into the middle of a tape. Disk storage created the concept of random access storage.

Modern digital computer systems use the idea of random access storage pioneered by early disk storage systems to create automated file systems managed by sophisticated operating systems. Modern operating system software provides utilitarian functions to define files and to allocate and record location of the file content on the physical storage device.

## CD and DVD Media

The alternative to using magnetic materials as the medium for storing bits was first developed in the late 1960's. The alternative was to use light and materials that were photosensitive to store data. This technology became important in the marketplace during the 1980's with the introduction of the compact disk (CD) format for storing digital audio materials.

Beginning during the 1990's, an alternative format for the optical disk emerged called digital versatile disc (DVD). This format was more versatile than the CD and allowed for more information to be stored on a single disk.

CD and DVD technologies are optical media. Digital information is stored as bits encoded by dark and light spots on the photosensitive surface of the disk. The bits are written and read using a laser. The advantages of the use of optical and the photosensitive medium are multiple. The density of bits that can be placed on the disk per unit of area is much higher with optical technologies than with magnetic technologies. Secondly, the span of time during which an optical disk remains readable is much longer. Lastly, the optical disk is not erased by the presence of magnetic fields.

A CD has capacity to store approximately 700 megabits. A DVD has capacity to store approximately 4.7 gigabits, which almost seven times more than a CD. However, the actual capacity of both technologies depends on the formatting used and other factors. Newer encoding technologies are emerging that will increase the capacity of these devices. One such technology is the Blu-Ray disk, which increases the capacity of a single disk to as high as 50 gigabits (Fast Guide to CD/DVD, 2009).

The major disadvantage of optical devices for storage is that the speed for reading and writing is much lower than with magnetic devices. Therefore, the use of optical devices is limited to applications and situations where the lower speed does not hinder the application.

## Flash Memory Devices

Both types of technology described above, magnetic and optical, require mechanical devices to move tape, to spin disks, and to move read/write mechanisms in order to work. During the 1980's a new type of memory device emerged. It is based on the use of semi-conductor devices capable of storing bits of information and that were non-volatile. Non-volatile means that the stored contents are preserved even when the device is without electrical power. This type of memory is often called flash memory.

There are two important advantages with storage devices based on flash memory. The first is that there are no moving parts. Secondly, the devices are much more tolerant of shocks and environmental hazards, which make them more idea for mobile uses. Modern mobile devices, such as cell phones, employ flash memories to store system programs and user data.

An important development using flash memory technologies has been the USB flash drive. USB, or universal serial bus, is a standard that has emerged to connect devices to digital computers in a manner that is recognized by current operation systems. USB memory devices combine the technologies of flash storage with the USB standard in a way that creates "disk drives" that can be plugged into any computer and used because they are recognized by all popular operating systems. They are called disk drives because they function in a manner which is the same as the legacy disk drives based on magnetic technologies. But there no disks or moving parts in flash memory devices.

USB flash memory technologies support data transfer rates that are much faster than are possible with optical technologies. Therefore, as prices of flash memory devices drop, their use as storage devices will more and more replace the use of optical devices. We could expect future mobile computers systems to replace the rotating magnetic disk drives with a flash memory store devices.

## Data Transmission

In this section we look at how information can be moved from one location to another. Historically, movement of information was accomplished by move the physical medium upon which the information was recorded. For example, books must be physically transported. Letters containing messages were carried from sender to receiver. Postal systems were established to move physical media with messages.

## Telegraph

One of the earliest systems that facilitated the movement of information from place to place was the telegraph. The basic ideas involved in the telegraph were evolving starting around 1800, but the first commercial successful device was invented by Samuel Morse in 1835. Morse introduced the idea of the Morse code and the use of electro-magnetic devices to transmit the codes using wires strung from one location to another. The Morse code defined a sequence of dots and dashes corresponding to each letter in the English alphabet. The electrical connection across the wires could then be used to transmit the sequences of dots and dashes. This enabled messages to be sent electronically rather than physically.

Even though the early telegraph system transmitted messages electronically, it was dependent on human operators are each end to create the codes to be sent and to convert the received codes back to letters, words, and messages. Later improvements include the development of the teletype in approximately 1870, which allowed messages to be keyed via a keyboard, sent across the electrical network, and typed on paper. Paper tape was also used by these devices to record messages that would be sent at times other than when keyed.

## Public Telephone Networks

In 1874, Alexander Graham Bell invented the telephone as an improvement to the existing telegraph system. The telephone system enabled the transmission of human voice using the wires connection places. Using a microphone to create modulated electrical signals to represent the sound of voice and a speaker to take the transmitted electrical signal back into sound, the telephone system allowed people to talk across distance.

As the telephone became a commercial success, more and more people had telephones that needed to be connected in some fashion to other telephones. This required switching networks to make these connections. Early implementations of switching networks involved the use of human operators that talked with callers to obtain information about the desired connection and then manually made the connection using a switchboard. Later, human operators were replaced with phone numbers, rotary dials, and electromechanical switching devices that established circuit connections between the caller and the party called.

## Circuit Switching

As the telephone system grew, networks of connections were developed to provide pathways to establish connections between telephones. These connections were established as calls were placed and were maintained for the duration of the call. When the call was completed, the connections were released. The concept of dedicating circuits for the connection during the phone call is circuit switching. Circuit switching required that the system be able to locate and to dedicate resources in the network to connect the phones. If the resources could not be located, then a busy signal resulted.

As the use of the telephone system grew, the demand for resources also grew. The demand grew for connections from one city to the next, commonly called "long distance" service. The network resources required for long distance connections were expensive, given the need to dedicated circuits for the duration of calls. This led to the development of digital technologies that could be used to transmit voice across distance. We have explained how voice can be converted to digits in an earlier section. Now we need to explore how digits can be transmitted.

## Analog vs. Digital Transmission

Early phone systems transmitted voice across distance by converting the sound waves into electrical waves using a microphone placing the electrical waves on wires connected to a speaker device at the other end of the wires. The information going across the wires was analog in the same way sound is analog. Using the technologies described above to convert audio signals into bits using a CODEC, newer networks were developed that transmitted bits across distance rather than analog signals. Transmitting bits is digital communication.

The simplest way to visualize how bits can be represented in an electrical transmission is to visualize two distinct levels of voltage. These two distinct voltage levels can then be used to encode ones and zeros. As the stream of bits is transmitted, the voltage level is changed as the bit values change. This example is simplistic in that modern encodings are much more sophisticated, the basic idea applies. If signal amplitudes are used to encode the bits, then the transmission is considered amplitude modulation (AM). However, a different type of modulations can be used. For example, if the frequency of the transmitted signal is varied to represent bits, then the transmission is considered frequency modulation (FM). Modern digital communication channels use various methods of encoding that represent bits that are beyond the scope of this discussion.

## Packet Switching

As described above, telephone systems use circuit switching to build connection between one telephone and the other. The process of converting voice signals to bits generates a stream of bits that then can be sent across the dedicated circuit. In order to provide a quality connection, the channel capacity must be reserved for the call. However, when computers are used in place of telephones at each end, the bits that must be transmitted are different: First, there may be relatively long periods of time during which no traffic is present. Secondly, computer to computer communication may be able to tolerate delays or variations in the effective transmission rate. These differences make using circuits designed for voice transmission for data transmission somewhat inefficient and expensive.

Rather than using circuit switching, digital communication channels use packet switching. A message is to be sent from one computer to another. The message is a string of bits. The process splits the message into smaller parts and each part is called a packet. Each packet is then sent across the network. Other users of the network are also sending packets. All packets contain addressing information so that they can be correctly carried across the network to the appropriate destination. When the packets arrive, they can be combined by the receiving computer to recreate the original message. Packets share the core communication channels. This allows the trunk lines, the core channels, to multiplex the transmission. This means that cost of the core channel connecting computers to computers is much lower than with circuit switched systems.

## The Internet

The purpose of this section is to explain how digital information and digital computers can interact via the world-wide connection called the Internet. We start with basic descriptions of digital networks and how connections occur. We then characterize in simple terms how the Internet works. The idea of protocol is developed. Finally, we brief describe typical applications used on the Internet.

A network is formed when two or more digital computers are connected. The connection may be wired or wireless. Wired connections use technologies commonly associated with local area networks (LAN) and interconnected local area networks, commonly called internets. Wireless connections use radio signals to replace wires, but the transmission of bits in wireless networks is fundamentally similar as with wired networks.

Any computer connected to a network is a host and must have an address that can be used to identify the computer in the network. The address used in internets is called an IP address. Any host can communicate with, that is exchange messages with, another host if the address of the other host is known. The Internet is the world-wide connection of networks made up of hosts identified by unique IP addresses.

An important resource in the Internet is the Domain Name System (DNS). The idea behind the DNS is to provide directory services to hosts that need to find IP addresses of services in the Internet. Applications normally use uniform resource locators (URL) to link to services in the Internet. DNS provides a distributed database of conversions from URL to IP addresses.

Most applications using the Internet typically operate as client/server applications. The client is a computer where a person is the user. The server is a computer that responds to messages from clients by returning messages or materials stored at the server. However, from the Internet point-of-view, any host can function as a client or as a server. This means messages arriving at a server may, in fact, originate at from another host that is really not a client, but a computer program.

The power and symmetry of the Internet in terms of hosts communicating and determining roles creates both an opportunity and a limitation. The opportunity is that applications systems can be devised with create architectures with little constraint to accomplish novel things. The limitation is that hosts operated by individuals with intent on disruption and exploitation can do the same thing.

## DIGITIZING THE OTHER SENSES: SMELL, TOUCH, AND TASTE

In this section, we consider what the possibilities might be for using digital technologies to interact with the human senses of smell, touch, and taste.

All of the discussion thus far has centered on the senses of sight and hearing. The sense of sight deals with the transmission of light and its interaction with the human eye. The sense of hearing deals with the transmission of sound in air and its interaction with the human ear. Both of these transmissions are propagated via the medium. That is, they travel through distance at a velocity that is relatively fast by human standards. The senses of smell, touch, and taste rely on delivery mechanisms that are very different.

In order to smell something, the human nose must encounter air containing molecules that stimulate the organs in the nose that smell. This means air must move into the nose. In order to taste something, the human taste buds, located in the mouth, must encounter molecules, in air or in other material, that stimulate the organs in the mouth that taste. Both of these senses depend on the delivery of molecules into the appropriate areas of the human body.

In order to touch or feel something, some part of the human body must physically be in contact with the thing felt. If we assume that a physical contact exists, then digital system could be created to exploit that contact. The digital system could create pressure, temperature, motion, vibration, or similar things that could then be sensed by the part of the body in physical contact. For example, a person without sight can read material by sensing the Braille codes placed on surfaces.

## CONCLUSION

This chapter has outlined what it means to use digital representations for things. We have examined representation of characters, words, and language.

We have examined the use of digits in number systems. We have examined how audio can be sampled and changed into digits. We have examined how images can be approximated by arrays of pixels, each with a set of digits encoding color information. We have examined digital memories and the transmission of digital information.

The most positive aspect of digital representation of information is that representations remain intact when stored and when transmitted. This is in contrast to non-digital representations where deterioration and distortions alter that representation. While deterioration and distortions do occur in digital systems, these systems are designed with sufficient redundancy to prevent loss of the basic unit of representation, the one and the zero. This means copies of original material are exactly the same in terms of quality.

There are two limitations in the continued development of digital technologies. The first has to do with the media used to store digital representations. The second has to do with the standards for representing digital materials.

The different types of media used to store digital representations have been described earlier. In all cases, these media have limited life expectancies. That is, after some length of time, the recorded bits will no longer be viable to be read by a digital input device. For example, the life span of magnetic tape, CD, and DVD technologies is estimated by various sources to be as short as a few years to as long, perhaps, as 100 years. However, these technologies are relatively recent and the actual life span may differ from estimates. When compared with information taken from artifacts thousands of years in age, the life span of digital media seems rather short.

However digital materials are stored, it must be encoded and organized using standards that comply with the software that records and reads the materials. Suppose we discover a CD material that last hundreds of years and we use this media to record our digital content. A problem will arise when, hundreds of years in the future, the standards have moved beyond what exists today and software cannot be located to read the materials.

Both of the above limitations mean that a fundamental limitation of digital technology is permanent retention of the information involved. In the real world, we have come to expect that great manuscripts and works of art can be retained indefinitely. This is due, in part, to the media used to represent the works. In the age of digital representations, the problem of indefinite retention becomes more difficult.

## REFERENCES

Begun, S. J. (1949). *Magnetic Recording*. New York: Rinehart & Company.

Fast Guide to CD/DVD. (2009). Retrieved October 15, 2009, from http://whatis.techtarget.com/definition/0,sid9_gci514667,00.html#

Gelatt, R. (1955). *The Fabulous Phonograph: From Tin Foil to High Fidelity*. Philadelphia: J. B. Lippincott Company.

Haralambous, Y. (2007). *Fonts & Encodings*. O'Reilly Media.

Herman Hollerith. (2009). Retrieved October 15, 2009, from http://www.columbia.edu/acis/history/hollerith.html

Jerri, A. (1977). The Shannon sampling theorem - its various extensions and applications: A tutorial review. *Proceedings of the IEEE*, 1567-1596.

Laudon, K. A. (2009). *Essentials of Management Information Systems*. Upper Saddle River, NJ: Prentice Hall.

Panko, R. (2009). *Business Data Networks and Telecommunications*. Upper Saddle River, NJ: Prentice Hall.

Unicode Consortium. (2009). Retrieved October 15, 2009, from http://www.unicode.org/

White, C. (2007). *Data Communications and Computer Networks*. Boston, MA: Thomson Course Technology.

## ADDITIONAL READING

Abelson, H., Ledeen, K., & Lewis, H. (2008). *Blown to Bits: Your Life, Liberty, and Happiness After the Digital Explosion*. Indianapolis, IN: Addison-Wesley Professional.

Baschab, J., Piot, J., & Carr, N. (2007). *The Executive's Guide to Information Technology*. Hoboken, NJ: Wiley.

Bergin, T. J., & Haigh, T. (2009). The commercialization of database management systems: 1969-1983. *IEEE Annals of the History of Computing*, *31*(4), 26–41. doi:10.1109/MAHC.2009.107

Brain, M. (2009). How analog and digital recording works. *HowStuffWorks*. Retrieved December, x, 2009, from http://communication.howstuffworks.com/analog-digital.htm

Brynjolfsson, E., & Saunders, A. (2009). *Wired for Innovation: How Information Technology is Reshaping the Economy*. Boston, MA: The MIT Press.

Cianci, P. J. (2007). *HDTV and the Transition to Digital Broadcasting: Understanding New Television Technologies*. Burlington, MA: Focal Press.

Downey, G. (2009). A complex history of the commercial Internet. *IEEE Annals of the History of Computing*, *31*(2), 80–81. doi:10.1109/MAHC.2009.18

Freyer, C., Noel, S., & Rucki, E. (2009). *Digital by Design: Crafting Technology for Products and Environments*. London: Thames & Hudson.

Hicks, M. (2009). Grace Hopper and the invention of the information age. *IEEE Annals of the History of Computing*, *31*(4), 116–117.

Laudon, K., & Laudon, J. (2009). *Management Information Systems*. Upper Saddle River, NJ: Prentice Hall.

Lowood, H. (2009). Videogames in computer space: The complex history of Pong. *IEEE Annals of the History of Computing*, *31*(3), 5–19. doi:10.1109/MAHC.2009.53

Messinger, P. R., Stroulia, E., Lyons, K., Bone, M., Niu, R. H., Smirnov, K., & Perelgut, S. (2009). Virtual worlds – past, present, and future: New directions in social computing. *Decision Support Systems*, *47*(3), 204–228. doi:10.1016/j.dss.2009.02.014

Montfort, N., & Bogost, I. (2009). Random and Raster: Display technologies and the development of videogames. *IEEE Annals of the History of Computing*, *31*(3), 34–43. doi:10.1109/MAHC.2009.50

Nice, K., Wilson, T. V., & Gurevich, G. (2009). How digital cameras work. *HowStuffWorks*. Retrieved October 15, 2009, from http://electronics.howstuffworks.com/digital-camera.htm

Pohlmann, K. (2005). *Principles of Digital Audio*. Columbus, OH: McGraw-Hill/TAB Electronics.

Rainer, K., & Turban, E. (2008). *Introduction to Information Systems: Supporting and Transforming Business*. Hoboken, NJ: Wiley.

Reid, K., & Dueck, R. (2007). *Introduction to Digital Electronics*. Boston, MA: Delmar Cengage Learning.

Reynolds, G. (2009). *Information Technology for Managers*. Boston, MA: Course Technology.

Schaefermeyer, S. (2007). *Digital Video Basics*. Boston, MA: Course Technology.

Snyder, L. (2010). *Fluency with Information Technology: Skills, Concepts, and Capabilities*. Upper Saddle River, NJ: Prentice Hall.

Stair, R., & Reynolds, G. (2009). *Principles of Information Systems*. Boston, MA: Course Technology.

Tocci, R., Widmer, N., & Moss, G. (2006). *Digital Systems: Principles and Applications*. Upper Saddle River, NJ: Prentice Hall.

Turban, E., & Volonino, L. (2009). *Information Technology for Management: Improving Performance in the Digital Economy*. Hoboken, NJ: Wiley.

Valacich, J., & Schneider, C. (2009). *Information Systems Today: Managing the Digital World*. Upper Saddle River, NJ: Prentice Hall.

Westcott, S., & Westcott, J. R. (2008). *Digitally Daunted: The Consumer's Guide to Taking Control of the Technology in Your Life*. Washington, DC: Capital Books.

White, R., & Downs, T. E. (2007). *How Computers Work*. Toronto: Que.

Williams, B., & Sawyer, S. (2009). *Using Information Technology*. Career Education.

Woodford, C. (2006). *Digital Technology*. London: Evans Brothers Ltd.

## KEY TERMS AND DEFINITONS

**Analog Representation:** A method where information can be transmitted as a continuous wave signal that is varied by changing the wave's amplitude and frequency. For example, sound is an analog signal that is transmitted by waves.

**American National Standard Code for Information Interchange (ASCII):** A coding scheme for representing characters in a digital computer. It is commonly used in microcomputers.

**Binary Number Base:** The base 2 number system that represents values using the digits 0 and 1. It is commonly used to represent data stored and processed in digital computers because of its relationship with the electromagnetic on-off switches used in digital technology.

**Digital Representation:** A method where information can be stored and transmitted using discrete on-off values to represent characters, numbers, images, sound, and moving images.

**Graphics Interchange Format (GIF):** An image format that uses 8-bit color representations to display up to 256 unique colors.

**Hexadecimal Number Base:** The base 16 number system that represents values using the sixteen unique digits 0-9 and A-F. It is commonly used as a simplified representation of binary values because it can represent four binary digits in one hexadecimal digit.

**Joint Photographic Experts Group (JPEG):** An image format that uses 24-bit color representations to display up to approximately 16.7 million unique colors.

**Moving Pictures Experts Group (MPEG):** A common format for storing and playing videos on digital devices.

**Uniform Resource Locator (URL):** The addressing scheme used to uniquely identify locations on the World Wide Web. For example, www.drake.edu is a URL used for Drake University's website.

# Chapter 2
# DRM Protection Technologies

**Gary Hackbarth**
*Northern Kentucky University, USA*

## ABSTRACT

*Digital Rights Management (DRM) is concerned with the ownership of digital information and access to that information. Organizations and individuals increasingly seek to prevent unauthorized or inadvertent release of owned, proprietary, or sensitive information. A variety of technologies are available to prevent the piracy and verify the true owners of digital content, unfortunately specifics of these technologies are often proprietary. Content can be protected by a variety of encryption techniques for the storage and transmission of digital information yet; these same techniques can limit access and usability of digital content. This chapter discusses the general state of digital security and technologies in use followed by a discussion of future directions for digital security research and practice.*

## INTRODUCTION

Information or data security is the general means of protecting information and information systems from unauthorized access, use, disclosure, disruption, modification or destruction. Organizations develop security policies to clearly articulate the specific rights and responsibilities of individual users, and to communicate these rights successfully to each employee so that there is an effective approach to information security across the organization (Doherty et al., 2009). The terms information security, computer security and information assurance are used interchangeably by the public because they share the common goals of protecting the confidentiality, integrity and availability of information. More specifically, (1) Information security is concerned with the confidentiality, integrity and availability of data regardless of the form the data may take: electronic, print, or other forms; (2) Computer security can focus on ensuring the availability and correct operation of a computer system without concern for the information stored or processed by the

DOI: 10.4018/978-1-61692-877-3.ch002

computer; and (3) Information Assurance (IA) is the practice of managing information-related risks (Alexei, 2006; D'Aubeterre, et al., 2008; Jean-Noel, et al., 2007).

It is important to understand that these fields overlap in that they confront the same issues but use different methodologies and techniques to address security issues from a different perspective. More specifically, a related issue to all three sub-disciplines is the issue of Digital Rights Management (DRM). DRM is a generic term for access control technologies that can be used by hardware manufacturers, publishers, copyright holders and individuals to try to impose limitations on the usage of digital content and devices (Fetscherin, (2002). The term is used to describe any technology which inhibits uses (legitimate or otherwise) of digital content that were not desired or foreseen by the content provider.

In reality, for most people, digital security just exists. There exists an expectation that digital content will arrive at a computer, be delivered to their TV, or heard on their radio. There is an expectation that personal information will be protected by financial institutions, business entities, educational institutions, the government, and others trusted with the responsibility of guarding intellectual and personal information. There is little thought given to the hackers who seek to capture and reuse digital content for their own profit. For most of us, safeguards that protect or help deliver digital content are something that happens in the background. Furthermore, many users assume the digital content downloaded to their personal device (TV, iPod, Computer, etc,) is free. The reasons for this assumption are complex but relevant to businesses trying to develop pricing schemes/strategies and product delivery models needed to sell digital content.

When we hear about identity theft, digital piracy, illegal copying of songs or other instances of digital abuse of protected information, we consider it an issue for law enforcement, security experts, or the businesses and people involved. Many IT professionals feel the same way. Digital security is about complicated algorithms, high-tech hardware, and complex communication configurations. The purpose of this chapter is not to convince you that digital security is important but rather to inform and instruct readers about the available technologies and issues required in managing digital rights.

## BACKGROUND

Digital media is replacing analog media as the primary technique in the way data or information is stored, transmitted, and used. The advantage of traditional analog information or other forms of traditional informational content (books, taped video, microfilm, records, etc.) is that it is relatively difficult and expensive to create high quality copies of the original materials. To this extent, traditional copyright law worked (Bates, 2008). As media shifted toward digital formats, the cost of reproduction declined and the capability to create exact high quality duplicates evolved. Under these circumstances, the protections given to authors under traditional copyright law begin to breakdown.

Copyright in the context of this chapter is intellectual property that gives the author of that intellectual property, exclusive right for a certain period of time to control publication, distribution, and adaptation of the original work, after which time the original work is released to enter the public domain (Crane, 2009). In general, copyright law applies to any expressible form of an idea or information that is substantial, is discreet in that it has a beginning and an end, and is complete in some final form. Complicating copyright law is that while there are some international standards, copyright law does vary by country. Internationally, copyright standards exist for the author between 50 and 100 years from the author's death or for a shorter period of time. Further, some international jurisdictions require administrative action to establish copyright, but most countries recognize any completed work. In general, copyright is a civil matter, although, copyright infringement may be a criminal action in some cases (Bates, 2008).

The key point is that in a digital society, it is relatively easy and inexpensive to make exact high quality copies of original intellectual property that exists or is transmitted in a digital format. The ease by which digital material can be copied has created opportunities for those who would pirate intellectual property. Piracy is the concept of copying digital material and using it or selling it without paying for it. Encryption has long been used to compress digital material into more efficient formats for transmission and storage but it has also been used to protect intellectual property from those who would pirate it. Hackers are those individuals who find ways to illegally decrypt digital information with the intent to use or sell the digital information. Copy protection helps prevent piracy by either making it impossible to duplicate a piece of content, or by inserting a watermark or other unique identifier, such as a digital fingerprint, that allows copyright holders to track down pirates (Glen, 2008).

## Protection Technologies

The protection of intellectual property and personal communication has long been an important consideration of the business community as well as the artistic community. It is not the protection of personal communications so much but rather being able to identify one's own intellectual property in digital form. In today's modern society, the growing use of digital media has made the security of digital media files of utmost concern against those users with malevolent intentions, especially on the Internet. In this context, security takes on a broader meaning. Intellectual property should be shared rather than being isolated and hidden away. The issue is to make intellectual property available yet be able to identify the originator so that proper acknowledgment and begin the originator.

To protect digital media files, industry and academic researchers propose and improve upon many data-hiding algorithms, which are known as steganographic algorithms, watermarking algorithms, and other data-embedding algorithms. There is much interest in these technologies. For instance, it is thought that the terrorists who planned the bombing of the World Trade Center in September 2001 used steganographic techniques to plan the bombing (Schmurr & Crawley, 2003). Steganography is meant to be a secure form of communication that hides objects in multimedia files and should not to be confused with watermarking techniques even though their main objectives are similar. Steganographic techniques have been used concurrently with anonymous remailers that strip off the headers of e-mail messages to hide the identity of the sender as well as padding the message with random data to conceal the true size of the transmitted file (Schmurr & Crawley, 2003).

Watermarking techniques are used to protect improper reproduction of digital media by protecting the integrity and copyright of images (Chang & Lin, 2008). Whereas steganography is used to transfer the hidden encrypted data to intended parties thereby preventing third parties from noticing the presence of hidden data and without anyone else except the intended recipient of the communication being able to decode the content of the hidden data. Both techniques are opposite sides in the battle between law enforcement and those individuals practicing cyber-terrorism and cyber-crime. On one hand you are trying to protect people and digital property, while on the other hand, you are trying to conceal illegal and harmful activities.

Industry uses two types of watermarking techniques: visible (perceptible) and invisible (imperceptible) (Samtani, 2009). Visible watermarking techniques are similar to the background printing of "DRAFT" in word processing documents when submitting documents for publication. One could also print the image of "Copyright protected" on case studies downloaded for instructor review prior to purchase. Watermarking techniques are used to protect multimedia contents over Internet trading so that ownership of the contents can be determined in copyright disputes (Cheung, Chiu, & Ho, 2008). Once the case studies for purchase are complete, then the visible watermark can be removed or a file without the watermark can be downloaded. There still could be an invisible watermark hidden in the document confirming the origin of the document and its successful purchase. Kodak (www.kodak.com) through its expertise in materials science and digital imaging provides an extensive portfolio of products and services designed to help companies protect against the growing problem of counterfeiting and piracy (Anonymous, 2008b).

Watermarking is not just for documents. Watermarking can be used for videos and other types of graphic material but, there are limitations. According to recent research from the United States, most video watermarking algorithms embed the watermark in I-frames, but refrain from embedding in P- and B-frames, which are highly compressed by motion compensation (Anonymous, 2008a). The problem is being able to find room within a complex algorithm to embed the watermark that does not interfere with the algorithm.

This sounds all well and nice but doesn't seem to be something practical that everyday people should be concerned with. As an example, witnesses see a crime or view an accident and take pictures to prove and document what they saw. Currently, digital camera images are not readily accepted as evidence because it is difficult for law enforcement, insurance, news, and other such agencies to authenticate the integrity, origin, and authorship of digital pictures (Blythe, 2005). The integrity of the evidence rests depends on the ability of the lawyer or witness to prove who did what when. Proof must exist and be verifiable that the digital camera image was not damaged or tampered with. Further, there must be confirmation when the picture was taken, what camera took the digital image, and who the digital photographer was (Blythe, 2005). One possible solution is to embed a biometric identifier (the photographer's iris), with cryptographic hashes, and other forensic data, concurrently into the original scene image (Blythe, 2005).

This problem affects the television industry. A variety of graphical content is now available on television or downloaded through the Internet. Movies, concerts, movies made-for-TV, home movies, etc. can be downloaded wirelessly or through cable to your TV or computer. As this diverse television content migrates to an increasing number of digital platforms like iPods, MP3 players, and the light, the need for content protection becomes paramount and it's an issue that will only get more important as high-value content continues to make its way onto the Web. Two major types of cross-platform content protection technology currently exist and are used together. Conditional access, or scrambling technology, blocks unauthorized viewers from watching premium content like HBO, pay-for-view TV etc (Glen, 2008).

Conditional Access (CA) protects content by requiring certain criteria to be met before granting access to premium content. CA is commonly used in relation to digital satellite television systems. Under the Digital Video Broadcasting Project (DVB), conditional access system standards are defined in the specification documents for DVB-CA (Conditional Access), DVB-CSA (the Common Scrambling Algorithm) and DVB-CI (the Common Interface). There are also standards for DVB-S2 for satellite networks, DVB-C2 for cable networks and DVB-T2 for terrestrial networks. In addition to these, a range of supporting standards exists such as service information (DVB-SI), subtitling (DVB-SUB), interfacing (e.g. DVB-ASI), etc. Further, Interactive TV, one of the key advances enabled by the switch from analogue to digital, required the creation of a set of return channel standards and the Multimedia

Home Platform (MHP), DVB's open middleware specification (www.dvb.org).

Currently, DVB embraces a broader network convergence by developing standards for innovative technologies that allow the delivery of services over fixed and wireless telecommunications networks (e.g. DVB-H and DVB-SH for mobile TV), content protection and copy management (DVB-CPCM), and looking at developing standards for IPTV, Internet TV and Home Networks. DVB is dedicated to constant innovation to keep up with both technological developments and market requirements.

The Digital Video Broadcasting Project (DVB) is an industry-led consortium of over 250 broadcasters, manufacturers, network operators, software developers, regulatory bodies and others in over 35 countries committed to designing open technical standards for the global delivery of more than 100 million DVB receivers for digital television and data services (www.dvb.org). These standards define an encryption scheme by which a digital television stream can be provided to those with valid decryption smart cards. This is achieved by a combination of scrambling and encryption where the data stream is scrambled with an 8-byte secret key, called the *control word*. Knowing the value of the control word at a given moment is of relatively little value, because it changes every few seconds and must be known slightly in advance of the data stream to prevent viewing interruption. The control word is generated automatically and randomly by a physical process, in such a way that successive values are not predictable. Encryption is used to protect the control word as an *entitlement control message* (ECM) during transmission to the receiver. The CA subsystem in the receiver will decrypt the control word when authorized to do so in the form of an *entitlement management message* (EMM). EMMs are specific to each subscriber smart card in each receiver and are issued about every 12 minutes or less. The contents of ECMs and EMMs are not standardized as they depend on the conditional access system being used. Several companies provide competing CA systems such as VideoGuard, Irdeto Access, Nagravision, Conax, Viaccess, Latens, Verimatrix and Mediaguard (also known as SECA) (Project, 2009).

The system is designed so that the control word can be transmitted through different ECMs at once allowing several conditional access systems access at the same time using a feature called *simulcrypt*, which saves bandwidth and encourages multiplex operators to cooperate. DVB Simulcrypt is widespread in Europe; some channels, like the CNN International Europe from the Hot Bird satellites, can use seven different CA systems in parallel.

There is a difference between Content Protection and Copy Management (CPCM) and CA. CPCM is concerned with content after it has been acquired and is separate from the CA or DRM (Digital Rights Management) systems that protected the content on its way to the consumer. The fundamental CPCM boundaries are the local environment and the Authorized Domain (AD). The AD is defined as a distinguishable set of DVB-CPCM compliant devices, which are owned, rented or otherwise controlled by each user. Each user specifies in their Usage State Information (USI) contract how the content is to be used. This concept is fundamentally different to today's CA and DRM techniques, which normally operate on a single device basis. Current mechanisms for protecting content within the home environment do not reflect the desires of consumer who use multifaceted devices in networked homes to download data so that the idea is that different devices exist to protect information rather than having one decryption device do it all (Project, 2009).

## Issues, Controversies, Problems

The primary problem of interest concerning digital media is economic. The information industry sells information content to users. They use cable, wireless, telephone lines, and other channels to deliver content to these same users. Users pay for the content as well as the secret decryption code or hardware device that decrypts the information prior to its use. This process can be timely, inconvenient, or costly to users. Thus, hackers and media pirates have incentive to circumvent communication channels, encryption codes, hardware devices, etc. to provide free or lower-cost content to users. The interesting part of this story is that the sellers of digital materials influence the degree of piracy.

The idea is that the more encryption that is imposed by the seller of digital media actually reduces or fails to create a viable market for the product. On one hand, software and hardware encryption technologies are expensive to build, maintain, and update and are costly to the consumer. The more costly the protection technologies are, the more incentive for the hacker or the video pirate to overcome these technologies. At some point, it may become prohibitively expensive to build, maintain and update these technologies. This is based on the assumption that hackers can ultimately circumvent any encryption (Dejean, 2009).

Another perspective is that as the quality of pirated goods increases, the seller of digital materials must lower prices and reduce profits as the total amount of informational product grows. The idea is that as price goes down there will be more buyers. As hackers make more products available, users will elect a less costly alternative because they would then be legal and have access to customer service. At this point the seller makes the most profit (Arun, 2004). That is, there is equilibrium point where spending to protect information versus the number of users is either too much or too little. If the digital rights management technologies weaken over time due to the underlying technology being weakened, the seller's choice is either to increase or decrease the level of technology based protection. The seller must consider the cost of effective protection technologies versus the value of the product in terms of profitability in order to align the effectiveness of the solution in preserving the ballot of the information provided to the legal users. For corporate customers the value of the information is a determining factor in growing profits. It is this reality that sellers of digital information must understand the technologies they're using both in terms of cost and user accessibility before they respond to threats from hackers were media pirates (Arun, 2004).

## SOLUTIONS, RECOMMENDATIONS, AND FUTURE DIRECTIONS

Grappling with how much to expand protecting your digital information is a complex issue. Vendors are hesitant to release details of encryption, watermarking, and other protective applications for fear of giving away trade secrets to hackers and pirates. Companies like Digimarc, Inc. (http://www.digimarc.com) provide general descriptions of the technologies they use but fail to provide real specifics. Even standards organizations are hesitant to discuss developing standards until they are agreed upon in the technology is mature and ranting information providers a window to optimize profits before actors force changes or adaptations to the technology. There have been some clues suggesting the direction that some researchers are following.

How to implement more successful watermarking techniques is a fruitful area of research. Some researchers have proposed the addition of computer code for which the topology of the control-flow graph encodes the watermark (Colberg, et al., 2009). Cheung et al. (2008) makes use of intelligenct user certificates to embed the identity of the users into the intelligence documents that are distributed. In particular, keeping the identity secrecy between document providers and users (but yet traceable upon disputes) is a key contribution of this protocol in order to support for intelligence applications (Cheung et al., 2008). El-Affendi (2008) suggests it is possible to design and build simple light watermarking protocols for placing authentic hand signatures and stamps on remote documents that is secure and reliable (El-Affendi, 2008). Another interesting process is the two-phase watermarking scheme extracts both the grayscale watermark and the binary one from the protected images to achieve the copyright protection goal (Ming-Chiang, et al., 2007).

Researchers looking at difficult three dimensional (3D) computer graphics models and digitally-controlled manufacturing have come together to enable the design, visualization, simulation, and automated creation of complex 3D objects. Aliaga and Atallah (2009) have proposed a framework for the designing and manufacturing of computer graphics objects such that no hacker or pirate can make imitations or counterfeit copies of the physical object, even if the adversary has a large number of original copies of the object, knowledge of the original object design, and has manufacturing precision that is comparable to or superior to that of the legitimate creator of the object (Aliaga & Atallah, 2009). The approach is to design and embed a signature on the surface of the object which acts as a certificate of genuinity of the object which is only detectable by a signature-reading device which contains some of the secret information that was used when marking the physical object. Further, the compromise of a signature-reading device by an adversary who is able to extract all its secrets, does not enable the adversary to create counterfeit objects that fool other readers, thereby still enabling reliable copy detection (Aliaga & Atallah, 2009).

In the United States, cable system operators use Cable Cards whose specifications were developed by the cable company consortium (CableLabs, http://www.cablelabs.com) and are mandated by the Federal Communications Commission but standards exist only for one way communication and do not apply to satellite television. The next generation approach is to eliminate physical cards develop hardware where downloadable software for conditional access is the norm. The main appeal of the software approach is that the access control may be upgraded dynamically in response to security breaches without requiring expensive exchanges of hardware. Another appeal is that it may be inexpensively incorporated into non-traditional media display devices such as Portable media players. Additionally, biometrics and Public key Infrastructure techniques may be used to identify and authenticate individuals accurately within a secure communication and transactional environment (Wagner, 2007).

Teaching ethics is another approach to protecting intellectual property. Unfortunately, ethics and morality differ widely across cultures and countries. This is proven to be largely ineffective. It had been thought that the public that used hacked or pirated digital media was costing jobs from those engaged in creating intellectual property (Ram & Sanders, 2000). Yet, many users felt slighted or overwhelmed by the high cost of intellectual property access. Thus they turned to lower-cost alternatives. Public libraries, universities, schools, and local governments pay for public access to digital information that further distorts how the creators of intellectual property intend to share, transmit, and sell their digital property (Dörte, 2008).

Times may be changing. The concept of digital rights management which this article has described as those technologies that control the copying and use of digital media has in fact infuriated consumers since its inception in the mid-1990s (Tekla, 2007). Consumer advocacy groups argue that preventing access to digital content not only prevents illegal use but legal use as well. The customary argument from the record and movie industries is that protection of an artists' intellectual property is vital to their creative efforts. Illegal coping of movies reduces income from that movie and prevents those in involved in the creation, distribution, and viewing of those movies from receiving the full compensation for their efforts. In fact, unlocking or circumventing digital locks on digital media is illegal and is enforceable through the 1998 passage of the Digital

Millennium Copyright Act, even if no copyright violation followed. Yet, DRM has been removed from some content.

Consumers who pay for content or download free content feel they have the right to shift content from one digital platform to another. Some business models are experimenting with alternate revenue models. Users can download movies and TV shows from websites like Hulu (http://www.hulu.com) and view them on iPods, computers and their TV's. Consumers are enticed with limited access which stimulates their desire to pay to see the TV or movie again in a higher quality definition or given access to previous episodes for a small fee. As these new business models evolve we may or may not see changes in DRM.

## FUTURE RESEARCH DIRECTIONS

Digital products take many forms. Newspapers, magazines, and book publishers have text and image content. Music and movie companies have multimedia files. And game and software companies have source code and an executable software product. Each of these products has unique characteristics, but the commonality is that they are all in a digital format which requires unique protection schemes. Future research on protecting digital products will involve an interdisciplinary approach involving the best solutions from scientists and engineers, business managers, and social scientists. Digital asset protection technologies continue to improve and there will be a continual battle to stay ahead of people who work to find ways around technical protections. Business management researchers will continue to work on answering questions about why people use digital products without authorization or without paying for it. Some relevant questions include: (1) What factors impact the behavior of individuals when they use digital products without authorization? (2) What can be done to change this behavior? and (3) What business strategies and models can be used to reduce this problem? These issues can also be addressed by the broader social sciences to look at the psychology, sociology, economics, and ethics issues involved in digital products and their protection. These are important issues for the future because the number of digital products and their impact on the global economy continues to increase.

## CONCLUSION

Protection of intellectual property can be a complex topic. We have avoided in this chapter article any discussion of the legal debate concerning intellectual property (Herman, 2008). This legal debate is a topic of its own and is global in its depth and breadth of content. And, there is the issue of the technology battle between those protecting intellectual property and those seeking to hack both the content and the means of communication for that content.

At a high level of understanding it is easy to understand that there is a trade-off between how much we spend to protect digital information versus accessibility to digital information. Companies and organizations who profit from digital information seek the highest profit margin based on the largest number of users of their content. Users seek digital information for free or at the lowest cost. One must remember that intellectual property has a single supplier. The creator of the intellectual property can sell to multiple buyers directly or for redistribution. It is control of the value chain from the creator to the user that is being fought over.

It is important to understand the process of delivering digital media and economic considerations of the parties involved more than the actual technologies involved for the typical reader. Individual organizations must calculate return on investment (ROI) based on the cost of digital protection versus the number of subscribers of their content (Miller & Wells, 2007). Government involvement in mandating standardization of hardware to protect consumers from high switching costs further complicates the issues for profit making firms but makes it easier for the user. The user must recognize that intellectual property is not free because if no one pays for it then who would invest the time and effort to create it (Cronan & Al-Rafee, 2008). Plus, users must have expectation of paying something for access to digital materials but at not too high a cost (Hoffman, 2009).

## REFERENCES

Alexei, N. (2006). Information assurance seals: How they impact consumer purchasing behavior. *Journal of Information Systems, 20*(1), 1. doi:10.2308/jis.2006.20.1.1

Aliaga, D., & Atallah, M. (2009). Genuinity signatures: Designing signatures for verifying 3D object genuinity. *Computer Graphics Forum, 28*(2), 437. doi:10.1111/j.1467-8659.2009.01383.x

Anonymous (2008a). Information forensics and security; New findings in information forensics and security described by M. Noorkami and co-researchers. *Computers, Networks & Communications*, 602.

Anonymous (2008b). Kodak; Kodak calls on businesses and industry to join the fight against counterfeiting. *Technology & Business Journal*, 59.

Arun, S. (2004). Managing digital piracy: Pricing and protection. *Information Systems Research, 15*(3), 287. doi:10.1287/isre.1040.0030

Bates, B. (2008). Commentary: Value and digital rights management - A social economics approach. *Journal of Media Economics, 21*(1), 53. doi:10.1080/08997760701806850

Blythe, P. A., Sr. (2005). *Biometric authentication system for secure digital cameras.* Unpublished Ph.D., State University of New York at Binghamton, United States -- New York.

Chang, C., & Lin, P. (2008). Adaptive watermark mechanism for rightful ownership protection. *Journal of Systems and Software, 81*(7), 1118. doi:10.1016/j.jss.2007.07.036

Cheung, S., Chiu, D., & Ho, C. (2008). The use of digital watermarking for intelligence multimedia document distribution. *Journal of Theoretical and Applied Electronic Commerce Research, 3*(3), 103. doi:10.4067/S0718-18762008000200008

Collberg, C., Huntwork, A., Carter, E., Townsend, G., & Stepp, M. (2009). More on graph theoretic software watermarks: Implementation, analysis, and attacks. *Information and Software Technology, 51*(1), 56. doi:10.1016/j.infsof.2008.09.016

Crane, D. (2009). Intellectual Liability. *Texas Law Review, 88*(2), 253.

Cronan, T., & Al-Rafee, S. (2008). Factors that influence the intention to pirate software and media. *Journal of Business Ethics, 78*(4), 527. doi:10.1007/s10551-007-9366-8

D'Aubeterre, F., Singh, R., & Iyer, L. (2008). Secure activity resource coordination: Empirical evidence of enhanced security awareness in designing secure business processes. *European Journal of Information Systems, 17*(5), 528. doi:10.1057/ejis.2008.42

Dejean, S. (2009). What can we learn from empirical studies about piracy? *CESifo Economic Studies, 55*(2), 326. doi:10.1093/cesifo/ifp006

Doherty, N. F., Anastasakis, L., & Fulford, H. (2009). The information security policy unpacked: A critical study of the content of university policies. *International Journal of Information Management, 29*(6), 449–457. doi:10.1016/j.ijinfomgt.2009.05.003

Dörte, B. (2008). Digital rights description as part of digital rights management: A challenge for libraries. *Library Hi Tech, 26*(4), 598. doi:10.1108/07378830810920923

El-Affendi, M. A. (2008). Completing the circuit in e-government process automation. *Business Process Management Journal, 14*(1), 96. doi:10.1108/14637150810849436

Fetscherin, M. (2002). Present state and emerging scenarios of digital rights management systems. *International Journal on Media Management, 4*(3), 164–171.

Glen, D. (2008). Safer digital information. *Broadcasting & Cable, 138*(27), 16.

Herman, B. (2008). Breaking and entering my own computer: The contest of copyright metaphors. *Communication Law and Policy, 13*(2), 231. doi:10.1080/10811680801941276

Hoffman, L. (2009). Content control. *Communications of the ACM, 52*(6), 16. doi:10.1145/1516046.1516052

Jean-Noël, E., Elspeth, M., & David, B. (2007). Mastering the art of corroboration. *Journal of Enterprise Information Management, 20*(1), 96.

Miller, C., & Wells, F. (2007). Balancing security and privacy in the digital workplace. *Journal of Change Management, 7*(3/4), 315. doi:10.1080/14697010701779181

Ming-Chiang, H., Der-Chyuan, L., & Ming-Chang, C. (2007). Dual-wrapped digital watermarking scheme for image copyright protection. *Computers & Security, 26*(4), 319. doi:10.1016/j.cose.2006.11.007

Project, D. V. B. (2009). DVB Fact Sheet – April 2009.

Ram, D. G., & Sanders, G. L. (2000). Global software piracy: You can't get blood out of a turnip. *Communications of the ACM, 43*(9), 82. doi:10.1145/348941.349002

Samtani, R. (2009, March). Ongoing innovation in digital watermarking. *Computer*, 111–113.

Schmurr, A., & Crawley. (2003). Cybercrime in the United States criminal justice system: Cryptography and steganography as tools of terrorism. *Journal of Security Administration, 26*(2), 51–76.

Tekla, S. P. (2007). Imagine there's no DRM... I wonder if you can. *IEEE Spectrum, 44*(7), 14. doi:10.1109/MSPEC.2007.4286549

Wagner, D. (2007). A comprehensive approach to security. *Sloan Management Review, 48*(4), 8.

## ADDITIONAL READING

Bhatt, S., Sion, R., & Carbunar, B. (2009). A personal mobile DRM manager for smartphones. *Computers & Security, 28*(6), 327–340. doi:10.1016/j.cose.2009.03.001

Bouganim, L., & Pucheral, P. (2007). Fairness concerns in digital right management models. *International Journal of Internet & Enterprise Management, 5*(1), 4.

Bradbury, D. (2007). Decoding digital rights management. *Computers & Security, 26*(1), 31–33. doi:10.1016/j.cose.2006.12.006

Burkart, P. (2008). Trends in digital music archiving. *The Information Society, 24*(4), 246–250. doi:10.1080/01972240802191621

Drossos, L., Tsolis, D., Sioutas, S., & Papatheodorou, T. (2008). *Digital Rights Management for E-Commerce Systems*. Hershey, PA: Information Science Reference.

Eckhardt, J., Lundborg, M., & Schlipp, C. (2007). Digital rights management (DRM) and the development of mobile content in Europe. *Computer Law & Security Report, 23*(6), 543–549. doi:10.1016/j.clsr.2007.09.008

García, R., & Gil, R. (2008). A Web ontology for copyright contract management. *International Journal of Electronic Commerce, 12*(4), 99–113. doi:10.2753/JEC1086-4415120404

Garcia, R., & Pariente, T. (2009). Interoperability of learning objects copyright in the LUISA semantic learning management system. *Information Systems Management, 26*(3), 252–261. doi:10.1080/10580530903018037

George, C., & Chandak, N. (2006). Issues and challenges in securing interoperability of DRM systems in the digital music market. *International Review of Law Computers & Technology, 20*(3), 271–285. doi:10.1080/13600860600852143

Harte, L. (2006). *Introduction to Digital Rights Management (DRM); Identifying, Tracking, Authorizing and Restricting Access to Digital Media.* Fuquay Varina, NC: Althos.

Hidalgo, A., Albors, J., & Lopez, V. (2009). Design and development challenges for an E2E DRM content integration business platform. *International Journal of Information Management, 29*(5), 389–396. doi:10.1016/j.ijinfomgt.2008.11.007

Jaisingh, J. (2007). Piracy on file-sharing networks: Strategies for recording companies. *Journal of Organizational Computing and Electronic Commerce, 17*(4), 329–348.

Korba, L., Song, R., & Yee, G. (2007). Privacy rights management: Implementation scenarios. *Information Resources Management Journal, 20*(1), 14–27.

Lee, S., Kim, J., & Hong, S. (2009). Redistributing time-based rights between consumer devices for content sharing in DRM system. *International Journal of Information Security, 8*(4), 263–273. doi:10.1007/s10207-009-0082-5

Lee, W.-B., Wu, W.-J., & Chang, C.-Y. (2007). A portable DRM scheme using smart cards. *Journal of Organizational Computing and Electronic Commerce, 17*(3), 247–258.

Leszczuk, M. (2007). Multimedia security technologies for digital rights management. *IEEE Communications Magazine, 45*(10), 16.

Lian, S. (2008). Digital rights management for the home TV based on scalable video coding. *IEEE Transactions on Consumer Electronics, 54*(3), 1287–1293. doi:10.1109/TCE.2008.4637619

Mohanty, S. P. (2009). A secure digital camera architecture for integrated real-time digital rights management. *Journal of Systems Architecture, 55*(10-12), 468–480. doi:10.1016/j.sysarc.2009.09.005

Muhlbauer, A., Safavi-Naini, R., Salim, F., Sheppard, N. P., & Surminen, M. (2008). Location constraints in digital rights management. *Computer Communications, 31*(6), 1173–1180.

Nishimoto, Y., Baba, A., Kimura, T., Imaizumi, H., & Fujita, Y. (2007). Advanced conditional access system for digital broadcasting receivers using metadata. *IEEE Transactions on Broadcasting, 53*(3), 697–702. doi:10.1109/TBC.2007.896972

Nishimoto, Y., Imaizumi, H., & Mita, N. (2009). Integrated digital rights management for mobile IPTV using broadcasting and communications. *IEEE Transactions on Broadcasting, 55*(2), 419–424. doi:10.1109/TBC.2009.2016496

Parthasarathy, A. K., & Kak, S. (2007). An improved method of content based image watermarking. *IEEE Transactions on Broadcasting, 53*(2), 468–479. doi:10.1109/TBC.2007.894947

Postigo, H. (2008). Capturing fair use for the Youtube generation: The digital rights movement, the Electronic Frontier Foundation and the user-centered framing of fair use. *Information Communication and Society, 11*(7), 1008–1027. doi:10.1080/13691180802109071

Samuel, S. (2009). *World of Watermarking: Digital Rights Management for JPEG Images.* Saarbrucken, Germany: VDM Verlag Dr. Müller.

Umeh, J. C. (2008). *The World beyond Digital Rights Management*. British Informatics Society Ltd.

Van Tassel, J. (2006). *Digital Rights Management: Protecting and Monetizing Content*. Burlington, MA: Focal Press.

Waterman, D., Sung, W. J., & Rochet, L. R. (2007). Enforcement and control of piracy, copying, and sharing in the movie industry. *Review of Industrial Organization*, *30*(4), 255–289. doi:10.1007/s11151-007-9136-x

Zeng, W., Yu, H., & Lin, C.-Y. (2006). *Multimedia Security Technologies for Digital Rights Management*. Burlington, MA: Academic Press.

## KEY TERMS AND DEFINITIONS

**Conditional Access:** Protects content by requiring certain criteria to be met before granting access to premium content in systems such as digital satellite television.

**Digital Millennium Copyright Act (DMCA):** A US copyright law, enacted in 1998, that criminalizes technologies or services that are meant to circumvent digital content protections such as digital rights management systems.

**Digital Rights Management (DRM):** Technologies used to control access to digital assets.

**Digital Video Broadcasting Project (DVB):** An industry-led consortium of over 250 organizations committed to designing open technical standards for digital television and data services.

**Digital Watermarking:** A security method that embeds unique identifying data in a multimedia file. For example, a digital watermark could be added to an image file to indicate who owns the copyright.

**Steganography:** A method used to protect secret digital communications by hiding objects inside multimedia files.

# Section 2
# Business Functions

# Chapter 3
# Legal Issues Facing Companies with Products in a Digital Format

**J. Royce Fichtner**
*Drake University, USA*

**Lou Ann Simpson**
*Drake University, USA*

## ABSTRACT

*Companies that deal in products in a digital format, such as magazines, newspapers, e-books, music, movies, games, or software, face unique legal challenges because they attempt to earn a profit by selling or licensing material that is easily copied and inexpensive to reproduce. This chapter discusses the four general categories of intellectual property law—patents, trade secrets, trademarks, and copyrights—and describes how each applies to products in a digital format. This chapter ends with a brief discussion of the changing societal norms toward copyright infringement for digital products and possible directions for future research.*

## INTRODUCTION

Digital products are content, multimedia, or software products that are in a digital format when possession is passed to the consumer. Examples include newspaper content, magazine content, e-books, music, movies, games, and all types of software. If the one who creates or develops a digital product cannot profit from his or her work by selling the product to someone else, then there is little incentive to invest the time and resources to develop new creative works. The framers of the

DOI: 10.4018/978-1-61692-877-3.ch003

U.S. Constitution recognized the importance of promoting this creative and expressive process and gave Congress the power to "promote the Progress of Science and useful Arts, by securing for limited Times to Authors and Inventors the exclusive right to their respective writings and discoveries" (U.S. Constitution, Art. I, § 8). The U.S. Congress and the individual state governments have used this power to create a series of intellectual property laws to protect inventive and artistic creativity.

Beyond the protections afforded by law, there has always been a second, more practical constraint on the unauthorized reproduction and distribution of creative works. This second constraint can be loosely classified as the technological and mechanical impediments to the physical process of reproducing the creative work (Menell, 2002). Stated more succinctly, unauthorized reproductions are encumbered because it is difficult to produce an identical copy of the original work in an economical fashion. However, recent advances in digital technology have virtually eliminated this constraint for many creative works.

Prior to the late 1980s, many creative works were reproduced using analog technologies. Analog technologies reproduce a creative work by deforming a physical object (such as paper or film) in such a manner to convey an image, audio frequency, or light intensity (Menell, 2002). Analog technology platforms impede unauthorized reproduction and distribution because the second comer must copy the creative work from an existing copy (a process which inevitably results in a work of lesser quality) and reproduce the copied work onto another physical object (which must be purchased or created). Digital technology eliminates both the cost of purchasing the physical media and the innate quality degradations that arise when copying from a physical media. By encoding creative works in binary form, digital computers allow for perfect reproductions across unlimited generations of reproductions (Menell, 2002). Also, because the work may be transmitted without transferring it to a physical media, there is little or no additional cost for each additional copy and such copies can be distributed without cost via the Internet.

Because there are no technological or mechanical impediments to the unauthorized reproduction and distribution of creative works which have been converted into digital products, there is increased pressure to limit such unauthorized reproductions through legal constraints. This chapter will discuss the existing legal constraints by introducing the four categories of intellectual property law—patents, trade secrets, trademarks, and copyrights. This chapter will describe each category of intellectual property law and describe how these laws apply to products in a digital format. This chapter also includes a brief discussion of the changing societal norms toward copyright infringement of digital products and possible directions for future research.

## Patents

A patent is a grant from the U.S. federal government that gives the inventor the exclusive right to make, use, or sell an invention for a limited period of time. A patent holder can prohibit others from using any product that is substantially similar and recover damages from anyone who uses the product without permission. Patents for inventions are valid for twenty years. Design patents (a patent granted on the ornamental design of a functional item) are valid for fourteen years. Once the patent period expires, the invention or design enters the public domain, which means that anyone can produce or sell the invention without paying the prior patent holder.

Unlike some other forms of intellectual property, patent rights do not arise once the invention is created. A patent is only acquired by filing an application and receiving approval from the U.S. Patent and Trademark Office. According to the U.S. Code, the patent application must "contain a written description of the invention, and of the manner and process of making and using it, in such full, clear, concise, and exact terms to enable any person skilled in the art to which it pertains... to make and use the same" (35 U.S.C. §112). Once approved, the patent owner can earn a profit by making, using, and selling the invention or by selling the patent or by licensing others to use the patent.

Patent holders own the exclusive rights to use and exploit their patents. A party who makes unauthorized use of a patented invention is liable for infringement. The holder of the patent must prove that the infringement occurred. If successful, the patent holder can recover monetary damages equal to a reasonable royalty on the sale of the infringed articles as well as other damages associated with a loss of customers. The successful plaintiff can also obtain a court order requiring the destruction of the infringing article and an injunction preventing the infringing party from infringing again. If the court determines the infringement was intentional, the court has the discretion to triple the amount of actual damages. Monetary damages in patent infringement lawsuits can be substantial. For example, in 1990, NTP Inc. developed and patented a method to deliver email wirelessly, but never released a product utilizing the technology. Later, when Research in Motion's (RIM) BlackBerry device utilized a similar technology, NTP notified RIM of its patent and requested a licensing fee. When RIM failed to meet NTP's demands, NTP filed suit to protect its patent rights. After years of litigation, RIM eventually paid NTP $612.5 million to settle the suit (Heinzel, 2006; Ewalt, 2005). In another case, Microsoft was sued for patent infringement for allegedly implementing another company's patented digital rights management technology. Microsoft eventually settled the case for approximately $440 million (Clark, 2004).

In order to be approved, a patent applicant must demonstrate that the subject matter of the proposed patent is novel, useful, and not obvious (35 U.S.C. §§ 102, 103). Patentable subject matters may include mechanical, electrical, or chemical inventions, processes, compositions of matter, or improvements to existing machines, processes, or compositions of matter. It may also include things such as designs for an article of manufacture, certain forms of plants, and living material invented by a person. Absent from the list of patentable matters are most of the more traditional forms of digital products such as newspaper content, magazine content, e-books, music, and movies. However, patents are available for one key digital product—software (Delta & Matsuura, 2009: § 7.02).

Years ago, it was difficult for developers and manufacturers of software to obtain patent protection because the basis for software was often mathematical equations or formulas, which are not patentable. Therefore, there was some doubt as to whether computer programs were novel and not obvious. However, in 1981 the U.S. Supreme Court decided that it was possible to obtain a patent for a process that incorporates a computer program (Diamond v. Diehr, 1981). As a result of this key decision, patents for software-related inventions are now allowed so long as the equations or formulas within the software are applied in a useful manner. As discussed below, software may also be protected under copyright law.

Recently, the rapid expansion of patentable subject matter has come under judicial scrutiny. In one recent case, an inventor filed an unsuccessful application to patent a method of encoding additional information on electronic signals emitted from digital audio files. Even though the process was very useful, the court ruled that it was not patentable because the signal was not a mechanical, electrical, or chemical invention or a process, a machine, or composition of matter (In re Nuijten, 2007). In another case, the Federal Circuit cut back on the scope of patentable subject matter by limiting process patents to either physical transformations or use tied to a particular machine. The U.S. Supreme Court recently agreed to review the Federal Circuit's decision in this case. If the Supreme Court decides to address the limits of patentable subject matter, it will be its first decision on the scope of patentable subject matter in more than 28 years. This decision could have profound implications for patent law, and in particular for process and software patents (Webster, 2009).

## Trade Secrets

The Uniform Trade Secrets Act defines a trade secret as information that derives actual or independent economic value because it is not generally known to others and is the subject of reasonable efforts to maintain its secrecy (National Conference of Commissioners on Uniform State Laws,

1985: § 1(4)). Stated another way, a trade secret is valuable secret information that makes an individual company unique and would have value to a competitor. Under this broad definition, a trade secret could be anything from a customer list, to a marketing technique, to the secret formula for Coca-Cola. Other common examples of trade secrets include processes, plans, or procedures that make a product more desirable to the consumer.

For digital technology companies, several different software techniques classify as trade secrets. This could include methods of data analysis and data storage as well as graphics display and encryption techniques. For example, some software companies operate under a proprietary business model in which their commercial products are provided only in binary form. They treat their source code as a valuable trade secret and resist disclosing it to any customer or outsider (Landy, 2008).

A company could chose to patent some types of trade secrets. However, the company may be reluctant to do so because patent registration requires that the trade secret be disclosed publicly. Also, patent protection expires after twenty years. Because there is no set time limit on the protection available for trade secrets and there is no requirement to reveal the secret to the public, computer software producers have an incentive to treat their works as trade secrets (Scott, 2007).

Trade secret law protects against wrongful appropriation, disclosure, or use of trade secrets. The owner of a trade secret can obtain a court order mandating that another party stop using the trade secret and receive civil damages if the trade secret was acquired by improper means. Examples of improper means could include blatant acquisitions through theft, trespass, wiretapping, or fraud. Acquisition by improper means could also include situations where the other party acquired it from a party who is known or should be known to have obtained it by improper means. This could include inducing an employee or former employee of another company to reveal trade secrets (Scott, 2007). Wrongful appropriation does not include the act of replicating a trade secret by independent invention or reverse engineering. Reverse engineering is the process where an individual takes a known product and works backward to determine how the product was made. Contracts for the sale or license of software and hardware often contain broad prohibitions against reverse engineering. Therefore, while technically permitted under trade secret law, reverse engineering could lead to civil liability under contract law.

Anyone who misappropriates a trade secret is liable to the owner for actual damages (lost profits) or a reasonable royalty for the improperly obtained profits. In addition, if the misappropriation was willful or malicious, the court may award attorney's fees and double the damages. In 1996, Congress passed the Economic Espionage Act (EEA) which makes it a criminal offense to steal trade secrets "related to or included in a product that is produced for or placed in interstate or foreign commerce" (18 U.S.C. §1832). The act broadly defines theft to include all types of conversion of trade secrets. This includes: (1) stealing, obtaining by fraud, or concealing such information; (2) copying, duplicating, sketching, drawing, photographing, downloading, uploading, photocopying, mailing, or conveying such information without authorization; (3) purchasing or possessing a trade secret with knowledge that it has been stolen.

Criminal punishment for thefts of trade secrets, attempts to steal trade secrets, or conspiracies to steal trades secrets includes fines and imprisonment for up to ten years. Organizations that violate the EEA are subject to fines of up to $5 million. In a recent case, a former employee of Coca-Cola was convicted and sentenced to eight years in prison for conspiring to steal Coca-Cola trade secrets and sell them to PepsiCo for $1.5 million. The employee was caught when PepsiCo notified Coca-Cola that it had received a letter offering to sell the trade secrets. Coca-Cola notified federal authorities, who eventually instituted a sting operation to catch the former employee and her accomplice (United States v. Williams, 2008). Absent such compelling evidence of a conspiracy to steal a trade secret, prosecutors are much more likely to pursue a violation of the EEA if the theft of trade secrets was coupled with other criminal activity such as an illegal wiretap, a break-in, or a theft of physical property (Landy, 2008).

## Trademarks

A trademark, according to the Lanham Trade-Mark Act, is any word, name, symbol, or device or combination thereof adopted and used by a manufacturer or merchant to identify his or her goods and distinguish them from goods manufactured or sold by others. A trademark could be, among other things, an advertising slogan, a jingle, a specific package design, or a logo. Similarly, the screen display and website containing a company's products in a digital format are covered by trademark law (Scott, 2007).

Trademarks are valuable assets for a business. A distinctive trademark allows a business to build a reputation in the goods and services it sells because trademarks enable the public to recognize that particular goods or services originated from a particular company. A distinctive trademark ensures that the trademark owner, and not an imitating competitor, will profit from the sale of a desirable product or service. If another party attempts to use confusingly similar elements and causes market confusion, a trademark infringement occurs.

Trademarks come into being through actual use, not necessarily registration. By using a "mark" on goods or displaying it in connection with offered goods or services, a company can acquire trademark rights. The first person to use a mark in commerce owns it. Once the seller of a product or service uses a mark in commerce or forms or has a bona fide intention to do so in the next six months, he or she may apply to register the mark with the U.S. Patents and Trademark Office (PTO). The PTO will review the application for distinctiveness. If the PTO denies the application, the denial may be challenged in a court of law. Likewise, the PTO's decision to grant an application may be challenged by a third party who believes that it would be injured by the registration of the mark (Scott, 2007: § 4.17).

Registered trademarks are placed on the Principal Register of the PTO. This registration provides the registrant with several benefits:

1. It gives the registrant the exclusive right to use the mark nationwide;
2. It provides constructive notice to the public of the registrant's claim of ownership of the mark and serves as a public record putting later users on notice of the registrant's superior rights;
3. The PTO will refuse to register closely related marks used in connection with similar goods or services;
4. The holder of a registered trademark generally has the right to use it as an Internet domain name;
5. It gives the holder the ability to prevent importation of infringing foreign goods.

Trademarks, like patents, should be registered in each country where protection is sought. In the U.S., priority is given to the first party to use the trademark. In most other countries, priority is given to the first party who files an application to register the trademark. Therefore, early registration is crucial for companies planning to offer products or services abroad.

A trademark registration is initially valid for ten years. The owner can renew the trademark for an unlimited number of ten year terms so long as the mark is still in use. Once a trademark has been obtained, the owner must take additional steps to ensure that the rights to the trademark are not lost. Trademarks may be "abandoned" if they are not used for three years.

Companies must be diligent to keep their trademarks from becoming generic because trademark protection will be lost if a trademark loses its distinctiveness and can no longer be used to distinguish one product from another. The terms "yo-yo," "escalator," "trampoline," "nylon," and "mimeograph" were once trademarks, but the widespread and generic use of these terms led to an environment where there was no way to distinguish one product from another. When people begin to use a trademark in a generic fashion, the trademark owner should take affirmative steps to retain exclusive rights to the trademark. For example, the Xerox Corporation has successfully protected its trademark of the word "Xerox" in the United States through an aggressive public relations campaign stressing that the term is an adjective describing its products and services, rather than a verb.

Courts have also extended trademark protections to include the packaging or dressing of a product. The "trade dress" of a product relates to the total image of the product and can include the color of the packaging, the configuration of the goods, and even the overall appearance of a business. In one case, the Supreme Court held that a Mexican restaurant was entitled to protection under the Lanham Act for the shape and general appearance of the exterior of its building as well as for the other features reflecting the total image of the restaurant such as its décor, menu, and server's uniforms (Two Peso, Inc. v. Taca Cabana, Inc., 1992).

A trademark is infringed when, without the owner's consent, another party uses the same or substantially similar mark in connection with the sale of goods or services. The test for trademark infringement asks whether an ordinary buyer would believe that both products or services came from the same source. The key element is the "likelihood of confusion" of the product or service. The court will determine whether the two marks are sufficiently alike to cause consumers to be confused as to their source or origin. The two marks do not have to be exactly alike, just close enough to be confusingly similar. For example, in 2009 a jury awarded Adidas nearly $305 million in damages for Payless Shoesource Inc.'s violation of Adidas's trademark for three-stripe shoes. Payless Shoesource Inc.'s claim that it did not violate the trademark because its shoes had two or four stripes, rather than three stripes, was unsuccessful (Ford & Ratoza, 2008).

Under the Federal Trademark Dilution Act of 1995 (FTDA), trademark owners can successfully sue to prevent others from using a trademark in a way that dilutes its value, even though consumers would not be confused about the origin of the product. The FTDA also prevents others from using a trademark in a way that tarnishes the trademark by association with unwholesome goods or services. For example, Toys "R" Us successfully sued a company that was selling sexual devices and clothing on the Internet under the domain name Adults "R" Us. The court found that Toys "R" Us family of marks were famous and distinctive and were established well before the defendants began identifying themselves as Adults "R" Us (Toys "R" Us, Inc. v. Akkaoui, 1996).

Courts protect trademarks against infringement by issuing injunctions, compensating the owner for damages, taking away the infringer's profits, and in some cases ordering the infringing party to pay the trademark owner's attorney fees. In the above mentioned case, the court awarded a preliminary injunction to stop the use of Adults "R" Us and ordered that the defendants immediately discontinue use of any Internet web site address or domain name that would infer any connection to Toys "R" Us.

## Trademarks as Domain Names

Domain names are not expressly reserved for a company that owns a trademark similar to the domain name. Instead, domain names are granted on a first-requested, first-served basis. Not surprisingly, conflicts arise between one party's claim of trademark rights and another party's claim of right over a domain name. Many of these conflicts are covered by the Anticybersquatting Consumer Protection Act of 1999 (ACPA).

The ACPA authorizes a trademark owner to institute civil action against any party who, having a "bad faith intent to profit" from the owner's mark, registers, sells, purchases, licenses, or otherwise uses a domain name that is identical or confusingly similar to the owner's mark. The ACPA does not define what constitutes "bad faith intent." Instead it lists several factors that courts can consider when deciding whether "bad faith intent" exists. Proof of this "bad faith intent" could be a defendant's intent to divert consumers from the trademark owner's online location to a site that could harm the mark owner's reputation. Proof of "bad faith intent" could also be that the defendant offered to sell the domain name to the mark owner without having used, or intended to use, the domain name in the offering of goods or services. If the trademark owner successfully proves its cybersquatting claim, the court may order the domain name transferred to the trademark owner. The court may also order the defendant to pay the trademark owner's actual damages plus profits attributable to the defendant's use of the domain name (Delta & Matsuura, 2009).

In one prominent case, the wine maker Ernest and Julio Gallo requested that Spider Webs Ltd., a domain-name speculator that owned numerous domain names consisting of famous company names, release or transfer the domain name "ernestandjuliogallo.com" to Gallo. When Spider Webs refused to do so, the winery filed suit under the ACPA claiming the domain name blurred or devalued its trademark in the name "Ernest & Julio Gallo." The court agreed, finding that Spider Webs operated with a "bad faith intent" to profit because it knew Gallo had a famous mark in which it had built up goodwill and Spider Webs hoped to profit from this by registering the domain name and waiting for Gallo to inquire about its use. The court ordered Spider Webs to transfer the domain name to the winemaker and enjoined Spider Webs from using any domain name containing the word "Gallo" or the words "Ernest" and "Julio" in combination. The court also ordered Spider Webs to pay Ernest and Julie Gallo $25,000 in statutory damages for its bad faith use of the domain name (E. & J. Gallo Winery v. Spider Webs Ltd., 2002).

## Copyrights

The area of intellectual property law that has the most pervasive impact on digital products is copyright law. Copyright law governs distributing and selling copies, preparing works based on earlier works, and public display and performance of works. The recording industry, publishers, broadcasters, video game developers, and the software industry rely on copyright law to protect their core products and services. The following section details the basic principles of copyright law and its impact on companies that sell, license, or distribute products in a digital format.

The Copyright Act of 1976 is the primary source of copyright law in the U.S. The Copyright Act protects the authors of original work from the unauthorized use of their copyrighted material. It protects a wide range of original works of authorship including books, articles, musical works, works of art, motion pictures, audiovisual works, architectural plans, sound recordings, lectures, and computer programs. This list is not exhaustive as the Copyright Act extends protection to "original works of authorship in any tangible medium of expression, now known or later developed" (Copyright Act, 1976: §102(a)). Because a motion picture, photograph, sound recording, or book can be distributed and copied in digital form, these digital products are controlled by copyright law. Copyright law also protects written words on a website and software programs running on a website. In 1980, Congress amended the original Copyright Act to specifically include computer programs in the list of items protected by copyright law. A computer program, whether in object code or source code, is protected from unauthorized copying whether from its object or source code version (Apple Computer Inc. v. Franklin Computer Corp., 1983).

The Copyright Act gives the owner of a copyright the exclusive right to do and to authorize others to (1) reproduce and make copies of the work, (2) distribute copies of the work to the public by sale, rental, lease, or lending, (3) perform the work publicly, display the work publicly, or, as in the case of sound recordings, to perform the work publicly by means of a digital audio transmission, and (4) prepare derivative works based upon the work (Landy, 2008: 13). A derivative work is a work based upon another work. Examples of derivative works include, among other things, a sequel to a movie, an updated version of an existing computer program, a port of a program into a different software-operating system (Landy, 2008), a translation of an existing work, an abridgement of an existing work, or a movie based on a novel. Processing works electronically also creates derivative works. For example, if you process a digital file from a music CD to create an MP3 file, you have created a derivative of the original file. If you want to create a work based upon someone else's work, you need the permission of the holder of the rights that your work is based on (Delta & Matsuura, 2009). If you create a derivative work without permission, you infringe the copyright in the underlying work. Likewise, if someone translates the original source language version of a program into a different source language, that translation would constitute an infringement of the exclusive right of translation (Synercom Tech., Inc. v. University Computing, 1979).

In essence, copyright law grants the copyright holder a legal time-limited monopoly on copying, distributing, and performing the work. Therefore, the owner of the copyright in a digital product has the right to exclude all others from copying, recording, adapting, publishing, and selling copies of that product. Likewise, the copyright owner has the right to transfer all or any one of these rights to another.

The Internet provides ample opportunities for copyright infringement beyond the traditional piracy of copyrighted material. For example, scanning a copyrighted work and displaying that work on a Web site violates the copyright owner's right to reproduction. Likewise, presenting copyrighted material on a Web site so that Internet users can view or listen to the copyrighted material is an example of infringing on the copyright holders' right to public display. In 2008, Senator John McCain's campaign staff created an Internet ad using Jackson Browne's signature 1977 hit "Running on Empty" to promote the Senator's bid for the presidency. Unfortunately, the campaign staff never received permission to use the song, so Browne filed a lawsuit to stop the infringement. The campaign quickly discontinued the use of the song and Senator McCain apologized for the unauthorized use. The lawsuit was eventually settled, but, as stated by Browne's attorney, "People like Jackson have the right to license the use of their songs for political campaigns, or to choose not to" (Neil, 2009).

In order to be protected by copyright, a work must be original, creative, and fixed in a tangible form. These requirements are minimal. To be original, the work must be the author's own work, not copied from somewhere else. The creative element is closely tied to the originality of the work. The level of creativity required for copyright is extremely low (Atari Games Corp. v. Oman, 1992). To be creative, the author must have used some slight degree of judgment or discretion to create the work. Copyright protection begins automatically as soon as an original work is fixed in a tangible form. A work is placed in a tangible form when it is written or typed on a piece of paper, saved on a computer hard drive (including ROM or RAM memory), or recorded in some fashion. Works that have not been fixed in a tangible form of expression would include improvisational speeches or performances that have not been written or recorded. For example, if someone creates a new song while singing in the shower, the song is not copyrighted. Anyone can use the song without permission. However, once that person records the song on an audiotape or a videotape or records it in an audio file, copyright attaches (Delta & Matsuura, 2009: § 6.01).

Registration with the U.S. Copyright Office is not necessary, though it is recommended. Registration involves payment of a fee and also involves filing a copy of the work with the U.S. Copyright Office. The following are some of the key benefits of registering a copyright:

- Registration establishes a public record of the copyright claim.
- Registration of a work is required before its author can sue for copyright infringement.
- If made before or within five years of publication, registration will establish prima facie evidence of the validity of the copyright.
- If registration is made before the infringement occurs, or within three months of first publication of the work, then, if the infringement was willful, the copyright owner can recover statutory damages and attorney's fees. Otherwise, only an award of actual damages and profits is available to the copyright owner.
- Registration allows the owner of the copyright to record the registration with the U.S. Customs Service for protection against the importation of infringing copies.

When a work is published, the creator should place a copyright notice on the work. The use of a copyright notice is no longer required under U.S. law; however, there are several reasons to use a copyright notice (Landy, 2008: 16). For example, if the work is later infringed, and a proper notice of copyright appears on the published copy, then the infringing defendant will be unable to claim that he or she did not realize the work was protected.

A work that was created (fixed in tangible form for the first time) on or after January 1, 1978, is automatically protected from the moment of its creation and is ordinarily given a term lasting for the author's life plus an additional 70 years after the author's death. In the case of a joint work prepared by two or more authors, the term lasts for 70 years after the last surviving author's death. For works made for hire, the duration of copyright will be 95 years from publication or 120 years from creation, whichever is shorter. A work originally created and published or registered before January 1, 1978 is subject to rules beyond the scope of this text.

While copyright law protects a broad range of works of authorship, it does not directly protect ideas, facts, procedures, processes, systems, methods of operation, concepts, principles, or discoveries. Instead, copyright law protects the ways in which they are expressed. For example, the facts in a newspaper article are not protected by copyright, but the story itself is protected by copyright because the author's creative decision as to how to arrange the presentation of those facts makes the story copyrightable. Therefore the newspaper, as the copyright holder of the story, has the exclusive right to sell or make copies of the article. However, the underlying facts or ideas in the article may be freely used by others.

This distinction highlights the differences between copyright law and other forms of legal protection. Patents protect the ideas themselves, not merely the expression of the ideas. For example, if a company invents a new method to encrypt data, copyright will not stop other parties from making programs that use the new method because inventions are protected by patent law. A copyright on a software program with patentable elements would only cover the program code itself; it would not cover the invention or any concepts or methods used in the invention (Landy, 2008: 15). Stated another way, copyright law would not keep a third party from independently developing a functionally equivalent software program. On the other hand, if the original program is protected by a patent, there may be an infringement issue based upon the equivalent functions of the two programs (Delta & Matsuura, 2009: § 7.02). Trade secret law might also be applicable because protection of trade secrets extends both to ideas and to their expression. Most importantly, copyright protection does not preclude other forms of protection, such as patent protection or trade secret protection, for computer programs and databases.

## Copyrights of Works Made for Hire

If the work was prepared by an employee within the scope of his or her employment (it was a part of his or her job duties to create the work), then the employer owns the copyright to the work. This means the employer company is both the legal author of the work and the legal owner of the copyright. For example, if a programmer is an employee of a software company, that company will own the copyright in the programs produced by that programmer, unless there was an express agreement to the contrary or the program was not written within the scope of his or her employment. However, if the employee creates a work on his or her own time that is later acquired by his or her employer, this is not sufficient to make it a "work made for hire" (Scott, 2007: § 2.16).

The fighting issue in work-made-for-hire cases is often whether the employee prepared the work within the scope of his or her employment. In order to establish the work was prepared within the scope of employment, the employer must prove (1) the work was the kind that the employee was employed to perform, (2) the work was created substantially within the authorized time and space limits of his or her job, and (3) the work was "actuated, at least in part, by a purpose to serve" the employer's interests (Restatement (Second) of Agency, 1958: § 228). In software development cases, it can be difficult to prove the work was created with the authorized time and space limits of the job because it is not uncommon for a programmer to work at home during times outside of normal work hours. Not surprisingly, a key question in these cases is whether the employed is salaried or paid on an hourly basis (Gezmer v. Public Health Trust of Miami-Cty, 2002).

A work made for hire may also exist if a customer and an independent contractor enter into a written agreement to prepare a copyrightable work. The work is considered to be a work made for hire if it was specially ordered and the parties agree in writing that it is a work made for hire. Stated another way, if an independent contractor develops a work while under contract with a customer, the independent contractor will own the copyright, unless the contract says otherwise. The customer will likely only receive a nonexclusive implied license for the intended use of the work. If an independent contractor is permitted to use subcontractors, the independent contractor should be required to have written work-made-for-hire agreements with each subcontractor.

## Copyright Infringement

Unless a valid exception applies, whenever the expression of an idea is copied in an unauthorized manner, an infringement of copyright occurs. The reproduction does not have to reproduce the original in its entirety; if a substantial portion of the original work is reproduced, there is copyright infringement. In order for a copyright owner to bring a suit for infringement, the copyright must be registered with the copyright office (unless the country of origin for the work is not the United States). At trial, the copyright owner must only prove it owns the copyright and the defendant violated one or more of the copyright owner's exclusive rights.

There are four basic types of copyright infringement:

1. *Direct* infringement occurs when one party exercises one of the exclusive rights of copyright without consent to do so from the

party that owns or controls the copyright. For example, one party puts copyrighted artwork on a new line of clothing without obtaining permission from the copyright owner. Likewise, if one person purchases a computer program from a computer store, burns a copy, and then sells that copy to their friend, this constitutes direct infringement.

2. *Contributory* infringement occurs when a party knowingly materially contributes to infringing activities. In its simplest terms, this occurs when one person helps another person make or sell infringing copies. Examples could include marketing a product that can be used to duplicate copyrighted software (Cable/Home Communications Corp. v. Network Prods., Inc., 1990), selling a special copier with a primary use of copying Sega video games (Sega Enter. Ltd. v. MAPHIA, 1994), or marketing a product that can override a software vendor's copy-protection system even though the product itself does not perform any copying (Scott, 2007). Internet search engines may also be liable for contributory infringement. In a recent case involving copyrighted images, the Ninth Circuit stated that a computer system operator could be held "contributorily liable if it has actual knowledge that specific infringing material is available using its system, and can take simple measures to prevent further damage to copyrighted works, yet continues to provide access to infringing works" (Perfect 10, Inc. v. Amazon.com, 2007). This case does not suggest that search engines are liable if they link to websites that offer infringing files. Instead, the fighting issue was whether search engines could be contributorily liable for providing direct means of accessing these files from the search results page (Boehm, 2009).

3. *Inducing* infringement occurs when a party knowingly induces (or even encourages) another party to engage in direct copyright infringement. This method of copyright infringement was established in the case Metro-Goldwyn-Mayer Studios Inc. v. Grokster, LTD (2005) where the defendants, Grokster, Ltd. and StreamCast Networks, Inc., distributed free software that allowed computer users to share electronic files through decentralized peer-to-peer networks. Hundreds of thousands of files were shared each month. MGM and other copyright owners sued Grokster and StreamCast alleging they were violating the copyright law by knowingly and intentionally distributing their software to users who would reproduce and distribute copyrighted works illegally. In review of a motion for summary judgment, the Supreme Court held that "one who distributes a device with the object of promoting its use to infringe copyright, as shown by clear expression or other affirmative steps taken to foster infringement, is liable for the results acts of infringement by third parties." The court pointed out that mere knowledge of infringing potential was not enough to attach liability. Instead, the inducement rule premises liability on purposeful, culpable expression and conduct.

4. If a party has the right and ability to supervise a direct copyright infringement and has a direct financial interest in the infringement, they may be liable for *vicarious* infringement. For example, a court held that the owners of a flea market, who had knowledge that infringing goods were being sold by vendors renting the booths, were vicariously liable for the underlying infringement because they had control over the direct infringers and had a direct financial interest in the infringing activity (Fonovisa, Inc. v. Cherry Auction, Inc., 1996).

The copyright owner need not provide direct proof that a work was copied. All that is needed is evidence that the alleged infringer had access to the copyrighted work and his or her work is substantially similar to the copyrighted work (E.F. Johnson Co. v. Uniden Corp. of America, 1985). To make it easier to prove that software was copied, useless pieces of software code may be embedded within a program. If the allegedly infringing copyright contains the useless code, this is powerful evidence that the original code was copied (Bagley and Dauchy, 2008).

The following remedies may be awarded in a copyright infringement case: (1) an injunction to make the infringer stop infringing, (2) impoundment and possible destruction of the infringing articles; (3) actual damages, plus profits made by the infringer that are additional to those damages, or statutory damages of at least $750 but no more than $30,000 ($150,000 if the infringement was willful); and (4) reasonable attorney fees. Because of the short market life of many software packages when compared to the lengthy litigation process, pre-trial relief by way of a temporary injunction is often very important. If the plaintiff receives a requested pre-trial injunction, he or she stops the defendant from selling, distributing, or using the software in question for the duration of the litigation. At that point the defendant may stop defending the suit because, win or lose, the software will likely have little value by the time the litigation has run its course.

Prior to 1997, criminal penalties under copyright law could only be imposed if unauthorized copies were exchanged for financial gain. However, much piracy of copyrighted digital products was not done for profit. Unauthorized copies were made and distributed simply to share with others. In 1997 Congress enacted the No Electronic Theft (NET) Act extending criminal liability for the piracy of copyrighted material to persons who exchange unauthorized copies of copyrighted material valued at more than $2500 even though they realized no profit from the exchange. The criminal penalties for violating the act include fines and up to five years of imprisonment.

The Family Entertainment and Copyright Act of 2005 established criminal penalties for willful copyright infringement by the distribution of a musical work, computer program, motion picture, or other audiovisual work, or sound recording being prepared for commercial distribution by making it available on a computer network accessible to members of the public, so long as the person knows or should have known the work was intended for commercial distribution. The criminal penalties are a fine and/or imprisonment for up to three years. If the infringement was done for commercial advantage or private financial gain, then imprisonment may be imposed for five years.

**Fair Use of Copyrights**

Not all unauthorized copying of copyrighted material constitutes an infringement. The doctrine of "fair use" is an affirmative defense to a copyright infringement claim. The rationale for this doctrine is that there is some copying where the social benefit of copying outweighs the benefit of copyright protection. The most influential court case addressing the doctrine of fair use involved whether the manufacturers of videotape recorders were contributory copyright infringers because customers bought the machines to record copyrighted television programs. The U.S. Supreme Court held that home recording for the purpose of "time shifting"—recording so that the television program could be viewed at a later time—was a fair use and the manufacturers of the videotape recorders were therefore not liable as contributory infringers because the machines had a "substantial non-infringing use" (Sony Corp. v. Universal City Studios, Inc., 1984).

The Copyright Act describes various purposes for which the reproduction of a particular work may be considered "fair use," such as criticism, comment, news reporting, teaching, scholarship, and research (Copyright Act, § 107). It also sets forth four factors to consider when determining whether or not a particular use is fair:

1. *The purpose and character of the use, including whether such use is of a commercial nature or is for nonprofit educational purposes.* If the use is for commercial purposes, it is less likely to be found to be fair use. If the copying is done for commentary or criticism, it is more likely to be found to be fair use. However, this factor "is not an all-or-nothing matter" (Nihon Keizai Shimbun, Inc. v. Comline Bus. Data, Inc., 1999). The commercial or non-commercial nature of the use is not determinative. It is but one of the four factors the court will consider when determining whether the use constituted fair use. (Campbell v. Acuff-Rose Music, 1994).
2. *The nature of the copyrighted work.* This factor relates to the kind of work the alleged infringer is copying. In general, works that are creative, such as songs, stories, and poems, are afforded more protection than items that are factual, such as new reports. Court have generally held that computer software and most online content are creative works entitled to strong protection (Scott, 2007).
3. *The amount and substantiality of the portion used in relation to the copyrighted work as a whole.* There is no specific number of words, lines, or notes that may safely be taken without permission. However, copying a small portion of a work is more likely to be deemed fair use. Similarly, a low-resolution production of a digitized picture (a thumbnail picture) is more likely to be fair use than a reproduction of the picture in its original resolution (Perfect 10, Inc. v. Amazon.com, Inc., 2007).
4. *The effect of the use upon the potential market for, or value of, the copyrighted work.* If actual or potential sales are not going to be hurt by the use, then the copying is more likely to be deemed fair use. As stated by the U.S. Supreme Court, this factor is "undoubtedly the single most important element of fair use" (Harper & Row Publishers v. Nation Enterprises, 1985).

In one case, the courts rejected a video maker's attempt to profit from the fair use doctrine. Passport video produced a sixteen hour video documentary of Elvis Presley's life and sold the videos commercially for profit. Approximately 5 to 10 percent of the videos were comprised of copyrighted music, movies, and television appearances owned by copyright holders other than Passport. Passport did not obtain permission to use those copyrighted works. Elvis Presley Enterprises, Inc. and various other companies and individuals owning these copyrights sued Passport for copyright infringement. Passport defended the lawsuit by arguing its use of the materials constituted fair use. The U.S. District Court disagreed and ordered Passport to stop distributing the videos. Passport appealed, claiming the videos consisted of scholarly research that would be protected as fair use. The Ninth Circuit Court of Appeals rejected this argument and affirmed the district court's decision. The court found that Passport's use of the copyrighted material caused market harm to the copyright holders because it would act as a substitute for the original copyrights and deny the copyright holders the value of their copyrights. The court of appeals held that the use of the copyrighted materials was not fair use, but instead was copyright infringement (Elvis Presley Enterprises, Inc. v. Passport Video, 2003).

Several courts have analyzed the application of the fair use doctrine to file-sharing software and services. Most courts have concluded that providing such file-sharing software, when coupled with evidence that the defendants encouraged the downloading of copyrighted materials, is not shielded by the fair use doctrine and therefore constitutes copyright infringement (Scott, 2007: 2-336).

The fair use doctrine should not be confused with the concept of public domain. Works that are not subject to copyright, or no longer subject to copyright, are referred to as works in the public domain. Because works in the public domain lack copyright protection, there is no need to get permission to use them. It is often difficult to discern whether a work is truly in the public domain. Public display of a work does not mean that the work is in the public domain. Likewise, creative materials available publicly on the Internet are not necessarily in the public domain. Generally, material is only in the public domain if the copyright for that material expired or if the material was not copyrightable in the first place.

## Transfers of Copyrights

The first sale doctrine permits a person who owns a lawfully made copy of a copyrighted work to sell the copy to another party. The purchaser cannot reproduce the work for resale because ownership of a copy does not include ownership of the copyright in the work itself. For example, if you purchase a music CD and then decide to resell it, that is permitted under the first sale doctrine. However, if you make a copy of the CD and then decide to sell either the original CD or copied CD, that constitutes copyright infringement. As explained below, producers of digital products seek to avoid the first sale doctrine by licensing, rather than selling, the digital product.

A license gives a person the right to do something he or she would not otherwise be permitted to do with someone else's property, without transferring the actual rights to the property. When an owner wishes to retain some rights to or control over its intellectual property, a license agreement is commonly used. Licensing agreements grant limited, specific rights to use intellectual property. For example, McDonalds own the rights to the many trademarks, such as the "Golden Arches." It licenses the rights to use this trademark to its franchisees, but the licensing agreement provides that McDonald's can take back the rights to use the trademark if the franchisee does not use the trademark properly.

The opposite of a license is a sale or an assignment. An assignment typically transfers all of one's interest in an item of intellectual property to a new owner. If the old owner tries to use the transferred rights in the intellectual property, the new owner can sue for infringement. For example, suppose that Cindy creates new software. Cindy then assigns the copyright in the software to a software company. After the assignment, the software company has the exclusive right to reproduce and distribute the software. If Cindy begins selling the software or creates an updated version of the software, she will be infringing upon the software company's rights as copyright owner.

Software developers will not typically sell their software to a customer. Instead, they will license the software to a customer. By doing so, the software developer can restrict how the software is used and, to some extent, who can use the software (Landy, 2008). Most software sold to a mass market is sold without a signed licensing agreement. Instead, software developers use a shrink-wrap license agreement or a click-wrap license agreement. With a "shrink-wrap agreement," the software contains a statement to the purchaser indicating that by opening the software packaging (tearing off the shrink wrap), the customer agrees to be bound by the terms written in the shrink-wrap agreement. With a "click-wrap agreement," the customer must indicate an acceptance of the licensing agreement before he or she is able to download or install the program. Some courts have thrown out such forms of licensing agreements, but the Uniform Computer Information Transactions Act, which has been gaining favor in a growing number of states, makes such agreements enforceable so long as they meet certain requirements (Delta & Matsuura, 2009: § 6.02).

## Digital Millennium Copyright Act

In 1998, Congress enacted the Digital Millennium Copyright Act ("DMCA") to both implement the United Nations World Intellectual Property

Organization Treaty and "strengthen copyright protection in the digital age" (Universal City Studios, Inc. v. Corley, 2001). The DMCA contains several provisions that affect digital products.

The DMCA provides copyright liability protection to Internet service providers who innocently store or transmit infringing materials posted by their users. To qualify for this protection, Internet service provider must take down a site (or the infringing material on the site) when a copyright holder informs the Internet service provider that a hosted site or individual user is infringing on its rights. The DMCA also specifically permits users to make a copy of a computer program for maintenance of the computer and allows libraries to make digital copies of works for their own internal use.

Most importantly, the DMCA prohibits the manufacture, distribution, and use of technology that can be used to circumvent systems that protect copyrighted material. This provision of the DMCA serves to protect the growing use of digital rights management to curb the rampant infringement of copyrighted material. Specifically, it makes it illegal to circumvent technological measures designed to protect works for either access or copying. This provision of the DMCA may be enforced through criminal prosecution or a civil lawsuit. For example, the U.S. Department of Justice used the DMCA to successful prosecute an individual for the sale of equipment used to obtain illegal access to satellite television transmissions (United States v. Whitehead, 2003), and the motion picture industry used the DMCA to stop an individual from selling a software application that enabled copying of DVDs protected by the Content Scramble System (Universal City Studios v. Reimerdes, 2000).

## FUTURE RESEARCH DIRECTIONS

Intellectual property law in the area of digital technology is constantly evolving through new legislation and judicial interpretations of existing statutory law. While the laws are constantly evolving, history shows that copyright laws are generally slow to react to technological innovations. Recently, Depoorter (2009) analyzed ten major innovative breakthroughs that gave rise to significant copyright issues during the past forty-five years. His research found that the average time it takes to ascertain an innovation's copyright status, from the time the technology is introduced to the time of a final resolution of the main copyright questions regarding this new technology through judicial decision or new legislation, is approximately seven years.

One recent lag between technological innovation and legal decision demonstrates how extended legal battles over the applicability of copyright law to innovative technologies may change societal norms and views toward copyright infringement for digital products. In the late 1990s, the introduction of file sharing software, alongside advanced compression technology and high-level bandwidth technology, led to effortless sharing of music on peer-to-peer networks. Napster first popularized this free transfer of MP3 files in 1999. However, it took nearly two years before the recording industry, through legal channels, successfully forced Napster to end its file sharing operations (Depoorter, 2009). Even when Napster ceased operations, numerous other file sharing websites, such as Grokster Ltd., operated similar file-sharing technologies that operated on a decentralized basis. Grokster's decentralized peer-to-peer file sharing technology presented such a unique legal challenge under existing law that both the trial court and the Ninth Circuit held that Grokster was not accountable under existing copyright law (Metro-Goldwyn-Mayer Studios Inc. v. Grokster, Ltd., 2004). Eventually, in 2005, the U.S. Supreme Court fashioned a new theory of copyright infringement when it decided that Grokster Ltd. was liable because it had induced the infringement (Metro-Goldwyn-Mayer Studios Inc. v. Grokster Ltd., 2005).

While the recording industry began litigation against Napster in 1999, it did not begin litigation against individual infringers until September of 2003 (Galluzzo, 2009). Many accused individual infringers settled their lawsuits, but others challenged the recording industry through proper legal channels. In 2009, ten years after Napster first facilitated the free transfer of MP3 files via the Internet, the first jury to consider such a case concluded that the individual infringer had willfully violated the copyrights on twenty-four songs and was liable for nearly two million dollars in damages (Des Moines Register, 2009).

Even though many of the key legal issues surrounding copyrights and peer-to-peer file sharing have been resolved, it is not known whether the years of rampant copyright infringement will have a lasting impact on society's views towards copyright law. During this time, many individuals grew accustom to peer-to-peer file sharing technology, and, not surprisingly, the years of unabashed music file sharing had a notable impact on societal norms. For example, in a 2003 study, only twenty percent of teenagers under the age of seventeen believed it was "wrong" to download a song without permission from its author (Press Release, 2003). Numerous popular commentators have also proclaimed that the era of copyright in the music industry is coming to a close (Mark F. Schultz, 2009) and some academic scholars have followed suit. One Harvard Law professor forecasts the end of the era of copyright as a means for supporting the production of music because peer-to-peer file sharing is "here to stay" (Benkler, 2004). He also claims that attempts to criminalize "one of our most basic social-cultural practices will, and ought to, fail" (Benkler, 2004).

Despite the recording industry's recent successes against individual infringers and companies promoting infringement through peer-to-peer file sharing, the sheer magnitude of illegal peer-to-peer downloading activity is astounding. It is estimated that there are 1.3 billion illegal downloads per year (Kaplan, 2008). It is also estimated that more than half of all college students engage in illegal peer-to-peer downloading activity (Kaplan, 2008). Future research should study whether societal norms and attitudes towards cyberspace copyright infringement have changed since the recording industry's recent legal triumphs against individual infringers. Future research could also explore whether the younger generation's views towards unauthorized reproductions of musical works also applies to other forms of digital products such as software and e-books. If large segments of society believe that a different set of rules should apply to digital products, there could eventually be enough political pressure to establish a new set of intellectual property laws for the cyberspace of the future, one that could be quite different from the system that currently governs owners and users of creative works in the analog world (Zimmerman, 2007).

## CONCLUSION

Companies that deal in products in a digital format face unique legal challenges because they attempt to earn a profit by selling or licensing material that is easily copied and inexpensive to reproduce. Because there are no inherent technological or mechanical impediments to the unauthorized reproduction and distribution of creative works which have been converted into digital products, digital product company managers must rely on a mixture of digital rights management technology and intellectual property law to limit such unauthorized reproductions. Of the four traditional categories of intellectual property law, copyright law is the most applicable to digital products.

This chapter also suggests that societal norms and views towards copyright infringement have been impacted by the extended legal battles over the applicability of copyright law to new digital technologies. For the foreseeable future, governments will continue to provide the means to successfully protect an owner's intellectual property right. However, it is not known whether shifting societal norms towards copyright infringement could eventually lead to a substantial overhaul of the current copyright scheme. Digital product managers must be cognizant that they may be required to shoulder more of the burden of protecting their interests in their own creative works.

## REFERENCES

"2nd jury rules against woman in music case." (2009, June 19) Associated Press. *Des Moines Register*, 11A.

Anticybersquatting Consumer Protection Act of 1999, § 15 U.S.C. § 1125(d) (2000).

Apple Computer Inc. v. Franklin Computer Corp., 714 F.2d 1240 (3d Cir. 1983).

Atari Games Corp. v. Oman, 979 F.2d 242 (D.C. Cir. 1992).

Benkler, Y. (2004). 'Sharing nicely': On shareable goods and the emergence of sharing as a modality of economic production. *The Yale Law Journal*, *114*, 273–358. doi:10.2307/4135731

Boehm, J. (2009). Copyright reform for the digital era: Protecting the future of recorded music through compulsory licensing and proper judicial analysis. *Texas Review of Entertainment & Sports Law*, *10*(2), 169–211.

Cable/Home Communications Corp. v. Network Prods. Inc., 902 F.2d 829 (11th Cir. 1990).

Campbell v. Acuff-Rose Music, 510 U.S. 569 (1994).

Clark, D. (2004, April 12). Microsoft strikes new patent accord. *Wall Street Journal*, p. A3.

Constance, E. B., & Dauch, C. E. (2008). *The Entrepreneur's Guide to Business Law*. Mason Ohio: South-Western.

Copyright Act of 1976, Pub. L. No. 95-553, 90 Stat. 2541 (1976).

Delta, G., & Matsuura, J. (2009). *Law of the Internet* (3rd ed.). Frederick, Maryland: Aspen.

Depoorter, B. (2009). Technology and uncertainty: The shaping effect on copyright law. *University of Pennsylvania Law Review*, *175*, 1831–1868.

Diamond v. Diehr, 450 U.S. 175 (1981).

Digital Millennium Copyright Act of 1998, Pub. L. No. 105-304, 112 Stat. 2860 (1998).

E. & J. Gallo Winery v. Spider Webs Ltd., 286 F.3d 270 (5th Cir. 2002).

Economic Espionage Act, Pub. L. No. 104-294, 110 Stat. 3488 (1996).

Elvis Presley Enterprises, Inc. v. Passport Video, 249 F.3d 622 (2003).

Ewalt, D. M. (2005, November 30). Judge sours BlackBerry settlement. *Forbes*. Retrieved January 11, 2010, from http://www.forbes.com/2005/11/30/rim-blackberry-lawsuit-cx_de_1130rimm.html.

Family Entertainment and Copyright Act of 2005, § 17 U.S.C. § 110(11).

Federal Trademark Dilution Act of 1995 (FTDA), Pub. L. No. 104-98, 109 Stat. 985 (1995).

Fonovisa, Inc. v. Cherry Auction, Inc., 76 F.3d. 259 (9th Cir. 1996).

Ford, S., & Ratoza, M. (2008). Landmark trademark infringement awarded to Adidas. Bullivant Houser Bailey PC. Retrieved January 12, 2010, from http://www.bullivant.com/Landmark-trademark-infringement-awarded-to-Adidas.

Galluzzo, V. (2009). When "now known or later developed" fails its purpose: How P2P litigation has turned the distribution right upside-down. *Florida Law Review*, *61*, 1165–1200.

Gezmer v. Public Health Trust of Miami-Cty, 219 F. Supp. 2d 1275, 1280 (S.D. Fla. 2002).

Harper & Row Publishers v. Nation Enterprises, 471 U.S. 539 (1985).

Heinzel, M. (2006, March 6). BlackBerry case could spur patent-revision efforts. *Wall Street Journal*, B4.

In re Nuijten, 500 F.3d 1346 (Fed. Cir. 2007).

Kaplan, T. (2008, July 20). Music industry zealous in tracking tune thieves. *St. Petersburg Times*, 1B.

Landy, G. K. (2008). *The IT/Digital Legal Companion-A Comprehensive Business Guide to Software, IT, Internet, Media and IP Law*. Burlington, MA: Syngress Publishing.

Menell, P. (2002). Can our current conception of copyright law survive the Internet age?: Envisioning copyright law's digital future. *New York Law School Law Review. New York Law School*, *46*, 63–199.

Metro-Goldwyn-Mayer Studios Inc. v. Grokster, Ltd., 259 F. Supp. 2d 1029 (C.D. Cal. 2003), aff'd, 380 F.3d 1154 (9th Cir. 2004).

Metro-Goldwyn-Mayer Studios Inc. v. Grokster, Ltd., 545 U.S. 913 (2005).

National Conference of Commissioners on Uniform State Laws. (1985). Uniform Trade Secrets Act (1985).

Neil, M. (2009). McCain says sorry for campaign use of signature Jackson Browne song. ABA Journal. Retrieved January 10, 2010, from http://www.abajournal.com/news/article/mccain_says_sorry_for_campaign_use_of_signature_jackson_browne_song/

Nihon Keizai Shimbun, Inc. v. Comline Bus. Data, Inc., 166 F.3d 65, 72 (2d Cir. 1999).

No Electronic Theft (NET) Act of 1997. Pub. L. No. 105-147, 111 Stat. 2678 (1997).

Perfect 10, Inc. v. Amazon.com, Inc., 508 F.3d 1146 (9th Cir. 2007).

Press Release. (Nov. 4, 2003). E-Poll, E-poll study looks at consumer's attitudes before and after RIAA lawsuits. Retrieved from http://www.prnewswire.com/cgi-bin/stories.pl?ACCT=104&STORY=/www/story/11-04-2003/0002050963&EDATE.

Restatement (Second) of Agency § 228 (1958).

Schultz, M. F. (2009, January). Live performance, copyright, and the future of the music business. *University of Richmond Law Review. University of Richmond*, *43*, 685–764.

Scott, M. D. (2007). *Scott on Information Technology Law* (3rd ed.). United States: Aspen Publishers.

Sega Enter. Ltd. v. MAPHIA, 857 F. Supp. 679 (N.D. Cal. 1994).

Sony Corp. v. Universal City Studios, Inc., 464 U.S. 417 (1984).

Synercom Tech., Inc. v. University Computing, 462 F. Supp. 1003 (N.D. 1979).

The Lanham Trade-Mark Act, 15 U.S.C. § 1125 (2000).

Toys "R" Us, Inc. v. Akkaoui, LEXIS 17090 (N.D. Cal. 1996).

Two Peso, Inc. v. Taca Cabana, Inc., 505 U.S. 763 (1992). E.F. Johnson Co. v. Uniden Corp. of America, 623 F. Supp. 1485 (D. Minn. 1985).

Uniform Trade Secrets Act (1985). 12 U.L.A. 433.

United States v. Whitehead. (C.D. Cal. 2003). Retrieved January 12, 2010, from http://www.usdoj.gov/criminal/cybercrime/whiteheadConviction.htm).

United States v. Williams, 526 F.3d 1312 (11th Cir. 2008).

Universal City Studios, Inc. v. Corley, 273 F.3d 429, 435 (2d Cir. 2001).

Universal City Studios v. Reimerdes, 111 F. Supp. 294 (S.D.N.Y. 2000).

U.S. Constitution, Art. I, § 8.

35U.S.C. §§ 102, 103, 112 (2000).

18U.S.C. § 1832 (2000).

Webster, B. (2009). In Re Bilski appealed to the Supreme Court. Retrieved January 13, 2010, from http://bfwa.com/2009/01/29/in-re-bilski-appealed-to-the-supreme-court/.

Zimmerman, D. (2007). Interdisciplinary living without copyright in a digital world. *Albany Law Review, 70*, 1375–1397.

## ADDITIONAL READING

Bagley, A., & Brown, J. (2007). Broadcast flag: Compatible with copyright law & incompatible with digital media consumers. *IDEA, 47*(5), 607–658.

Bambauer, D. (2008). Faulty math: The economics of legalizing the grey album. *Alabama Law Review, 59*, 345–406.

Bartley, M. (2008). Slinging television: A new battleground for technology and content holders. *IDEA, 48*(4), 535–560.

Burk, D. (2008). The mereology of digital copyright. *Fordham Intellectual Property. Media & Entertainment Law Journal, 18*, 711–739.

Cacovean, C. (2009). Is free riding aided by parody to sneak between the cracks of the Trademark Dilution Revision Act. *Hastings Communication and Entertainment Law Journal, 31*(3), 441–462.

Crawford, J., & Strasser, R. (2008). Management of infringement risk of intellectual property assets. *Intellectual Property & Technology Law Journal, 20*(12), 7–10.

Cundiff, V. (2009). Reasonable measures to protect trade secrets in a digital environment. *IDEA, 49*(3), 359–410.

Darrow, J., & Ferrera, G. (2007). Social networking Web sites and the DMCA: A safe-harbor from copyright infringement liability or the perfect storm? *Northwestern Journal of Technology and Intellectual Property, 6*(1), 1–35.

Gatto, J., Blaise, B., & Esplin, D. (2009). Worlds.com Saber-rattling portends a trend in virtual world and video game patents. *Intellectual Property & Technology Law Journal, 21*(5), 8–12.

Gore, K. N. (2009). Trademark battles in a Barbie-cyber world: Trademark protection of website domain names and the Anticybersquatting Consumer Protection Act. *Hastings Communication and Entertainment Law Journal, 31*(2), 193–222.

Handel, J. (2009). Uneasy lies the head that wears the crown: Why content's kingdom is slipping away. *Vanderbilt Journal of Entertainment and Technology Law, 11*, 597–636.

Heald, P. (2008). Property rights and the efficient exploitation of copyrighted works: An empirical analysis of public domain and copyrighted fiction bestsellers. *Minnesota Law Review, 92*, 1031–1063.

Helberger, N., & Hugenholtz, P. B. (2007). No place like home for making a copy: Private copying in European copyright law and consumer law. *Berkeley Technology Journal, 22*, 1061–1098.

Hetcher, S. (2009). Using social norms to regulate fan fiction and remix culture. *University of Pennsylvania Law Review, 157*, 1869–1935.

Jacobson, S. (2007). Now playing on an iPod near you: Rip, mix, burn - it's your music: But is it your video. *Tulane Journal of Technology and Intellectual Property, 9*, 349–363.

Kappos, D. J., Thomas, J. R., & Bluestone, R. (2008). A technological contribution requirement for patentable subject matter: Supreme Court precedent and policy. *Northwestern Journal of Technology and Intellectual Property, 6*(2), 152–170.

Koransky, J. (2009). Magazine publishers exhale: Exploiting collective works after Greenberg. *The John Marshall Law School Review of Intellectual Property Law, 9*, 161–183.

Latham, R., Butzer, C., & Brown, J. (2008). Legal implications of user-generated content: YouTube, MySpace, Facebook. *Intellectual Property & Technology Law Journal, 20*(5), 1–11.

Lunceford, B., & Lunceford, S. (2008). The irrelevance of copyright in the public mind. *Northwestern Journal of Technology and Intellectual Property, 7*(1), 32–49.

Madison, M. J. (2008). Intellectual property and Americana, or why IP gets the blues. *Fordham Intellectual Property. Media & Entertainment Law Journal, 18*, 677–710.

Mahesh, G., & Mittal, R. (2009). Digital content creation and copyright issues. *The Electronic Library, 27*(4), 676–683. doi:10.1108/02640470910979615

Masters, R., & Weber, B. (2009). Intellectual property cases and trends to follow. *Intellectual Property & Technology Law Journal, 21*(6), 14–24.

Ng, A. (2008). Authors and readers: Conceptualizing authorship in copyright law. *Hastings Communication and Entertainment Law Journal, 30*(3), 377–417.

Peitz, M. (2006). Piracy of digital products: A critical review of the theoretical literature. *Information Economics and Policy, 18*(4), 449–476. doi:10.1016/j.infoecopol.2006.06.005

Reilly, T. (2008). Debunking the top three myths of digital sampling: An endorsement of the Bridgeport Music Court's attempt to afford "sound" copyright protection to sound recordings. *The Columbia Journal of Law & the Arts, 31*, 355–408.

Rosenblatt, B. (2008). Rights management and its role in social media markets. *Journal of Digital Asset Management, 4*(2), 112–122. doi:10.1057/dam.2008.13

Sawyer, M. (2009). Filters, fair use & feedback: User-generated content principles and the DMCA. *Berkeley Technology Law Journal, 24*, 363–404.

Smith, A. D. (2009). The future of digital music sales among Web-enabled professionals: An empirical investigation. *International Journal of Business Information Systems, 4*(3), 263–289. doi:10.1504/IJBIS.2009.024096

Sween, G. (2009). Who's your daddy? A psychoanalytic exegesis of the Supreme Court's recent patent jurisprudence. *Northwestern Journal of Technology and Intellectual Property, 7*(2), 204–223.

Tao, H. (2010). Performance and its legal protection in the digital era. *Intellectual Property & Technology Law Journal, 22*(1), 20–27.

## KEY TERMS AND DEFINITIONS

**Digital Products:** Content or multimedia products that are in a digital format when possession is passed to the consumer. Examples include newspaper content, magazine content, e-books, music, movies, games, and software.

**Copyright:** A form of legal protection provided to the authors of original works of authorship fixed in a tangible form of expression, including literary, dramatic, musical, artistic, and certain other intellectual works.

**Copyright Infringement:** Unless a valid exception applies, whenever the expression of an idea is copied in an unauthorized manner, an infringement of copyright occurs.

**Patent:** A governmental grant that gives the inventor the exclusive right to make, use, or sell an invention for a limited period of time.

**Trademark:** A protected word, name, symbol, or device which is adopted and used in the trade of goods to indicate the source of the goods and to distinguish them from the goods of others.

**Trade Secret:** Information that, when used in business, has actual or independent economic value because it is not generally known by competitors.

**Work Made for Hire:** An employer-employee relationship exception to the general rule that the person who actually creates a work is the legally-recognized author of that work.

# Chapter 4
# Pricing in the Digital Age

**Chip E. Miller**
*Drake University, USA*

## ABSTRACT

*Consumers have greater ability than ever to compare prices on products using the Internet. Also, information goods can be sold at much lower prices because of greatly reduced or almost non-existent costs of production. However, because of the ease of pirating information goods, company pricing strategy must take steps to offset losses from unauthorized copies of digital goods. An overview of traditional pricing strategy is presented, followed with research findings of specific actions to undertake for optimal pricing strategy in various scenarios. Discussions of versioning, windowing, bundling and unbundling, with recommendations for use of each, follow. This chapter explores the pressures placed on prices, the strategies companies use when setting price, and provides examples and discussion of sales methods on the Internet for both physical and digital goods.*

## INTRODUCTION

Pricing items for sale in the Digital Age, whether they be information goods, services, or physical products is far more complex than its corresponding number in traditional marketing settings. Nonetheless, as the population moves inexorably toward more interactive transactions and global customers become the order of the day for most businesses, learning to adapt to this relatively new medium is essential.

The Internet provides a variety of sources for products and services unparalleled in the history of commerce. Whereas before a customer in a large city might compare prices by going from one store to another in town, and then choosing the best deal, customers can now easily check multiple websites offering identical products and choose which one has the best price. The ordeal of comparison shopping with one's feet has an alternative. In the case of rural buyers, or those in

DOI: 10.4018/978-1-61692-877-3.ch004

foreign markets with limited choices, having no options at all except the local dealer or a catalog has been supplanted. These developments all point to a greater degree of price transparency, wherein the customer need not wonder whether or not they are getting the best deal possible. With a few clicks of the mouse, they can be relatively assured of an optimal exchange.

Comparisons of physical goods for best price are very easy on the Web. Model numbers and specifications are identical across suppliers and retailers, so the only question remaining is what the item costs and what the additional charges are for such ancillary services as shipping and handling. Similarly, shopping for some commodity services such as airline tickets or concert tickets from Ticketmaster or its competitors is simplified on the price front. Knowing one's destination, price comparisons can enable the consumer to choose the best option for their budget, in many cases eliminating a middleman's fees.

Why is identifying and setting an optimal price so important to the company? Revenues from sales provide the capital essential to operate the firm. Making less money than possible will lead to disgruntled stockholders and a drop in stock prices. Declining investment generally results in the company being less competitive, less able to invest in research and development or innovations that ensure its continued success in the market. Proper pricing strategy is at the heart of financial success for the firm.

The chapter will proceed as follows. It will define what information goods are to better understand the need for a fresh approach to their pricing strategy. Next will be a discussion of the underpinnings of pricing—corporate objectives and their effect on pricing policy, constraints on pricing, influential factors and the actual methods chosen to set price. Within each of these functional areas of pricing strategy will be a discussion of how pricing on the Web differs from traditional models and what academic researchers have found to be optimal strategies in the digital realm.

The chapter concludes by presenting some novel digital product pricing strategies that could be used in the future.

## BACKGROUND

### Information Goods

The terms information goods and digital goods will be used interchangeably throughout this chapter. Such products are items that can be transmitted electronically to the consumer, have few or no variable costs related to their reproduction, and minimal to no distribution costs depending on their channel. The variety of products fitting this description is fairly wide. It includes the following items: online content for news, stock quotes or research; software; music; entertainment such as movies or videogames; electronic teaching cases; e-books; graphics or clip art; ringtones. This list is expected to expand rapidly as more and more products are converted to digital form (Sundararajan, 2004).

Information goods are special in economic terms. While the initial cost of creation may be quite large—millions of dollars for a movie or videogame—duplication costs are extremely low. Also, it is quite difficult in many cases to ascertain the quality of a particular product. Whether one will like a song by an unknown artist or a movie can only be judged after obtaining the work for inspection.

Lal and Sarvary (1999) note that goods have both digital and non-digital attributes. The former are those characteristics that can be conveyed through the Internet and, for the most part, are features that can be evaluated through visual inspection. The Web has the ability to enlarge the number of digital attributes over those found in catalogs or even in traditional stores. For example, samples of music can be listened to over one's computer before making a music purchase. Non-digital attributes are such things as fit (e.g. of clothing), smell, taste or freshness. While generally the competitive environment on the Internet results in lower prices, these researchers highlight circumstances when the opposite is true.

## Pricing Objectives

The first step in discussing price is to look at the objectives of the corporation. Firms have traditionally emphasized one of three objectives in determining pricing strategy. These are profit maximization, increasing sales, and increasing market share.

## Profit Maximization

Profit maximization involves gaining the most profit possible for each unit sold. Generally a short term phenomenon, it functions best when there are few or no substitutes for one's product. Examples of this would be the pharmaceutical industry, where high prices are the norm while a product is under patent protection. Similar examples can be found with digital information products such as video games. New games are priced to maximize their return while the novelty still exists. This objective will, of course, suffer in those circumstances when other competitors can easily enter the market, reducing or eliminating the opportunity for excess profits. Because of the global nature of the Internet, wherein competitors can immediately see your price and challenge it almost as quickly, profit maximization would be a difficult strategy to pursue.

## Increasing Sales

Increasing sales can be useful in that it sends positive signals to investors. Given the vast number of start-ups in Internet commerce that are using venture capital to get established, it is important to the firm to be able to demonstrate that its product or service is indeed viable. Being able to repay some of the very high start-up costs is important if the firm is to survive. That said, the competition on the Internet is especially fierce, and an offering that is priced higher than alternatives may have slim chances of being selected. Thus, pursuing this objective often involves penetration pricing, discussed below, so that repeat sales in the future cover the minimal profits or even losses of early stages. The risk involved with this strategy is mitigated somewhat by the fact that most digital products have very low or non-existent variable costs. This approach is the model followed by perhaps a majority of Internet commerce firms. For example, AOL early on instituted a flat rate fee that was minimal, but designed to attract the greatest number of users. AOL became attractive to advertisers and others who sought access to their user base, so the minimal or non-existent profits from the access fees were offset by other revenue streams.

The same model can be followed in a B2B environment. Amazon, in its earlier stages of development, charged lower commissions to companies that were in its partner network. As its marketing power increased, it was able to raise those fees and still retain its connections with independent booksellers and others.

Success under this scenario is built on the supposition that, in the future, earnings will increase to cover early losses. This may prove elusive or impossible in the Internet environment because of the global competition one faces and the ability of other competitors to offer, very swiftly, similar or superior offerings to that of the first mover. One method to enhance the likelihood of success is to engage in versioning—product line pricing in a more traditional sense—where future offerings are superior to (and more expensive than) the entry level or early versions. Taking advantage of the tendency of people to stay with existing items and not seek change because of potential risks involved, this approach can lead to substantial future revenue streams. As Lal and Sarvary (1999) point out, with the advent of the Internet, many consumers will merely continue to buy what they are familiar with instead of comparison shopping. Because the Internet obviates the need to even go to the store, the likelihood of enhanced brand loyalty is increased because the cost of making a trip to the store to perhaps compare other goods with one's current choice is greater than merely buying the same item again on the Internet. In this instance, the price charged by the Internet retailer can even be higher than that in a traditional store.

## Increasing Market Share

Sales alone are not a definitive indicator of the success of the company vis-à-vis its competition. Sales can rise merely because the entire market has grown, and the company may actually be shrinking in relation to competitors. Similarly, an economic downturn may cause a drop in sales that has nothing to do with the product or its desirability, and very little in the way of incentives or price reductions will attract the audience. An alternative objective, then, is seeking to increase one's market share. This is determined by the proportion of the firm's sales against total market sales.

One way information goods firms seek to create market share is by offering goods to the public for free or at very low prices. This benefits the firm in several ways. Such wide circulation brings favorable network effects that stimulate later sales and, if sheer audience size is the goal, attracts advertisers who wish to be seen by the audience. Also, the very low variable costs of information goods permit price setting that leads to higher volumes of sales through penetration pricing (Linde, 2009).

Increased market share brings many advantages to the firm that possesses it. First, economies of scale hand the firm a cost advantage. In e-commerce environments, this may not be as obvious as in traditional firms, especially if the company is selling digital goods. The manufacturing economies of scale would not pertain in this instance. Nonetheless, larger firms with more market share have greater pricing flexibility. The axiom about a million customers each generating a dollar of profit or one customer generating a million dollars in profit holds here. Customers can come and go, but the firm is buttressed against severe losses because of the sheer size of its client base. Moreover, in an environment where customers expect free products or services, it is important to the Internet firm—blogger, search engine or other entity—to be able to provide a substantial "customer base" for use by advertisers or companies that wish to sponsor a slot on a search engine to enhance their visibility.

Similarly, firms with greater market share have increased bargaining power with suppliers, partners and channel members. Amazon.com is able to charge greater fees to its syndicated selling partners. Internet stores with global audiences have the potential to negotiate favorable terms with their suppliers, benefiting the former with increased sales to a larger customer base and themselves with lower prices.

Finally, when industries are stagnant, it becomes important for a firm to increase its market share. In that manner, sales can still increase. Once again, Internet firms have an inherent advantage in this regard. Because sales declines are not generally universal across the globe, slow sales in one country can be swiftly offset by new sales gained in another market outside the home country. No need for setting up new factories or stores abroad.

## Price Constraints

There is a wide range of factors that modify how much a firm can charge for its goods. Among these are consumer demand, stage of the product life cycle, product line depth, economies of scale, legal restrictions and piracy. Most researchers adopt the stricture in their models of a monopolist, and we will continue that convention here.

The fundamental constraint on price is the demand for the product. According to basic economic theory, as demand for a product or service increases, price will also rise. Lower demand for a product results in lower prices, all else being equal. The degree to which a firm can raise its price is, to some extent, determined by the number of substitutes available to the consumer. When many substitutes exist, prices tend to be held down because a rise in the price of a good will drive consumers to seek readily available alternatives. This has been advanced as a fundamental expectation of lower prices resulting from Internet sales because consumers can so easily compare prices and substitute goods. Exceptions to the expected rule can be found in Lal and Sarvary (1999).

Where the product or service is in its life cycle also has an effect on what the company can charge. In the early stages of the product life cycle, prices are generally higher. Costs to the firm are substantial at this point—recovery of research and development, extensive promotion costs to make the public aware of the product, fees to retailers to induce them to carry the product—all contribute to a higher unit cost. Also a factor is that sales are comparatively low at this stage, therefore the variable costs are higher per unit and the fixed costs are spread over fewer units as well. Finally, when a product is new and has few or no substitutes, the public must pay what the manufacturer charges or do without.

As the product life cycle progresses, prices tend to drop as competitors enter the market and more substitutes are created. In the maturity stage, discounting is common in an effort to keep sales up. When the product reaches a decline stage, pricing may either rise or fall. In the former case, there is a persistent niche market that desires the good and will pay a premium to obtain it from the few remaining suppliers. For example, phonograph needles are in far less demand since the advent of digital music. Audiophiles seeking such needles must needs pay the asking price or do without. An example from the video games industry might be an old version of "Donkey Kong". It has been supplanted by superior games, but may have some nostalgic value and therefore limited demand. Given that the product is digital, there is virtually no cost to reproducing another copy for the consumer beyond the packaging, so some profits can still be made.

Price may also fall in the decline stage. Demand declines as products become obsolete. In this case, the company may seek to dispose of its remaining inventory and lowers price to get rid of remaining product before its value falls to zero. What value would antivirus software from 1999 have in the current era? Thus, following a standard economic model, producers will seek to sell their goods as long as marginal revenue exceeds marginal cost. Given that digital products have extremely low marginal costs, digital entertainment goods that keep some value over time, for example e-books, music, movies, or games, can continue to be sold for longer periods of time than most physical goods.

Having a product line to show the customer introduces the idea of cross-elasticity of demand. A consumer can compare one product of a company's line with others immediately above and/or below it in price to get a better sense of its value. Thus, having a very highly priced product may increase the sales of items directly below it as the consumer believes the lower priced item to be a better value compared to the highly priced place-holder. Digital examples would be the standard movie on DVD and a supreme version with out-takes, additional scenes, interviews with the actors and so forth.

Legal constraints exist for many goods as well, restricting them to a more narrow range of prices. Companies could in the past easily charge different prices across international boundaries because the markets were physically isolated. With customers being able to shop globally, some firms are attempting to restrict what products are sold where in order to sustain revenues. In some cases, such as the EU, governments are forbidding firms to penalize buyers for purchasing outside their home territories. The ease with which such shopping can be accomplished on the Web is a cause for concern among some firms that lack the most cutting-edge goods or the lowest prices.

A special point of interest that is more prominent in information goods than traditional physical goods is piracy or unauthorized copying. Because digital products such as software and music are relatively easy to copy and, in many cases, the pirate can produce perfect copies for little or no cost, piracy has a profound effect on demand for the product and its ultimate price. Production costs will remain higher if piracy reaches substantial proportions, and companies may be tempted to raise the price of the good even further to offset predicted losses due to unauthorized copying. This approach, however, seems only to exacerbate the problem of piracy.

Cheng, Sims and Teegen (1997) investigated some of the underlying factors of piracy and discuss how it affects pricing. They note that products that have a short but intense usage life are more likely to be pirated. Hence, video games become boring after relatively limited but intense use and aficionados are prone to merely copy new games to save money.

Some consumers will obviously presume that software or digital goods are too expensive, and thus justify in their own minds that copying is permissible. This makes less sense in the case of free music downloads, however. Again, Cheng, Sims and Teegen (1997) address both of these cases in their study. Users seeking to test out a new product or who will only use it temporarily will be more likely to copy rather than purchase. Because entertainment software has a short lifespan, it will remain vulnerable to piracy even if priced cheaply.

Khouja and Park (2008) and Sundararajan (2004) offer suggestions on pricing strategies in the face of piracy. It is a given that revenues and pricing power will decline in the face of piracy, so methods of defeating it are especially sought after. Sundararajan sets forth guidelines for pricing under the threat of digital piracy. Under low threats, one should price as if the threat is not present and adjust downward based on the value the lowest customer type would get if they pirated the good.

As piracy becomes more likely, more careful value differentiation is called for because lowering the price will actually draw in some customers that might otherwise have been inclined to pirate the good. In the maturity stage of the product's life cycle, desired usage levels for the product rise and the price is able to be increased as well. Khouja and Park note further that when pirates dominate the market and variable costs are low, price cutting is in order. This would appear to be the case in the music industry, where high quality reproduction of songs is easily accomplished among peers or from download sites. Hence, we see the 99 cent singles available from many distributors because younger users do not suffer ethical angst due to downloading music for free. Conversely, if ethical consumers dominate the market (as might be the case when businesses purchasing a site license are the target), there is less incentive to cut prices.

## Influences on Pricing

There are many forces that result in prices on the Internet being driven downward. The greatest and most obvious of these is consumer search capability. It is easier than ever for consumers to shop nationally or even globally for the best price. Aiding in this endeavor are sites like that of PriceScan (www.pricescan.com) where prices of products are listed in ascending order. This helpful site not only provides prices, but will also include vendor ratings, shipping charges and a calculator to enable the customer to determine the total cost of the good in question. Companies that cannot compete successfully on price are likely to be immediately removed from the consumer's selection set. Other shopping agents like CNET Shopper (http://shopper.cnet.com/) list vendors by sponsor rather than price in order to relieve price pressure.

A second factor that depresses Internet prices is the speed with which such prices are visible to one's competition. Instead of the old method of sending an employee to check the prices of the store across town, one now need only visit the website to see what is transpiring. Moreover, competitors are global, not local, intensifying the pressure to keep one's prices in line with others in the trade. Skimming strategies, which allow one to set a high price initially for unique offerings, are generally less useful in this setting because of the ability of the buyer to compare prices swiftly across many vendors. The information good adaptation of skimming is windowing, which will be discussed in more detail further on. Penetration pricing, in which one lowers price dramatically at the outset in order to wall out contenders and attract the widest range of customers, seems more prudent if the resources to sustain such an approach exist. With many digital goods, the penetration pricing model is quite viable, as production costs are minimal.

The majority of Internet sales take place across state boundaries. At least presently, that enables the buyer to avoid paying sales tax on the purchase. This is an obvious advantage that reduces the overall cost of the purchase by anywhere from 3% to more than 10%. For the time being at least, the U.S. government has chosen to uphold this tax haven for buyers.

Many Internet companies, either because of the attitude of their investors or because of corporate strategy, have taken a long term view regarding corporate growth. With that aim in mind, low prices or even free offerings early on enable the firm to gain a great deal of buzz and the potential to raise their share of the market. While the average Internet user has been conditioned to receive offerings for free, whether it be information or use of the website's service, there is a risk with this strategy in the long term. The assumption is that once brand equity and consumer satisfaction have been established, retention will occur even when the previously free items now have a price tag attached. A common outcome, however, is that once a free item is no longer free, customers tend to leave, undermining the idea of building share by the company.

The cost of goods sold online also faces upward pressure as well, with the outcome of higher prices passed on to the consumer. Lal and Sarvary (1999) model, in detail, circumstances when prices on the Internet tend to be higher. One factor is situation where there are few or no channel members whose services increase value and price. Channel members indeed raise prices by offering the value of storage close to the consumer and product display, among other things. Consequently, when those channel members are not present, the Internet firm selling a physical product must still get that product to the consumer. When the channel member performs this service, it does so from stocks held on site, generally bought in large quantities and shipped in a single order to save costs. The Internet retailer usually ends up sending out each order individually to customers, eliminating any economies of scale in shipping and incurring additional charges of preparing the order for shipping. If the retailer passes this cost on to the buyer, some resentment may ensue as the buyer can find other sites that will ship for free. If the product is identical, and no other mitigating circumstances exist, the first seller loses an order. In the case of international orders, absorbing the cost of shipping may be ruinous, especially for products such as books with their relatively low value to weight ratio. This concern will not exist, of course, for digital products such as downloaded software, information such as hotel prices or airline tickets.

Many websites are not readily available to the consumer as they are likely to get lost in the sea of online options for a product. Consider these booksellers from Amazon.com—Jerods, LeapYear Books and Bugs8997. While they all are highly rated by customers that Amazon has referred to them, the likelihood that any of them would surface in a search of book dealers that have a specific title is infinitesimal. Hence they rely on syndicated selling, in which Amazon charges a commission whenever a referral from Amazon purchases a book from the dealer. These fees run from 7% to 15% and may significantly inflate the cost of buying the object in question. Nonetheless, for smaller and/or more obscure retailers, this is a cost that is well worth paying. The exposure gained from links on Amazon or other major shopping sites is immense, and may determine the difference between marginal success and a comfortable income for the small business owner.

Some experts argue that advertising on the Internet is cheaper than its traditional counterparts of print and broadcast. This is due to the ability of the Internet to more precisely target customers such that the advertiser only pays for those who actually click through—hence, no wasted reach.

Brandt Dainow of Think Metrics provides this example. An ad in the *New York Times* costs $1000 per column inch. A $10,000, one-quarter page ad reaches 2 million people, of whom 100 buy your goods and generate $10,000. Your cost per sale would be $100. Google, by contrast, charges 54 cents per visitor. With a standard two percent conversion rate for your website, your cost per acquisition is a mere $27.00. Advantage, Google.

Other elements exert pressure to keep prices up in Internet sales. Website development and maintenance is substantial. For even a small online retailer of airguns, annual expenses for a website were $45,000 on just over $1 million in sales. Forrester indicates that development costs may range up to $100,000, exclusive of the ongoing maintenance costs for hardware upgrades, software and other costs. Because some consumer desertion will occur if sites are not constantly updated, and changes in competitor prices and product offerings are more frequent on the web, maintenance costs for the site must be passed on to the consumer.

Online consumers were conditioned from the beginning to treat the Web as a source of free information. This attitude has carried over into other aspects of e-commerce, sometimes forcing companies to perform services at no cost in order to gain customers. Obviously, revenues must be generated elsewhere to offset these giveaways, driving overall prices upward.

## Pricing Models

There are a number of standard pricing models used in traditional marketing. Variations of these standard approaches can also be seen in the Web environment. The most commonly seen pricing approaches are prestige pricing, price lining, versioning, windowing, captive product pricing, bundling, loss leader pricing and demand-based pricing. Demand-based pricing, in turn, has different strategies that can be pursued. These include fixed fee vs. usage-based pricing, discounting and the dynamic pricing approach of yield management pricing.

## Prestige Pricing

Prestige pricing centers on the fact that the consumer will pay a premium price for the product or service because the quality of the item is very high or because the psychological value of ownership is quite desirable. In the latter case, there are many reasons why the customer may wish to possess the item. One would be the status conferred by ownership. Possessing a rare piece of art, fine jewelry, leather-bound first edition books or an expensive Italian sports car definitely raises one's standing. This may be especially important to those aspiring to be part of the next higher socioeconomic group, hence they acquire at least some of the trappings of those they wish to join or emulate.

Other reasons exist to pay high prices for expensive goods. Some individuals may reason that having better quality equipment may improve their game. Hence, less able players may have the most expensive equipment available, assuming that using such will at least partially offset their lack of practice or innate skill.

Some individuals will not purchase the finest equipment available merely for looks, however. Olympic athletes, world class surgeons, Formula One racers and others with consummate skill can benefit from the best equipment there is. Their abilities are such that they can wring the last bit of performance out of the item.

Finally, paying more can result in better use of one's time. For example, business executives that purchase first class tickets on airlines are able to make the most use of their time. They can work more easily while on the aircraft, gain swift entrance and egress instead of wasting time waiting in line, and arrive at their destination relatively rested and able to perform their duties.

Prestige pricing has the same place in Internet sales that it does in conventional stores. There is always a segment of the population that is willing to pay a premium price for a good. The major advantage of the Internet is the global reach it possesses, giving customers worldwide the chance to purchase products. Coupled with this can be an auction format, wherein buyers will likely bid up the price on a particularly desirable object, perhaps even beyond market value.

In the case of digital goods, prestige pricing might be feasible in more than one way. For example, the new video game Halo 3 ODST comes in both the conventional and the collector's edition. The latter is $99.99 compared to the $59.99 standard version. Additional features are included with the deluxe model, making it more desirable despite the higher price. Too, hard-core gamers are more likely to be attracted to the high-end version and willing to pay the price premium. The consumer surplus generated by the deluxe model sales helps to subsidize the lower priced version and offset some potential losses from piracy.

A combination of good and service can also fetch a premium price. Publishers of journals can provide the abstract alone for free, while subscribers get the full article and additional search options to help with their research. Data providers such as Mediamark Research or Hoover's allow free use of slightly older data or limited information, but provide paid subscribers with a full range of additional information and supplementary services (Venkatesh & Chatterjee, 2006).

## Price Lining and Versioning

Price lining occurs when products or services are priced at discreet levels, with gaps between each level. The implication is that, with each step up in price, the quality or number of features increases and hence justifies the greater expense to the consumer. For example, a manufacturer sells shirts under its label at $34.95, $49.95 and $109.95. There is an obvious gap in the line that could be filled with a $79.95 shirt. Utilizing the principle of cross-elasticity of demand, the most expensive shirt increases the demand of the new shirt beyond what it would be if the $79.95 model were the top of the line. Even though the most expensive shirt may sell low volumes, the cost of producing it is more than covered by the increased sales of the lower priced items. Similarly, Corvette serves as a desirable car to GM mostly in that it greatly increases the sales of Camaros. A 2009 Z06 is priced at $64,000 and gets few buyers. However, the certainty that some of the engineering skill and product features such as the engine may be had in slightly less exotic form in a Camaro for no more than $27,000 is appealing and makes it worthwhile to offer the high end model, even as a loss leader.

Versioning is digital goods' equivalent of price lining. Newer versions of software, games or information are continuously being produced. When several of these items are available at the same time, price lining can take place with the oldest (or least sophisticated) version being available for the lowest price. In this environment, many buyers will see the comparison table for, say, McAfee Antivirus and decide that the least expensive version carries with it too many potential hazards. Thus the step up to the next level of protection is not as great as the perceived price would be if the intermediate version were all that were offered.

Versioning plays an important role in addressing the needs of various customer segments and maximizing profits by matching customer value with price. TurboTax has a free version for those doing 1040 E-Z, Deluxe ($29.95) for personal use, Premier ($49.95) for those with investments or rental property, as well as higher versions for business users (http://turbotax.intuit.com/). Each subsequent version has more features available to provide more utility to the consumer in exchange for a higher price.

Linde (2009) notes that at least three versions of the product are optimal. If only two versions exist, customers often choose the less expensive version. If three are present, customers will tend to purchase the middle product. More discriminating customers will gravitate toward the highest priced item, while introducing a high quality mid-range option will encourage low-end buyers to trade up to a more profitable version. Types of versioning that are used in information goods sales include up-to-dateness, availability of the information, scope of work, perceived friendliness and processing speed.

Versioning can also be helpful in maintaining relationships with one's traditional retailers as well. For instance, in order not to alienate magazine distributors such as newsstands or bookstores, it is common for the Internet source to have special content restricted to it that is not replicated by the paper version. Pricing the Internet version can take many forms in this case. It can be bundled with the paper subscription, either at the regular price or at a premium price. It can also be sold in separate packages, with the Internet version being priced at a slight premium over the paper version because of its greater depth of content, ease of retention for future use, and other features that cannot be found in the paper version, such as links to related articles on the Web or access to archived material on demand (Venkatesh & Chatterjee 2006).

Wu and Chen (2008) look at versioning as a tool to combat piracy. They address the issue of how to design versioning strategy, how many different quality levels to offer, and at what price. They note that selling a significantly lower priced and limited capability version of products can capture low end markets that otherwise would be inclined to pirate software. Their suggestion is to design for the high end of the market first and then downgrade to pursue other segments. Reinforcing the findings of other research, they argue for a single version policy when piracy is not a threat, and dual versions (one high for price insensitive customers and one low for those inclined to pirate) when piracy is likely.

## Windowing

This concept comes from TV and film media and was described by Owen and Wildman (1992). As explained by Linde (2009), windowing involves the staged introduction of a product in different forms, at different times. The goal of this strategy is to extract the full profit potential of the product by optimally pricing each of the various forms and divining the timing for each new introduction of the good. To avoid cannibalization, the offers are staged.

For example, a movie first comes out to be seen in theaters. Later, the movie appears in video form for rent or purchase. Finally, the movie makes its way to pay-TV or free TV. Each of these levels of introduction has a slightly different form, and is priced for the price elasticity of the audience in mind. Movie goers presumably have a keen interest to see the movie at its first showing and will pay the higher price commanded by a theater. Usually the subsequent DVD will surface quite some time after the initial theater run to ensure all demand at the theater is exhausted. The parallel to this in the traditional marketing literature would be a skimming strategy, although in the case of windowing, the product takes on different forms throughout the price decline process instead of merely keeping the same item and lowering price over time.

## Captive Product Pricing

Captive product pricing refers to a situation where a product can only function with dedicated components. An example from traditional marketing would be a printer cartridge that is specially designed for one printer only, and for which there are no after-market substitutes. In this case, the manufacturer can practically give away the printer itself, while charging a premium for the replacement cartridges. Similar examples of digital goods would be when Apple software was not fully compatible with Microsoft products, or video games could only be played on a single platform. With regard to copyright material on the Internet—journal articles, for example—one can easily search for references to old articles or books and perhaps even find abstracts of the information. However, to access the full text, a hefty fee may be charged. A subscription to a paper journal might cost the subscriber only $55 a year for four issues, but a single article from a back-issue, available only from the journal via download, may run as high as $35. At the very least, access to the information will require that one be a member of the society, for an annual fee. The scarcity of the content, and its copyright status, permit the owner to charge whatever they feel the market will bear.

## Bundle Pricing

Many customers like to get what they perceive as a bargain. Assembling a variety of related products or services into a bundle, then selling the bundle for less than the combined total of the individual components, represents a good deal even if some of the items aren't really desired or needed by the customer. Such bundle pricing has many uses for both manufacturers and retailers. Bundle pricing can aid in moving merchandise faster to improve cash flow; rid the seller of slower moving, outdated or otherwise less desirable goods when combined with a popular item; or attract customers with favorable prices who will in turn buy other goods at full price.

Bundle pricing has two variations. One is the equivalent of a quantity discount, in which the consumer buys a large amount of a single item. The other alternative is a mix of various goods and/or services in a bundle with the price dependent on what combination is selected (Kannan & Kopalle, 2001). Mobshop.com and Mercata.com represent the first option, where consumers aggregate to form what is in essence a buying co-op to garner better prices. While Mercata did not survive beyond 2001, Mobshop is still intact (Cook, 2001). Prices for combination bundles vary with the content of the bundle and customer type. For example, individual consumers, businesses and students may all pay different prices for a bundled subscription to the online *Wall Street Journal* and its paper form.

Let us look to music and books as typical online examples of bundle pricing. Amazon.com has music downloads for albums ranging from less than $5.99 new to $16.99 for some oldies albums. This is in line with typical demand-based pricing. The unique nature of the Internet and digital format of the music allows Amazon to pursue either bundle pricing or single item pricing. When the songs are combined into an album, the cost per song is lower. For example, a CD of Frankie Valli and the 4 Seasons has 20 songs for $11.49. The customer gets each song for less than 58 cents per song. The variable costs to Amazon are trivial, in that the expense involved in burning a CD is minute. The cost of the box and its accompanying artwork are the major expense, along with musician royalties of less than 25%. The customer pays shipping, and Amazon even makes a profit there. The only drawback to the customer is that, given the nature of music, some of the songs are likely ones that they will not want. Nonetheless, at 58 cents each, who will complain? This contention is supported by the findings of Bakos and Brynjolfsson (1999) as well.

Other users will be more selective and only purchase favorite songs. Amazon can accommodate that wish as well with solos for 99 cents each. Because the music is a download, Amazon incurs virtually no expense here. The flexibility offered with single songs attracts the market segments that either cannot or will not spend the money for an entire CD. Rather than lose the business altogether, Amazon is able to make money at 99 cents per song because the variable costs are so low.

So how did 99 cents get chosen? One suggested answer is that, as long as the item is below a dollar, it is an impulse purchase with little financial risk. Also, since the items are not bundled, most customers would accept the fact that individual products have always cost more than larger packages on a per unit basis. Amazon has no storage costs on the MP3 downloads, and very little for the CDs themselves—presumably they can be prepared on a just-in-time basis to reduce warehousing.

The literature on bundling is extensive and shows numerous well developed models on how to determine optimal bundling strategy. For example, Bakos and Brynjolfsson (1999) look extensively at the concept of bundling and the optimal approach to the strategy. They assume that most information providers are multiproduct monopolists, which reflects the state of most offerings such as software, news and research reports. Their model can be used to analyze bundling of complements and substitutes, contend with budget constraints and identify optimal sets to be offered. Their model shows that multiproduct monopolists can gain substantially higher profits with bundles compared to selling individual goods. Another important distinction for information goods is that, because distribution costs are much lower than for physical goods, even unbundling can be a profitable approach. This can be seen in the example above regarding the unbundling of individual songs for sale. Finally, the authors note that several bundles at different price points should be offered if consumer segments have different values for a set of information goods.

Linde (2009) argues that mixed bundling produces optimal profits for the firm. In this case, customers are free to choose either individual components or a package. For example, individual elements within Microsoft Office can be purchased separately, or they can be bought in a various suites of products at different price points bundled for consumer convenience.

Venkatesh and Chatterjee (2006) looked at print magazines with an online presence. They found that a wider product line resulted in higher prices than for either product alone. Interestingly, over time, the optimal price of the print magazine increased as online readership grew. In the circumstance when the print and online versions are substitutes, the more profitable product should be sold at full price and the other priced as an add-on to avoid cannibalization.

Looking at Amazon's book offers, we see many deals wherein the buyer is tempted with such deals as super saver shipping (buy over $25 worth of books and Amazon will ship them for free) or discounts when other books within the genre, by the same author, or that other buyers liked and purchased at the same time, show up as pop-up screens. If the additional item is slow moving, Amazon benefits by disposing of inventory and also can reduce its shipping costs and processing by getting the customer to buy several items in a single order.

ISPs have readily available information to compare as well, but are very difficult to sort through for the best deal with the bewildering variety of single service prices, installation fees, termination fees and other charges. Verizon charged $179 for early termination and $49.99 to activate. Mediacom charged $29.95 to activate as did Comcast. Comparing just the normal bundle of phone/Internet/TV from three major providers, we see the following in Table 1.

Those lured by advertised price alone would probably take the Verizon package. Careful read-

*Pricing in the Digital Age*

*Table 1. ISP pricing comparison*

| ISP | Bundle | Duration | Comments |
|---|---|---|---|
| Mediacom | $89.85 per month | First year | |
| Comcast | $114.99 per month | First year | HD |
| Verizon | $79.99 per month | First year | Online coupon for $60 for bundle for 1 year |

*Table 2. Wii software sales, 2006 to 2009*

| Sale Dates | Sales Volume | Price |
|---|---|---|
| November 19 to Dec. 31, 2006 | 1.2 million units | $250 |
| 2007 | 7.4 million units | |
| 2008 | 11.4 million units | |
| January 1 to Sept. 30, 2009 | 4.7 million units | $200 |

ing of the terms and conditions leads to other desirable choices. A hefty early termination fee, higher than average activation fees and perhaps other negatives await the unwary.

## Loss Leader Pricing

The concept of loss leader pricing was very commonplace on the Internet, especially since many digital goods had such low variable costs. Being free to produce, one could give them away to customers for nothing. The goal, of course, was to attract a loyal following that, in the future, would be willing to pay for products offered by the site or, if the site were for information only such as a blog, attract enough visitors that advertisers would pay to promote their wares on the site. Traditionally, stores used this method to attract customers with a commonly purchased, desirable item such as milk in the hope that the customer would buy other, full priced goods at the store during the trip. Smaller profits would be recouped by eventual customer loyalty.

The Internet attempted the same approach, but suffered more from innumerable freeloaders who only showed up for the free samples. To offset this problem, one solution would be to have a very limited free option, augmented by a limited trial period for a very low price. The incremental attraction to the customer should, over time, generate more business for the site. Venkatesh and Chatterjee (2006) argue that very limited circumstances exist for free online versions of products.

What are some other examples from the Internet and how do they fit the methods described above? Let's look first at video games, a high demand item with very high fixed costs of development. According to neoseeker.com, next generation video games in 2006 were expected to cost between $15 million and $20 million to develop. Wii, on the other hand, was brought to market for between $5 million and $8 million. The tremendous difference in cost allows Wii to engage in discounted pricing that, in addition to its breakthrough technology, gives it a significant pricing edge. Although both Wii and competing video games have the same variable costs once produced, the difference in fixed costs is substantial.

Wii, the amazingly popular game that involves player motion for some versions, was launched in November, 2006, and sold 1.2 million units in 6 weeks at a cost of $250 per unit (www.vgchartz.com). Their volume subsequent to that, and benchmark prices over time, are seen in Table 2.

At a projected $6 profit per unit (www.joystiq.com, 2008), Nintendo tidily recouped its investment, ostensibly in the first six weeks. This is a primary example of skimming strategy in marketing. The new Wii was a breakthrough product and priced accordingly. In addition, its lower development costs enabled it to be priced lower than competing games, giving Nintendo a double advantage. Because the nature of video games is such that sales drop precipitously after the first week—50-65%—the ability to maintain price for an extended period of time is unusual.

A more recent example of the pricing strategies noted above is Halo 3 ODST from Microsoft. Released on September 22, 2009, it set unprecedented sales records by moving 2.2 million units in its first week. Moreover, its sales volume only dropped 40% after the first week, giving its developers at Microsoft a greater return for a longer period of time than normal for such a good. This item comes in two versions, pursuing the lead of the prestige pricing and price lining models above. The standard edition is $59.99, while the collector's edition with a special controller is $99.99.

The example of Halo 3 incorporates several common pricing strategies at the same time. Skimming strategy is used in that the product is priced at a "maximum" when first launched to capture the excitement of a new game and its relative price inelasticity. As its popularity wanes, it will be reduced in price until it is replaced by another, newer model (versioning). Also used immediately is versioning, with the standard edition and a separate collector's edition available at launch. Finally, there is odd-even pricing, using ending digits slightly below the full dollar amount. This can trick the buyer into thinking that the product is less expensive than it first appears. Many prospective buyers will state that the purchase price for the standard edition is "around $50", when indeed it is nearer $60. In the case of the collector's edition, it is "less than $100", which is psychologically smaller by a notable margin than the actual $100 price.

Halo 3 is not sold online by the company itself to avoid competition with its various retailers. Microsoft does have other items on the website to enhance sales of the game, both gratis and for a fee. Game videos, for example, are free, while add-ons and gamer pictures are not. This effective versioning permits Microsoft to extend the product life cycle while generating revenue for itself that does not interfere with channel members' income streams.

## Demand Based Pricing

Given technology's advances, it is possible to tailor demand in real time, maximizing the return for an individual sale. One unusual example would be when Coca-Cola tested "smart" vending machines that could set up its prices to change with the temperature and demand for its vending machine products, as well as manipulate demand by offering discounts to help move product in less-than-optimal conditions (Cortese & Stepanek, 1998).

A common Internet example would be the price paid by advertisers on websites. Advertisers can pay rates based on the cost per thousand impressions. More likely, advertisers will pay a negotiated fee by the cost per click or cost per action. These options ensure that the viewer actually paid attention to the ad, reducing wasted reach that is prevalent in traditional advertising.

Up to this point, we have been discussing fixed price strategies. There are also those situations where dynamic pricing is optimal. Some items like airline seats drop to zero value if left unfilled. Therefore it is in the interest of the marketer to find customers to buy these items, because some revenue is better than none. We will discuss discounts as an elementary form of dynamic pricing, followed by yield management pricing.

## Fixed-Fee vs. Usage Based Pricing

Which is optimal and when? Sundararajan (2004) tackles this problem for managers. Both have desirable aspects for specific consumers. The fixed-fee approach makes planning budgets easier and users will tend to pay a higher price for fixed-fee, unlimited usage. Because production costs are near zero, such pricing is a viable alternative for the provider. Moreover, usage-based systems have substantial implementation and monitoring costs that militate against their use.

Most new technology markets have a high percentage of experimenters with low usage rates and a handful of innovators with high usage rates.

This suggests that a preferred pricing scheme would be to enter the market with a low fixed fee—in essence, penetration pricing. Over time, the fee should be raised to induce consumers to adopt usage-based plans.

## Dynamic Pricing

Coupons and other sales promotion tools have always been used to reduce the risk involved in purchasing new products, encourage purchases of larger quantities than normal or accelerate the sale of slow-moving items. The main contribution of the Internet to this pricing tool has been the reduction in cost for creating and distributing coupons. Consumers visiting the website of the store can find readily available coupons, encouraging immediate purchase and reducing the likelihood of comparison shopping because a price reduction is present. The cost of printing and distributing the coupon is virtually zero, and waste is further reduced because only those customers actually interested in receiving the coupon print it.

Coupons can, of course, be dynamic as well. Consumers entering a website can receive, based on their browsing habits, e-coupons of different values. This in turn allows e-tailers to set the most effective price for each group of consumers. Posted prices remain the same, but the price paid by each consumer varies. This strategy can even go so far as to match price of other competitors, changing the price listed on the current site to match the lowest option among all alternatives (Kannan & Kopalle, 2001).

Distribution of free copies of product (e.g. music singles) or free services in the hopes of either gift generating wide interest in the supplier is commonplace in information goods marketing. Generally, the assumption was that one could make up for the free disbursements through charging more for advertisers. However, both Gallaway and Kinnear (2001) and Venkatesh and Chatterjee (2006) point out that this is not an optimal approach.

Another means of discounting unique to digital goods is that of P2P (peer to peer) networks. Generally content is transferred free from one individual to another in this setting. Land and Vragov (2005) describe methods whereby these networks can be utilized by the company to its advantage. Prices are set in such a manner as to provide financial incentives for the peers to move the information goods to a wider scope of their fellows, similar to the home party plans of companies such as the Pampered Chef and Tupperware.

Dynamic pricing has several caveats based on consumer perceptions and type of product being sold. Consumers can take offense if the dynamic pricing policy of the firm thwarts their efforts to find or obtain the best price. This is most likely to happen for frequently purchased goods (where the possibility of remembering previous prices is higher) and for less price-sensitive, more loyal consumers. These effects are less likely to be a problem in yield management situations such as the airline tickets discussed below (Kannan & Kopalle, 2001).

Yield management pricing—found in the airline seat example—has been made more viable by the use of sites such as Travelocity and Orbitz. Operating in real time, the airline can maximize the value of the seat by not reducing the price unless demand appears to be so low that only a reduction will cause the seat to sell. To stimulate demand, some prices for empty seats can be adjusted to be sold at bargain prices, perhaps even below marginal cost. The possibility of being able to pick up such a bargain entices customers to return to the site and raises the likelihood that all seats on a given flight will be sold.

## FUTURE DIGITAL PRODUCT PRICING STRATEGIES

Many different pricing strategies have been discussed up to this point, but most of them involve a fairly straightforward application of traditional product pricing strategies applied to these unique digital products. Differentiating a digital product from its competitors in the future may require a unique or novel strategy that has not been tried before. The two examples discussed below include (1) variable or auction pricing, and (2) mixed subscription and sales pricing.

## Variable or Auction Pricing

One potential strategy would be to use some form of variable pricing based on the amount each individual is willing to pay. This may involve variable pricing at different points in the product's life cycle, or some form of auction mechanism for setting a price. The common auction method used online is a progressive, or English, auction equivalent where prices start low and bids are made at higher prices for some time period or until no more bids occur. This would produce some level of price discrimination where people that are willing to pay more would be charged more, and others would be charged less. For example, if someone pre-orders an e-book months ahead of its release day, they may pay a low price. At the time of release the price may be higher. And as time passes the price may decline to the point where, years after release, an e-book may be sold for a very low price, perhaps only a dollar or two. The online market is capable of handling variable pricing or auction pricing and given that the marginal cost of an additional copy of an e-book is near zero, any revenue received years after release would be mostly profit. The main cost involved would be to maintain a copy on the e-book file on a server. Data storage costs continue to decline so this cost would be inconsequential. This same model could be used for new music, movies, or games. The downside for a digital product auction pricing model would be that the buyers know that there is a virtually unlimited supply of digital product units, a very different situation from the physical product world.

## Mixed Subscription and Sales Pricing

Another approach would be to receive revenue from a combination of subscription fees and product sales prices. This would be similar to the shopping club model. For example, a customer could pay a monthly subscription fee to an e-book seller that would allow them to pay very low prices for all of their purchases for that time period. They would not be allowed to purchase any e-books without paying the subscription fee. This would provide some steady revenue for the seller while still receiving additional revenue when popular products become available. This model could be used in any of the other digital product industries. People may actually buy more products because the sale prices would be very low. The consumer may view the subscription fee as a sunk cost. The buyer would only take a small risk because they could drop their subscription at any time.

## MANAGERIAL IMPLICATIONS AND CONCLUSION

While many of the pricing models from traditional products are applicable to information goods, some modification of their use is often called for. Managers seeking insights for their own strategies should take these ideas from the research literature into consideration.

Bakos and Brynjolfsson (1999) and Venkatesh and Chatterjee (2006) provide valuable insights into the concept of bundling. For instance, selling a low quality good in a bundle will lead to greater profit than selling a high quality good outside the bundle and an unprofitable item sold alone can become profitable if included in a bundle.

More products are not necessarily better. Seeing the same sports scores from several services or very similar articles in multiple online magazines does not enhance customer value. Also, unique items such as special events on TV are pay-per-view because a small fraction of consumers will pay a premium to see them, but the general public will not watch at all, even if part of a bundle. The potential surplus should be realized outside the bundle as long as some knowledge of specific segments' price elasticity is known.

Managers should offer a menu of bundles at different prices to capture the maximum consumer surplus. Some customers will wait for information or entertainment in exchange for a lower price while those requiring instant gratification can be offered a bundle of higher priced items. Also, offer "economy" and "premium" bundles, either in terms of total items in the bundle or relative quality of the contents. Even if the cost of creating the bundles is the same, the lower priced option allows firms to pursue low-demand customers without losing revenue from high demand consumers. In those situations where the firm cannot, on its own, create bundles because of trade restrictions or sparse product offerings, it can still improve profits by selling to a broker that does the bundling.

Versioning is a well-accepted approach to product offerings and maximizing profits through price discrimination. The recommendation here is that at least three versions of a product should be offered to take advantage of consumer behavior and cross-elasticity of demand (Linde 2009).

Demand for a product has characteristics of both duration and intensity (Cheng et al., 1997). For those goods with short duration and high intensity of use, like entertainment software, the pricing strategy needs to adjust to the fact that a significant amount of piracy is likely to occur. A cheaper trial version is one option, but charging more for all units sold to cover piracy losses will only exacerbate the problem.

Using intertemporal price discrimination, as suggested by Gallaway and Kinnear (2001), allows firms to maximize their returns on both aging and new content. Using songs as the example, older music is charged a successively lower price the older the music gets. Consumers of new music showed a willingness to pay amounts equal to the per-song average of a new CD, so price inelasticity for current hits should allow companies to recoup their investments and avoid free downloads. A similar approach might be feasible in video games and similar products.

Under circumstances where the importance of non-digital attributes in determining choice is not overwhelming, products on the Internet can command a higher price than in-store equivalents. This is most often true for goods purchased regularly but not too frequently. Customers will accept higher prices on the Internet in exchange for not making a trip to a physical store and in the assurance that replacing the product they currently use meets their needs without the necessity of comparing new items with the old (Lal & Sarvary, 1999).

Obviously, the digital pricing landscape is both more complex and more simple than the world of traditional stores and products. Nevertheless, the combination of real world experience and the insights provided by academic research should light the way to a successful pricing strategy.

## REFERENCES

Bakos, Y., & Brynjolfsson, E. (1999). Bundling information goods: Pricing, profits, and efficiency. *Management Science*, *45*(12), 1613–1630. doi:10.1287/mnsc.45.12.1613

Cheng, H. K., Sims, R. R., & Teegen, H. (1997). To purchase or to pirate software: An empirical study. *Journal of Management Information Systems*, *13*(4), 49–60.

Cook, J. (2001, January 12). Venture capital: Where Mercata led, consumers were unwilling to follow. *Seattle Post-Intelligencer*. Retrieved from http://www.seattlepi.com/business/vc122.shtml.

Cortese, A., & Stepanek, M. (1998, May 4). Good-bye to fixed pricing? *Business Week*, 70-84.

Dainow, B. (2009). Comparing offline with online advertising. Retrieved October 8, 2009 from http://www.visibilitymagazine.com/think-metrics/brandt-dainow/comparing-offline-with-online-advertising.

Gallaway, T., & Kinnear, D. (2001). Unchained melody: A price discrimination-based policy proposal for addressing the MP3 revolution. *Journal of Economic Issues*, *35*(2), 279–287.

Joystiq. Retrieved October 1, 2009, from http://www.joystiq.com/

Kannan, P. K., & Kopalle, P. K. (2001). Dynamic pricing on the Internet: Importance and implications for consumer behavior. *International Journal of Electronic Commerce*, *5*(3), 63–83.

Khouja, M., & Park, S.(2007-8). Optimal pricing of digital experience goods under piracy. *Journal of Management Information Systems*, *24*(3), 109–141. doi:10.2753/MIS0742-1222240304

Lal, R., & Sarvary, M. (1999). When and how is the Internet likely to decrease price competition? *Marketing Science*, *18*(4), 485–503. doi:10.1287/mksc.18.4.485

Lang, K. R., & Vragov, R. (2005). A pricing mechanism for digital content distribution over computer networks. *Journal of Management Information Systems*, *22*(2), 121–139.

Laudon, K., & Traver, C. (2004). *E-Commerce: Business, Technology, Society* (2nd ed.). Boston, MA: Addison-Wesley.

Linde, F. (2009). Pricing information goods. *Journal of Product and Brand Management*, *18*(5), 379–384. doi:10.1108/10610420910981864

Neoseeker. Retrieved October 6, 2009, from http://www.neoseeker.com/.

Owen, B., & Wildman, S. (1992). *Video Economics*. Cambridge, MA: Harvard University Press.

Sreenivasan, S. (2009). Web retailers finding allies at sites with nothing to sell. Retrieved October 8, 2009 from http://www.nytimes.com/1997/04/14/business/web-retailers-finding-allies-at-sites-with-nothing-to-sell.

Strauss, J., & Frost, R. (1999). *Marketing on the Internet—Principles of Online Marketing*. Upper Saddle River, NJ: Prentice-Hall.

Sundararajan, A. (2004). Managing digital piracy: Pricing and protection. *Information Systems Research*, *15*(3), 287–308. doi:10.1287/isre.1040.0030

Sundararajan, A. (2004). Nonlinear pricing of information goods. *Management Science*, *50*(12), 1660–1673. doi:10.1287/mnsc.1040.0291

Turbotax. Retrieved January 26, 2010, from http://turbotax.intuit.com/

Venkatesh, R., & Chatterjee, R. (2006). Bundling, unbundling, and pricing of multiform products: The case of magazine content. *Journal of Direct and Interactive Marketing*, *20*(2), 21–40. doi:10.1002/dir.20059

Vgchartz. Retrieved October 6, 2009, from http://www.vgchartz.com/

Wu, S.-Y., & Chen, P.-Y. (2008). Versioning and piracy control for digital information goods. *Operations Research*, *56*(1), 157–172. doi:10.1287/opre.1070.0414

## ADDITIONAL READING

Angwin, J., & Wingfield, N. (2000, October 16). Priceline offshoot ate millions in costs to subsidize customers. *Wall Street Journal*, 1.

Armstrong, M. (1996). Multiproduct nonlinear pricing. *Econometrica*, *64*(1), 51–75. doi:10.2307/2171924

Bakos, Y., & Brynjolfsson, E. (2000). Bundling and competition on the Internet. *Marketing Science, 19*(1), 63–82. doi:10.1287/mksc.19.1.63.15182

Belleflamme, P. (2003). Pricing information goods in the presence of copying. In Gordon, W., & Watt, R. (Eds.), *The Economics of Copyright Developments in Research and Analysis*. Cheltenham, UK: Edward Elgar Publishers.

Bhargava, H., & Choudhary, V. (2001). Information goods and vertical differentiation. *Journal of Management Information Systems, 18*(2), 89–106.

Bhargava, H., & Choudhary, V. (2008). When is versioning optimal for information goods? *Management Science, 54*(5), 1029–1035. doi:10.1287/mnsc.1070.0773

Brynjolfsson, E., & Smith, M. (2000). Frictionless commerce? A comparison of Internet and conventional retailers. *Management Science, 46*(4), 563–585. doi:10.1287/mnsc.46.4.563.12061

Dickson, P. R., & Sawyer, A. G. (1990). The price knowledge and search of supermarket shoppers. *Journal of Marketing, 45*, 116–129.

Hanson, W., & Martin, K. (1990). Optimal bundle pricing. *Management Science, 36*(2), 155–174. doi:10.1287/mnsc.36.2.155

Kannan, P. K., & Biehal, G. (2000). The impact of dynamic e-coupons on consumers' reference prices and purchase behavior. Working paper, Smith School of Business, University of Maryland, College Park.

Lynch, J. G. Jr, & Ariely, D. (2000). Wine online: Search costs affect competition on price, quality, and distribution. *Marketing Science, 19*(1), 83–103. doi:10.1287/mksc.19.1.83.15183

Monroe, K. (2003). *Pricing: Making Profitable Decisions* (3rd ed.). Boston, MA: McGraw-Hill/Irwin.

Mui, K., Monroe, K., & Cox, J. (2002). Pricing on the Internet. *Journal of Product and Brand Management, 11*(4), 274–287.

Nascimento, F., & Vanhonacker, W. R. (1988). Optimal strategic pricing of reproducible consumer products. *Management Science, 34*(8), 921–937. doi:10.1287/mnsc.34.8.921

Shapiro, C., & Varian, H. (1998). Versioning: The smart way to sell information. *Harvard Business Review, 76*(6), 106–114.

Shapiro, C., & Varian, H. (1999). *Information Rules: A Strategic Guide to the Network Economy*. Boston, MA: Harvard Business School Press.

Varian, H. (2000). Buying, sharing and renting information goods. *The Journal of Industrial Economics, 48*(4), 473–488. doi:10.1111/1467-6451.00133

Venkatesh, R., & Mahajan, V. (1997). Products with branded components: an approach for premium pricing and partner selection. *Marketing Science, 16*(2), 146–165. doi:10.1287/mksc.16.2.146

Wilson, R. B. (1993). *Nonlinear Pricing*. New York, NY: Oxford University Press.

Zagorsky, J. (1990, July 2). The upside of software piracy. *Computerworld*, •••, 21.

Zentner, A. (2006). Measuring the effect of file sharing on music purchases. *The Journal of Law & Economics, 49*(1), 63–90. doi:10.1086/501082

# KEY TERMS AND DEFINITIONS

**Bundling:** Combining goods or content in a package for sale at a single price. For instance, the articles in a magazine or a compilation of songs on an album constitute a bundle. Bundles may consist of very popular and less popular items (e.g. hit songs and less well liked numbers) or merely differing content (e.g. news articles).

**Captive Product:** A product that can only be utilized in the presence of a dedicated component. Digital examples would be video games that must be played on a specific platform.

**Digital and Non-Digital Attributes:** Digital attributes are product features that can be evaluated through visual inspection and do not need the product to be physically present. Non-digital attributes such as fit, freshness and flavor require the product to be present for evaluation.

**Dynamic Pricing:** Prices that fluctuate with changes in demand or other factors. Examples would be prices for airline seats or Coca-cola's flexible soft drink prices that moved with the temperature and availability of soda in the vending machine.

**Digital Goods and Information Goods:** Goods such as research reports, software, music, video footage or other items that are in electronic form. Such goods are easily reproducible in perfect or near perfect form at little or no cost to the purchasing copying the original.

**Internet Marketing:** Conducting marketing functions (information exchange, promotion, sales) of products through the medium of the Internet via computer linkages.

**Piracy:** Unauthorized copying of digital goods.

**Prestige Pricing:** Setting a high price to attract quality-conscious or status-conscious consumers.

**Versioning:** Different editions of particular digital product. The various editions may vary in quality, number of features available, or enhancements. Versioning is used to help more fully capture various market segments with different demands.

**Windowing:** Similar to versioning. Releasing the same product in slightly different form periodically to capture all consumer surplus. An example would be movies in the theater, followed by rental/purchase DVDs, available for viewing on cable TV, and finally on free TV.

**Yield Management Pricing:** Varying the price of a good or service in real time to reflect current demand. The aim is to maximize profits for perishable goods such as airline seats or hotel reservations.

# Chapter 5
# Financing Digital Product Companies

**Richard B. Carter**
*Iowa State University, USA*

**Frederick H. Dark**
*Iowa State University, USA*

## ABSTRACT

*Faced with the prospect of positive and negative network externalities and the all-or-nothing phenomenon, digital product (DP) firms must choose the timing of their capital acquisitions carefully. Moreover, with typically high fixed-to-variable cost ratios, the risk to recovering the initial investment is critical. In this chapter the authors discuss various forms of financing for the DP firm, both short-term and long-term, with these issues in mind. But our primary focus is the initial public offering of equity (IPO) and particularly its timing. Through empirical analysis and case studies we show that if DP firms issue too early in their life cycle they may receive a price for their shares that is not commensurate with long-term prospects. However, issuing too late may mean that they either cannot sell shares or are unable to recover their initial investment.*

## INTRODUCTION

Digital product (DP) firms include software developers, e-books, e-newspapers, digital music and movies, and games. Depending on their size and maturity, all firms have a myriad of choices for raising capital and DP firms are no exception. Many DP firms have pursued a strategic growth strategy to take advantage of newly available opportunities like personal computers, electronic mail, the Internet, and electronic commerce. In some instances there have been barriers to introduction because larger successful companies already existed or were already producing a competing product that would dominate and create an all-or-nothing market share environment. There is also a large amount of fixed cost investment for the typical DP firm that must ultimately be recovered. But if the firm is successful, and a critical mass of users reached, potential profits and returns on investment can be very large. However, with growth comes an increasing need for capital.

DOI: 10.4018/978-1-61692-877-3.ch005

The typical forms of capital include short-term or long-term debt or equity. There are also hybrids like convertible bonds where the debt instrument can be exchanged for proportional shares of equity following a waiting period. The appropriate form of capital depends on many factors but for many DP firms, organized in the past few decades, the ultimate form is the initial public offering of equity (IPO). With the IPO DP firms can recover fixed costs and finance growth into the future, as well as provide extensive wealth for their founders.

For the DP firm the timing of their IPO is critical. Up to a certain point in the firm's life cycle prospects are uncertain. The digital product must be compatible with existing hardware and other software platforms – an objective that is often elusive given the rapidly changing industry environment. There is also the issue of whether the ultimate user will accept the product as the premier platform version if they accept it at all. If the DP firm issues their IPO too early the discount to its shares could be extreme because of the impounded risk – and not commensurate with their ultimate potential. But should they wait too long they may fail and be unable to sell shares at any price.

The objective of this chapter is to present the problems faced by DP firms as they make the capital acquisition decision and how various forms capital available to the firm may provide solutions. We focus on the IPO as a good source of financing for the DP firm and the timing decision discussed above. With empirical analysis and brief case studies we show how IPO timing – a point in the firm's life cycle – can determine the success of the IPO in recovering the original owners' investment and finance future growth.

## BACKGROUND

What was not understood well in the 1980s and early 1990s when the growth in DP firms escalated was that these firms, and their products that are in a digital format, are unique in comparison to traditional product firms. One of these unique characteristics is the importance of network externalities where a software development firm would have an advantage if there was product compatibility across many potential users. While network externalities can be a feature of traditional firms (e.g., railroads and trucking), it appears to be more important for DP firms. For example, externalities can decrease survivability for pioneers but increase survivability for technologically intense products and larger firms, and those with an installed base of customers (Srinivasan, Lilien, & Rangaswamy, 2004).

A potential consequence of markets with network externalities is an expression of the winner-take-all phenomenon (Yamamoto et al., 2002). For various reasons only a limited number of firms provide the product that becomes the dominant design while the others are left to whither. Microsoft Windows, for example, became the consumer favorite for operating systems and graphical user interfaces while others, like Linux, have not. And once one product leads its competitors its success accelerates as new users are more compelled to choose it because of its greater perceived utility. Though a number of firms may enter the market, only a limited number will succeed - relegating losers to technology lockout (Schilling, 1998). Schilling (2002) shows that in such markets failure to invest in learning, or poor market entry timing, can be detrimental. "Firms now (post-Internet) have to compete not only within, but also across differentiated channels, with some of the firms competing in multiple channels and transferring competition across them (Viswanathan, 2005)." As research shows, with differences in cost structures and/or externalities, the probability of success or failure can hang on the vagaries of consumers (Srinivasan, Lilien & Rangaswamy, 2004).

Another unique feature of DP firms is their heavy reliance on fixed costs. Unlike traditional firms that manage both ongoing fixed and variable costs such as materials, logistics and labor, DP firms are more fixed-cost driven. For example, the fixed costs involved in developing the first copy of a software package are very high and typically require large capital investments (Bakos, 1991). Low variable costs as in the development of software for business and individuals also produce substantial economies of scale (Bakos, 1991; Rayport & Sviokla, 1995). For traditional firms variable costs can be significant and managed throughout the growth and maturity phases of the firm life cycle. While there may be a wide range of fixed-to-variable cost ratios, the preponderance of traditional firms have appreciable variable costs that can be controlled. Moreover, these controllable variable costs can also be a source of recovery for traditional firms when sales wane.

As part of our analysis for this chapter we examined a sample of 528 DP firms. We compared the DP sample to a sample of 4,281 traditional, non-DP firms including firms from several industries including manufacturing, mining, transportation, etc. All of the firms went public over the period 1981 through 2005. We find fixed-to-variable cost ratios of 5.62 and 2.47 for DP firms and non-DP firms, respectively. The association between high fixed-to-variable cost ratios, operating leverage, and risk has long been established (Lev, 1974). Hence, risk for these DP firms is very high early during start-up because they must recover this large fixed cost from product sales in a timely manner or risk failure.

These unique characteristics of the DP firm require different managerial decisions at different points in the firm's life cycle than for traditional, non-DP firms. Moreover, there is evidence that early decisions by a firm's management will have a lasting impact (Wilbon, 2002). One of these early decisions relates to financing strategy – how and when to raise capital.

Studies of equity capital acquisition generally treat firms with different operating cost structures alike (for a survey of related research see Ritter & Welch, 2002). Moreover, much of the research concerning DP firms treats them more as technology firms than in a more fundamental cost-structure framework (see for example Schill & Zhou, 2001). In these studies, a computer hardware manufacturer would be considered in the same class as a software developer or online service provider. Yet the combination of the high fixed-to-variable cost ratio (FCVC) and importance of network externalities for the DP firm may have a profound effect on when to acquire capital.

## FINANCING THE DIGITAL PRODUCT FIRM AND THE TIMING OF THE IPO

### Sources of Capital

Early sources of capital are generally from the private resources of the firm's founders - personal savings, life insurance, credit cards, etc. – often referred to as bootstrap financing (Carter & Van Auken, 2005). Once these private funds have been exhausted, management turns to outside sources. Because the growth of DP firms can be substantial the choices are limited.

Traditional pecking order theory suggests that after depleting private funds, firms first go to long-term debt financing from banks and private lenders; then to public debt like notes and bonds - saving outside equity for last (Myers, 1984). Contradicting theory, Smith & Watts (1992) and Barclay, Morellec & Smith (2001) argue that high growth firms, like DP firms, have relatively less long-term debt in their capital structure. This may be due to an incentive problem for risky firms where owners are less likely to take potentially lucrative projects if they know long-term debt holders will take a disproportionate share of the return (Myers, 1977).

While short-term debt may be a temporary solution to the incentive problem, external equity provides a long-term solution. A major external source for early-stage capital is private investors or angels. These are wealthy individuals interested in the high risk and high return opportunities that new business enterprises offer. Angels invest $30-$40 billion annually in entrepreneurial companies (Sohl, 1999). Advantages of angel capital are the size and accessibility of the investment pool as well as a positive reputation many wealthy angels bring to the table. The major disadvantage is the matching of the right angel with the right firm. In some cases, investors and entrepreneurs have resorted to paying costly finders fees in an effort to link ideas and capital. Moreover, the angel investor may not have expertise in the selected firm's industry but still interfere with the management of the firm in an effort to protect their capital. Another potential problem arises when the original angel does not have enough money to reinvest should the need for additional capital develop.

A second source of outside equity, venture capital funds, represents wealthy investors who have pooled their resources and hired professionals to make investment and other related decisions. They consist of about 500 funds that manage $35-$45 billion and make $3-$4 billion in new investments annually (Sohl, 1999). Venture capitalists have become intensely stratified and generally specialize within industries and sectors, in the stage of the prospective company, in the type of ownership structure, or in a particular locale. They are aware of the high risk to their investment and often look for target rates of return of 500% on their investment within three years and 1000% within five years (Sohl, 1999). In addition, they conduct extensive research of the company and its operations, commonly referred to as due diligence, before committing funds and subject the current owners and management to intense scrutiny. Unlike angel capital, venture capitalists are often able to bring additional money to the table if needed. Major problems include the lack of understanding of the venture capital market by potential users and the matching of investors with entrepreneurs when the venture capitalists vary so dramatically in their tastes for investment and requirements for participation.

The definitive source of equity capital, in terms of the size needed for DP firm growth and wealth potential for its owners, is an IPO. The firm generally contracts with an investment bank to help price the stock and distribute the shares. Choosing the right investment bank can be critical for obtaining a fair price for the firm's shares (Carter & Manaster, 1990). But there is evidence to suggest that the choice is not always in the hands of the firm (Fernando, Gatchev, & Spindt, 2005). While the value of the firm can increase considerably following the offering, it is a costly endeavor from finding the right investment bank to paying the bank's commission and myriad expenses. The proprietary nature of the investment banking industry may also mean that the firm is at the mercy of the bank's pricing and distribution practices. Among these practices is under-pricing, where the bank deliberately discounts the stock price from its estimated value, and spinning, where shares of hot IPOs -- those that are expected to escalate in price -- are reserved for the bank's best clients. While the IPO adds tremendous value to the firm, under-pricing and spinning appear to add more cost to the offering process as a rapid increase of the stock price in the after market suggests the firm was sold for considerably less than its market value.

## The Timing of the Digital Product Firm IPO

Theory suggests that firms issue equity when they believe their stock prices are relatively high (Myers & Majluf, 1984). The theory has support both in the apparent valuation at the offering and the performance in the after-market and beyond (see for example, Loughran and Ritter, 1995). For IPOs, Alti (2005) argues that the timing decision is based on a signal of investor interest generated by the recent relative offer prices of other IPOs. An increase (decrease) in IPO offer prices reflects a decrease (increase) in uncertainty of a common valuation factor which in turn elicits (inhibits) the marketing of further IPOs. Schultz (2003) suggests that market timing is not driven by insiders' ability to predict future returns but what he refers to as pseudo market timing. Schultz argues that managers are simply trying to offer equity when they believe their stock is at its highest value. As a result offerings are at the peak market value for their firm in most circumstances. The explanation fits well with the poor post-offering performance IPOs (Ritter & Welch, 2002).

The idea that managers are trying to sell stock at its peak may be a way to understand the differences in timing of IPOs for DP and non-DP firms – especially considering the importance of timing in markets where network externalities exist (Schilling 2002). For DP firms that face markets where consumer interest has a self-promoting effect, management must access capital markets prior to resolving the winner-take-all dilemma. If they wait too long investors can identify losers - in which case their shares may be worthless.

Why might a DP firm issue too early or too late? Carter, Strader and Dark (2009) argue that digital product and service firms that go public early do so because of their own insecurity regarding eventual success. With the all-or-nothing phenomenon that DP firms face, insecurity and apprehension about timing decisions is high, or at least higher than non-DP firms that know there is always a chance to recover – through controlling variable costs for example. Hence, some DP firms may jump the process before they fail and are unable to recover private capital – even if they must sacrifice a potentially higher price for their stock should they wait. However, if DP firm management waits too long it is likely they have information that the firm is a winner – with little apprehension about timing. This leads to the supposition that the later a DP firm issues stock the higher their offer price and the less likely they are to fail. This is not to imply that market timing is not important, but rather that developmental timing is more critical for DP firms.

By looking at the timing of IPOs in a company's life cycle we can get an idea of a reasonable time for going public given the nature of their firm: their idiosyncratic risk and developmental maturity. We compare our sample of 528 digital product firms with initial public offerings between April of 1981 and September of 2005 to the 4,281 non-DP firms issuing IPOs over the same time period. Descriptive statistics for both the DP firms and the non-DP firms along with difference tests are presented in Table 1.

Among the significant differences between DP and non–DP firms are beta, the after-market return variance, pre-IPO sales and net income. These differences suggest DP firms are riskier than non-DP firms. But while the higher initial return for DP firms likely reflects their riskier nature, the average long-run returns for both firm types are similar. Other interesting differences are ownership retention and underwriter reputation, where DP firms appear to have more reputable IPO underwriters and their owners tend to retain a larger fraction of the public firm than non-DP firms. Both of these differences may suggest confidence by DP firm owners in their long-term prospects – yet a larger portion of DP firms than non-DP firms fail in their first five years.

To examine outcomes should a firm go public early or late we must first determine a normal time frame for IPOs. Using an ordinary least squares regression model, we begin with the following independent variables: size of the offering, the after-market return variance (a measure of risk) and the level of revenues at the time of the offering. These variables have been used in IPO studies to estimate risk and/or maturity (Ritter, 1984; Carter & Manaster, 1990; Ritter & Welch, 2002). We include the fixed-to-variable cost ratio for reasons discussed above. Using the actual age the firm went public as the dependent variable and the observations from a benchmark period, we estimate an equation for DP and non-DP firms to estimate their normal age to go public and compare it to their actual IPO age during a test period.

Table 1. Characteristic variables for digital product vs. non-digital product firms at the IPO

|  | Digital Product Firms (N=528) | | Non-Digital Product Firms (N=4,281) | | |
|---|---|---|---|---|---|
| Variable | Mean | Std Dev | Mean | Std Dev | Difference Tests |
| Offer Price ($) | 12.90 | 5.01 | 12.09 | 5.00 | *** |
| Age of Firm at IPO(Years) | 8.63 | 8.85 | 14.66 | 22.71 | *** |
| Carter & Manaster Reputation Rank | 8.24 | 1.41 | 7.75 | 1.71 | *** |
| Beta | 2.29 | 1.34 | 1.48 | 1.12 | *** |
| After-market Variance (%) | 5.48 | 2.34 | 3.89 | 1.90 | *** |
| Pre-IPO Sales (Mill $) | 58.07 | 212.09 | 236.17 | 1270.89 | *** |
| Pre-IPO Net Income (Mill $) | -11.61 | 74.76 | 6.01 | 93.75 | *** |
| Market Value: IPO Day (Mill$) | 549.93 | 2,364.49 | 301.78 | 1,215.52 | ** |
| Initial Return First Day (%) | 37.96 | 66.03 | 14.11 | 30.39 | *** |
| Owners' Shares in IPO (%) | 14.84 | 18.65 | 13.24 | 21.70 | * |
| Owners' Firm Retention (%) | 73.35 | 13.78 | 63.42 | 25.87 | *** |
| Fixed Cost/Variable Cost | 5.62 | 15.21 | 2.47 | 28.48 | *** |
| 5-Year Raw Return (%) | 32.81 | 251.02 | 44.93 | 269.05 | |
| % Failed or Failing in 5 Years | 25.19 | 57.38 | 17.72 | 49.50 | *** |

*p <.05 *** p <.01

While the total sample involves a span of 25 years, the first time period (1981-1992) is only used to form the IPO age benchmarks. Certainly there has been a dramatic evolution in information technology use and e-commerce from 1981 through 2005. The period we chose for testing is 1993 through 2005 to incorporate the more recent surge in digital product firms. It is difficult to identify exactly when the digital product and service revolution began but it is likely that it was in the early 1990s with the advent of the World Wide Web when digital products could be advertised and delivered through various electronic means. For our sample there are three times as many DP firm IPOs per year in our test period than prior – this is compared to only a 29% increase for non-DP firms.

Using our benchmark period and the equation discussed above, we estimate the normal issue age for the test period as 8.40 years and 12.01 years for the DP and non-DP firms, respectively. This difference is significant at better than the 0.01% level. Comparing the expected to the actual age firms issued their IPOs in the test period, DP firms issued in 8.48 years after founding and non-DP firms 15.05 years after founding. This difference is also significantly different at the 0.01% level.

In Table 2 we provide some performance results for DP and non-DP firms as a function of IPO timing during the test period - if they went public sooner than expected (early) or later (late). To highlight differences we define early and late as those IPOs where actual age less the normal age is outside the 25$^{th}$ percentile from the mean – although we obtained similar results when we varied these operational definitions to some degree. The difference test referenced is a Wilcoxon rank sums test – as population distribution characteristics are uncertain.

For DP firms, early IPOs have significantly greater initial return, worse long-run performance and are significantly more likely to have failed or be failing than late firms. For firms going public earlier than expected there is a signifi-

*Table 2. Initial returns, failure rates & long-run return for IPOs issued early vs. late*

|  | Actual Age < Expected Age | | | Actual Age > Expected Age | | | Difference |
|---|---|---|---|---|---|---|---|
| Variable | Mean | Std Dev | N | Mean | Std Dev | N | Test |
|  |  |  | 820 |  |  | 237 |  |
| Initial Return (%) | 456.89 | 751.47 |  | 8.23 | 11.37 |  | *** |
|  | Total | % of N |  | Total | % of N |  |  |
| Failed & Failing | 158 | 19.27 |  | 12 | 5.06 |  | *** |
| 5-year Return (%) | 40.34 | 222.32 |  | 48.53 | 267.25 |  |  |
| *Panel A: Non Digital Product Firms* | | | | | | | |
|  | Actual Age < Expected Age | | | Actual Age > Expected Age | | | Difference |
| Variable | Mean | Std Dev | N | Mean | Std Dev | N | t stat |
|  |  |  | 96 |  |  | 45 |  |
| Initial Return (%) | 529.64 | 597.29 |  | 22.03 | 35.16 |  | *** |
|  | Total | % of N |  | Total | % of N |  |  |
| Failed & Failing | 27 | 28.28 |  | 3 | 6.67 |  | *** |
| 5-year Return (%) | -21.50 | 104.46 |  | 31.39 | 126.72 |  | *** |
| *Panel B: Digital Product Firms* | | | | | | | |

*** p<.01

cantly larger percent (at the 0.01% level) of DP firms either having failed or failing – the latter defined as losing 95% of their value - than for the non-DP firms. These results are supportive of the Carter, Strader & Dark (2009) digital product and service firm IPO timing theory. Also in support, the five-year market-adjusted return is substantially and significantly worse (at the 0.1% level) for the DP firms than for non-DP firm early issuers. A contemporaneous Nasdaq market return was subtracted from each five-year IPO return. However, we obtained similar results using a sample matched to book-to-market and market value. Moreover, in their study, where a number of control variables are included in the analysis, Carter, Strader and Dark (2009) find that early digital product and service firm IPOs are significantly more likely to fail than early non-DP IPOs. While our sample period covers a number of years, the independent variables provided in Carter, Strader and Dark (2009) likely control for systematic differences across periods. For both firm types the late issuers have better survival rates than early issuers but only the DP firms have a significantly lower initial return and therefore better relative offering price than for the early issuers. These results demonstrate the critical nature of timing for all firms but especially those with digital products.

## Three Cases of Timing for the Digital Product Firm IPO

We may gain some further insight by looking at a few individual digital product firm cases that went public during our test period and where the firm issued equity early or late according to our estimates. The cases were selected from among those that fell into the early and late categories in order to illustrate the importance of IPO timing. One of the late DP firms was Blackbaud Incorporated. Blackbaud provides software and related services for nonprofit organizations worldwide. Its solutions include The Raiser's Edge, a software application to manage nonprofit organizations. They went public on July 22, 2004 using J. P.

Morgan as their managing underwriter. At the time, J. P. Morgan was considered among the elite of investment banking houses. According to our calculation they should have gone public 7.31 years after their founding but waited 15 years. At the time of their offering their return variance in early after market trading was roughly 12.25 and their beta was 1.72. These figures are substantially lower than the 28.52 variance and 2.23 beta of the overall DP firm sample. Revenues were $92 million with a net loss of $1.4 million in the year prior to the IPO but Blackbaud achieved several very profitable years previously. They offered their stock at $8 per share and it rose to $8.53 by the end of trading on the first day. An increase of only 7% compared to nearly 115% increase in first day return for other firms that went public around that time. As of mid 2009 Blackbaud was trading at roughly $15 per share, representing an annualized return of 12% since going public. Overall, these results are representative of late issuers.

A firm that appeared to have issued stock too early was Software Net Corporation who went public on June 17, 1998. They produced commercial off-the-shelf computer software for consumers, small business and large-enterprise markets. Much of their marketing was done online. The company was founded in 1994 but only emerged in its final form one year prior to marketing their IPO. According to our calculations they should have gone public roughly 6.5 years after founding but chose to issue their IPO within 4 years of incorporating and within one year of an important organizational change. Their managing underwriter was Deutsche Bank Securities, a prestigious investment bank. Revenues were fairly strong but reported a substantial net loss for the year prior to the IPO. Their return variance was 79.2% and their beta was 2.85. Both of these imply a very risky firm and considerably more risky than our average DP firm. As they wrote in their offering prospectus: "Since inception, the Company has incurred significant losses (Software Net Corporation Form 424B1, 1998, p. 5)." As predicted by Carter, Strader and Dark (2009) the stock price rose dramatically on the first day ... from $9 to $13.25 a 47% increase, suggesting the firm and its owners did not receive a good price for their shares. Within three months Software Net had lost nearly all of their value and were delisted from Nasdaq.

A second firm that, according to our estimates, issued stock early was CyberCash. Using physical points of sale and integrated enterprise payment solutions, CyberCash initially took a commanding position in Internet payment software. While we calculated that they should have gone public in their seventh year, CyberCash issued their IPO in February of 1996 - less than two years after their founding. W.R. Hambrecht, a prestigious investment bank, was their book-runner. Like Software Net, CyberCash reported a net loss prior to going public and their beta in after-market trading was a very risky 3.86. As predicted by Carter, Strader and Dark (2009), the stock price went from $17 per share to $28 on the first day of trading - a 70% increase and was selling for as much as $60 within a short period of time. As with Software Net, the run-up suggests that the firm and its owners did not receive a fair price for their shares. But in five years CyberCash's stock price had fallen to less than a dollar and they declared bankruptcy one month later.

Incidentally, all three of the firms outlined above issued their stock in roughly the same market environment. In each case the stock market had increased an average of about 1.5% in the two months prior to their offerings. The other interesting similarity is that after the offering the original owners retained about the same amount of the firm – roughly 80%.

## FUTURE RESEARCH DIRECTIONS

We have provided evidence that there is likely an ultimate window of opportunity for issuing public equity for the digital product firm. However, we have not examined those firms that waited too long and failed. Obviously, if a firm is trying to wait until their expected stock market price reaches a peak to issue their IPO but in the meantime their product loses favor with the consumer the IPO is likely to fail. In this research we confined our sample to those firms that completed a successful offering. An important follow-on study is an examination of this group of firms to determine if there is evidence of imminent failure in enough time to issue public equity and recover initial capital. A survey of managers of these firms may provide timing information that may be useful to the management of private digital product firms. Future studies could also address these issues for each of the digital product industries separately to see if there are significant differences for IPOs offered by online content companies (magazines and newspapers), e-book publishers, multimedia product companies (music and movies), or software companies.

## CONCLUSION

Digital product firms are faced with some unique circumstances when attempting to find financing to support their growth. First, there is the prospect of positive network externalities where the value of a product or service increases as more people use it. Second, there is the all-or-nothing phenomenon where one particular platform may become the standard to the exclusion of all others. Finally, typically high fixed-to-variable cost ratios increase the risk to recovering the initial investment.

The lessons learned from the evidence presented here, as well as the brief case studies, is that the initial public offering provides a viable form of capital acquisition especially for firms financing rapid growth. But timing is important. Digital product firms should be cognizant of the value to investors of the firm's record knowing full well that the original owners are trying to recover at least some of their own invested capital. Selling shares prior to having a sustained record of success, at least on some level, might mean receiving a price for their shares that is not commensurate with the eventual outcome for the firm. These early issuers also have poorer long-term performance than non-DP firms. On the other hand, unless the owner is certain they will succeed, waiting too long may mean no issue at all for the digital product firm. As we pointed out above we did not determine how many DP firms waited too long and failed before being able to issue their IPO.

To conduct our analysis we had to estimate the normal age a DP firm goes public using such measures as the market value of the firm at the time of the IPO and the relative prestige of the underwriter. While we are not recommending the use of these measures alone to estimate IPO timing, we do believe the DP firm managers/owners, in conjunction with an investment bank with experience in marketing DP firm IPOs, can determine if they have a sufficient track record to issue public equity. But as they make their IPO decision they must bear in mind that if they wait to find the ultimate time they may miss their window of opportunity.

## REFERENCES

Alti, A. (2005). IPO market timing. *Review of Financial Studies*, *18*, 1105–1138. doi:10.1093/rfs/hhi022

Bakos, J. Y. (1991). A strategic analysis of electronic marketplaces. *Management Information Systems Quarterly*, *15*, 294–310. doi:10.2307/249641

Barclay, M. J., Morellec, E., & Smith, C. W. (2001). On the debt capacity of growth options. Unpublished manuscript, University of Rochester, New York.

Carter, R., & Manaster, S. (1990). Initial public offerings and underwriter reputation. *The Journal of Finance*, *45*, 1045–1068. doi:10.2307/2328714

Carter, R., & Van Auken, H. (2005). Bootstrap financing and the characteristics of small business owners and their business environment. *Entrepreneurship and Regional Development, 17*, 129–144. doi:10.1080/08985620500067548

Carter, R. B., Strader, T. J., & Dark, F. H. (2009). *Digital product & service firms: fixed-to-variable costs and the timing of IPOs*. Unpublished manuscript, Iowa State University.

Fernando, C., Gatchev, V., & Spindt, P. (2005). Wanna dance? How firms and underwriters choose each other. *The Journal of Finance, 60*, 2437–2469. doi:10.1111/j.1540-6261.2005.00804.x

Lev, B. (1974). On the association between operating leverage and risk. *Journal of Financial and Quantitative Analysis, 9*, 627–641. doi:10.2307/2329764

Loughran, T., & Ritter, J. (1995). The new issues puzzle. *The Journal of Finance, 50*, 23–51. doi:10.2307/2329238

Myers, S., & Majluf, S. (1984). Corporate financing and investment decisions when firms have information that investors do not have. *Journal of Financial Economics, 13*, 187–221. doi:10.1016/0304-405X(84)90023-0

Myers, S. C. (1977). Determinants of corporate borrowing. *Journal of Financial Economics, 5*, 147–175. doi:10.1016/0304-405X(77)90015-0

Myers, S. C. (1984). The capital structure puzzle. *The Journal of Finance, 39*, 575–592. doi:10.2307/2327916

Rayport, J. F., & Sviokla, J. J. (1995). Exploiting the virtual value chain. *Harvard Business Review, 75*, 75–85.

Ritter, J. (1984). The 'Hot' issue market of 1980. *The Journal of Business, 57*, 23–51. doi:10.1086/296260

Ritter, J., & Welch, J. (2002). A review of IPO activity, pricing, and allocations. *The Journal of Finance, 57*, 1795–1827. doi:10.1111/1540-6261.00478

Schill, M., & Zhjou, C. (2001). Pricing an emerging industry: Evidence from Internet subsidiary carve-outs. *Financial Management, 29*, 5–33. doi:10.2307/3666374

Schilling, M. (1998). Technological lockout: An integrative model of the economic and strategic factors driving technology success and failure. *Academy of Management Review, 98*, 267–284. doi:10.2307/259374

Schilling, M. (2002). Technology success and failure in winner-take-all markets: The impact of learning orientation, timing and network externalities. *Academy of Management Journal, 45*, 387–398. doi:10.2307/3069353

Schultz, P. (2003). Pseudo market timing and the long-run underperformance of IPOs. *The Journal of Finance, 58*, 483–518. doi:10.1111/1540-6261.00535

Smith, C. W., & Watts, R. L. (1992). The investment opportunity set and corporate financing, dividend and compensation policies. *Journal of Financial Economics, 32*, 263–292. doi:10.1016/0304-405X(92)90029-W

*Software Net Corporation Form 424B1*. (1998, June 18). Retrieved September 14, 2009 from http://www.sec.gov/Archives/edgar/data/1060531/0000891618-98-002957.txt

Sohl, J. (1999). The early stage equity market in the USA. *Venture Capital: An International Journal of Entrepreneurial Finance, 1*, 101–1120. doi:10.1080/136910699295929

Srinivasan, R., Lilien, G., & Rangaswamy, A. (2004). First in, first out? The effects of network externalities on pioneer survival. *Journal of Marketing, 68*, 41–58. doi:10.1509/jmkg.68.1.41.24026

Viswanathan, S. (2005). Competing across technology-differentiated channels: The impact of network externalities and switching costs. *Management Science, 51*, 483–496. doi:10.1287/mnsc.1040.0338

Wilbon, A. (2002). Predicting survival of high-technology initial public offering firms. *The Journal of High Technology Management Research, 13*, 127–141. doi:10.1016/S1047-8310(01)00052-9

Yamamoto, H., Okada, I., Kobayashi, N., & Ohta, T. (2002). The information channel effect in the winner-take-all: a multi-agent simulation. *Proceedings of the 6th World Multi-Conference on Systemics, Cybernetics and Informatics, 2*, 510-513.

## ADDITIONAL READING

Advani, A. (2009). The angel in your pocket. *Entrepreneur, 37*(11), 83.

Alti, A. (2006). How persistent is the impact of market timing on capital structure? *The Journal of Finance, 61*(4), 1681–1710. doi:10.1111/j.1540-6261.2006.00886.x

Benninga, S., Helmantel, M., & Sarig, O. (2005). The timing of initial public offerings. *Journal of Financial Economics, 75*(1), 115–132. doi:10.1016/j.jfineco.2003.04.002

Brau, J., & Fawcett, S. (2006). Initial public offerings: An analysis of theory and practice. *The Journal of Finance, 61*(1), 399–436. doi:10.1111/j.1540-6261.2006.00840.x

Coakley, J., Hadass, L., & Wood, A. (2007). Post-IPO operating performance, venture capital and the bubble years. *Journal of Business Finance & Accounting, 34*, 1423–1446.

Cockburn, I., & MacGarvie, M. (2009). Patents, thickets and the financing of early-stage firms: Evidence from the software industry. *Journal of Economics & Management Strategy, 18*(3), 729–773. doi:10.1111/j.1530-9134.2009.00228.x

Gubbins, E. (2008, November 3). VC chill could slow innovation. *Telephony*, 5.

Hale, G., & Santos, J. (2008). The decision to first enter the public bond market: The role of firm reputation, funding choices, and bank relationships. *Journal of Banking & Finance, 32*(8), 1928–1940. doi:10.1016/j.jbankfin.2007.12.016

Ibrahim, D. (2008). The (not so) puzzling behavior of angel investors. *Vanderbilt Law Review, 61*(5), 1405–1452.

Inderst, R., & Mueller, H. (2009). Early-stage financing and firm growth in new industries. *Journal of Financial Economics, 93*(2), 276–291. doi:10.1016/j.jfineco.2008.07.004

Jain, B., Jayaraman, N., & Kini, O. (2008). The path-to-profitability of Internet IPO firms. *Journal of Business Venturing, 23*(2), 165–194. doi:10.1016/j.jbusvent.2007.02.004

Klein, D., & Mingsheng, L. (2009). Factors affecting secondary share offerings in the IPO process. *The Quarterly Review of Economics and Finance, 49*(3), 1194–1212. doi:10.1016/j.qref.2009.04.002

Lee, Y., & Lee, J. (2008). Strategy of start-ups for IPO timing across high technology industries. *Applied Economics Letters, 15*, 869–877. doi:10.1080/13504850600820650

Lerner, J. (1994). Venture capitalists and the decision to go public. *Journal of Financial Economics, 35*(3), 293–316. doi:10.1016/0304-405X(94)90035-3

Levisohn, B. (2008, September 19). Corporate financing in a pinch. *BusinessWeek Online*, 24.

Lucchetti, A., & Bauman, L. (1998). Underwriters and nice timing boost IPO. *Wall Street Journal – Eastern Edition, 232*(92), C1.

Luna, L. (2001). It's all in the timing. *Telephony, 240*(2), 44.

Magri, S. (2009). The financing of small innovative firms: The Italian case. *Economics of Innovation and New Technology, 18*(2), 181–204. doi:10.1080/10438590701738016

Maksimovic, V., & Pichler, P. (2001). Technological innovation and initial public offerings. *Review of Financial Studies, 14*(2), 459–494. doi:10.1093/rfs/14.2.459

Minola, T., & Giorgino, M. (2008). Who's going to provide the funding for high tech start-ups? A model for the analysis of determinants with a fuzzy approach. *R & D Management, 38*(3), 335–351. doi:10.1111/j.1467-9310.2008.00518.x

Neus, W., & Walz, U. (2005). Exit timing of venture capitalists in the course of an initial public offering. *Journal of Financial Intermediation, 14*(2), 253–277. doi:10.1016/j.jfi.2004.02.003

Shen, Y., & Wei, P. (2007). Why do companies choose to go IPOs? New results using data from Taiwan. *Journal of Economics and Finance, 31*(3), 359–367. doi:10.1007/BF02885725

The website is the business. *Fortune – Winter 2001 Technology Guide, 142*(12), 144-150.

Vong, A., & Zhao, N. (2008). An examination of IPO underpricing in the growth enterprise market of Hong Kong. *Applied Financial Economics, 18*, 1539–1547. doi:10.1080/09603100701704256

Wang, C., Wang, K., & Lu, Q. (2002). Do venture capitalists add value? A comparative study between Singapore and US. *Applied Financial Economics, 12*, 581–588. doi:10.1080/09603100010011963

## KEY TERMS AND DEFINITIONS

**Going Public:** The process by which a privately held company sells a portion of its ownership to the general public through a stock offering. Owners generally take their firms public because they need additional large sums of equity funding that they are unable or unwilling to contribute themselves.

**Initial Public Offering (IPO):** The first sale of stock by a company to the public. The most common reason for a company to initiate an IPO is in order to raise more capital.

**Initial Return:** The relative difference between the IPO offer price and the closing bid price on the first day of trading.

**Network Externality:** The effect that one user of a good or service has on the value of that product to other people.

**Operating Leverage:** The percentage of fixed costs in a company's cost structure. Generally, the higher the operating leverage, the more the company's income is affected by fluctuation in sales volume and the greater the risk to the firm.

**Technology Lockout:** The all-or-nothing phenomenon where consumers reject a vendor-specific digital product in favor of a preferred variety.

**Underwriter:** An intermediary between an issuer of a security and the investing public, usually an investment bank.

# Chapter 6
# Accounting for Digital Products

**Yasemin Zengin Karaibrahimoğlu**
*Izmir University of Economics, Turkey*

## ABSTRACT

*Digital products are "content" goods such as software, books, music, or movies which can be digitized and traded on a digital market place. With the increase in trade and ownership of digital products, several important management issues have arisen. Accounting treatment for digital products is one important management issue. It is argued that digital products require regulations in terms of recognition, measurement, valuation, reporting and taxation. Therefore, the purpose of this chapter is to discuss the accounting problems that arise as a result of the growing importance of digital products in the business environment and to propose suggestions based on the accounting concepts and standards. For this purpose, first, the increasing importance of digital products is briefly explained. Then, the challenge created as a result of expanding trading volume of digital products are discussed in terms of accounting with suggestions for the appropriate accounting for digital products.*

## INTRODUCTION

Electronic commerce (e-commerce) has been the main subject to emerge in recent years. With the use of the Internet and digital devices there has been a significant alteration in traditional business models (Luxem, 2000). E-commerce falls into two categories: (1) trade in physical goods such as computers, televisions or books with a physical substance and (2) the trade in digital products such as software, music and video products, and several other different services that lack any physical substance (Vincent, 2004). Therefore, e-commerce not only involves a digital process of selling or buying (as a trade process) of physical products, it also involves content based products with and without physical substance. The development of e-commerce has caused a change in traditional business methods. Companies operating as publishers, software developers or computer

DOI: 10.4018/978-1-61692-877-3.ch006

programmers, banks and insurance agents have started to offer a wide range of digital products (Hui & Chau, 2002).

Recently, as a result of technological developments, the way of doing business has changed through e-commerce. Rapid development of digital technologies and use of the Internet facilitate the communication and the transfer of payment around the world eliminates the borders of trade, transportation, and marketing. However, in addition to its advantages such as simplifying the trade among different countries or improving competition among companies, e-commerce has brought uncertainty about the nature of trade and legal treatments.

According to Rayport and Sviokla (1995), products are either classified as tangible (physical) or intangible (service related). Poon and Joseph (2000) argued that goods are classified based on "tangibility"- how tangible the goods are. According to them, while highly tangible goods can only be traded or used in a physical form, low tangibility goods are content-based goods used preferably in a physical form or a digital form. For example, while a refrigerator is a highly tangible good, a software package used through physical means might be classified as low tangibility. Even though software is a content-based product, it is used or traded through other means. However, there are also intangible goods either content-based goods or services downloadable or used only in digital form. This kind of product may be consumed over an infinite life time, or it may be restricted by the seller, or produced in limited numbers for a limited time, or limited in some other respects. Depending on their characteristics, digital products may be considered as intangible according to United States (US) and International Accounting codifications.

For physical products, the case is simple. All digitally traded physical products are recognized, measured, valued, and reported in accordance with accounting standards. Similarly, as these products have a physical substance, whether domestically or internationally traded, they are taxed in accordance with legal rules prevailing in the domestic country or across borders, respectively.

However, in the case of digital products, accounting problems arise, especially in the valuation, recognition, reporting and taxation of digital products. As they are intangible, it is difficult to measure the initial or subsequent value of the products. In addition, the revenue recognition related to any digital products or services is becoming increasingly complicated as the physical delivery and the transfer of control of the products cannot be effectively observed. Similarly, goods and services traded digitally escape import duties and value added taxes (Vincent, 2004). In sum, the following questions on the accounting of digital products have been raised.

- What is the value of digital products? For the initial measurement which price is used; purchase price or subscription fee? For subsequent measurement, which valuation method will be used? Historical-based or fair value?
- What is the accounting treatment for digital products? Do they need to be expensed directly in the period of purchase or to be capitalized as an asset?
- How can the revenues generated from the sale of the digital products be recognized? How can the revenues be matched with expenses for the services provided and resources consumed as digital products?
- How can the information produced be stored and reported in financial statements?
- How can the taxation issues associated with digital products be solved?

The answers to the above questions depend on the classification of digital products as either goods or services. In this chapter, first the classification of the digital products will be explained and then the management and control of digital products will be discussed, in terms of accounting, by considering the recognition, measurement, valuation, reporting and taxation problems.

## BACKGROUND

Digital products are "content" goods or services such as software, books, music, movies or financial services that can be digitized and traded on a digital market place. The new form of business, e-commerce possesses several challenges on how to manage the digital products.

A number of different definitions are used in the literature. Whinston, Stahl and Choi (1997) described digital products as products that can be digitized and transferred over a digital network. Seddon (1998) defined digital products as a combination of products, processes either physical or digital and players either physical (buyers, sellers, third parties) or digital (agents). In other words, a product is categorized as a digital product depending on its form, trading terms or players (buyers, sellers, third parties or agents) engaged in the trade of the products. According to Shaw, Gardner and Thomas (1997), compared to traditional products, the delivery of digital products is radically different. In addition to the differences in delivery, as soon as digital products have been produced, their inventory level is infinite (Tao et al., 2006).

The rapidly expanding digital economy offers a wide range of digital products which differ from each other in terms of characteristics and nature. Digital products were traditionally traded as part of a physical carrier medium and classified as either goods or services. For tax purposes, their valuation was based on the value of the carrier medium (Vincent, 2004). However, with the use of the Internet, the physical borders of, and constraints on, digital products have been eliminated. Today, digital products are traded all over the world via the Internet with no physical substance. This change in the form of delivery of digital products has increased the debate on the classification of digital products, as goods or services, which is important for their legal taxation and accounting treatments.

In accordance with the explanation above, Hui and Chau (2002) developed a comprehensive framework for the classification of digital products. According to this framework, based on their attributes, delivery processes and serving purposes digital products are divided into three categories; tools and utilities, content-based digital products, and online services. *Tools and Utilities* are products that are used as an assistant tool to meet specific purposes of customers. Software programs are one of the most frequently traded digital products categorized under tools and utilities. *Content-based Digital Products* are products that contain information. E-books or databases are examples of content-based digital products. *Online Services* refers to other digitally traded services including online resources, consultancy services, banking services, subscription services.

In literature, studies on "digital products" and "digital economy" started in the 1960s and there has been a rapid growth in the studies especially since the 1990s. A vast amount of studies on digital products have been conducted on their marketing, distribution and management models (Wang et al., 2001; Tao et al. 2006; Shaw et al., 1997; Park & Scotchmer, 2005; Gayer & Shy, 2003; Kuan, 2006; Luxem, 2000; Hui & Chau, 2002). However, studies specifically concerning the accounting treatments for digital products are limited. Mullins (2005) provided insights into the emergence on accounting of digital products, although failed to present any suggestions. As David (2002, p.67) and Mullins (2005) argued, uncertainty over the accounting of digital products and accounting treatments for digital products should be investigated, because this is important for the development of the regulations and standards.

## ACCOUNTING TREATMENT FOR DIGITAL PRODUCTS

Mullins (2005) explained the absence of the accounting rules, standards or principles for software, as an example. According to his study, the costs of software are either written-off in the period incurred or amortized. Even it does not reflect the true pattern, there is a general accounting treatment for software purchased or acquired through other means. On the other hand, there is no accounting treatment for the digital products created internally or purchased digitally. According to current accounting practices, assets internally created do not correlate with revenue performance (Mullins, 2005).

Accounting is reactive, developing in parallel to changes in the business environment. As a result of these developments in the digital products market, debate on accounting treatment for digital products has increased. Traditional accounting methods do not completely meet the requirements of the emerging digital economy. In traditional accounting, it was simple to measure and record many cases. Revenue is recognized when the goods or services are delivered to customers (DeMark, 2004). However, as a result of the characteristics of digital products, their measurement and valuation process and trading terms are complicated.

In accounting, the classifications of assets are made based on liquidity, current assets and long term assets. In addition, in order to differentiate the assets with no physical substance, another classification has been used and non-monetary assets (non-financial instruments) with no physical substance are defined as intangible assets. Although at first, digital products seems like intangible assets, Kuan (2005, p.69) argued that a digital content product should be defined as neither tangible nor intangible. Therefore, it may not be possible to apply completely the accounting standards for intangible assets to digital products. It would be appropriate to classify digital products held for sale in the ordinary course of business as inventory. Therefore, the classification of digital products is important to understanding possible accounting treatments and interpreting the applications of accounting standards for digital products.

In this chapter, accounting treatments for digital products will be discussed under the following main topics; recognition, measurement and valuation, reporting and taxation. In Figure 1, accounting treatment for digital products is summarized.

## Recognition

The first issue associated with the accounting of digital products is to determine whether to capitalize or expense the digital products.

In accounting, assets are defined as resources controlled by the entity, resulting from past events and from which a future benefit is expected (FASB, 2006; IASB, 2009). A digital product, either self created or acquired, can be classified as an asset and capitalized if a future economic benefit is expected to flow into the entity through its use. Otherwise, any other digital products that fail to meet the asset definition would be recognized as an expense when incurred. Using this base of classification, tools and utilities and content-based digital products may be examples of assets that should be capitalized, while most online services should be expensed in the period incurred. If the payment of the online services has been made in advance, it should be expensed in accordance with the proportionate usage of the digital services.

In the sale of digital products, another issue that has arisen is how to match the revenues with expenses resulting from trading digital products. As the transfer of control among sellers, buyers and distributors cannot be accurately determined, a revenue recognition problem exists. Besides the risks associated with the sale or transfer of control and ownership of the digital product, warranty and after sale obligations related to products cause serious problems (Mullins, 2005). According to International Accounting Standards Board (IASB), IAS 18 and Financial Accounting Standards Boards (FASB, 2009) revenue is recognized if (1) it is probable that any future economic benefit associated with the item of revenue will flow to the entity, and (2) the amount of revenue can be reliably measured.

**Accounting for Digital Products**

*Figure 1. Summary of accounting treatment for digital products*

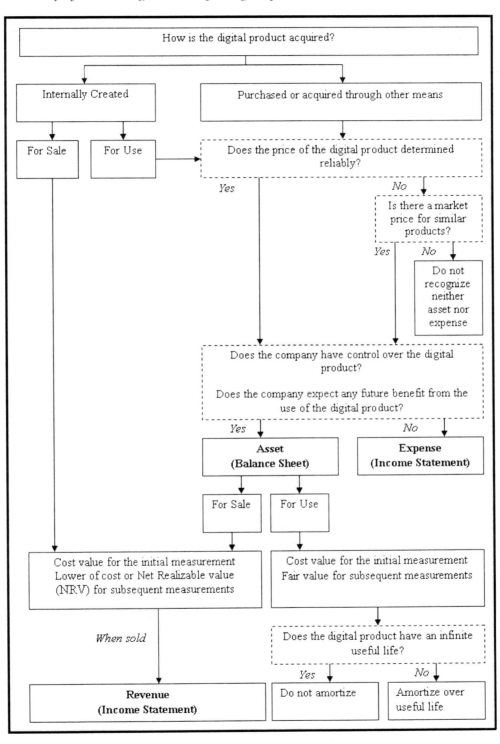

Thus, in the accounting of digital products, revenue should be recognized in accordance with the delivery of goods or services. For warranty and after sale obligation, there is a requirement for a sale agreement which clearly explains the responsibilities of both buyers and sellers with sales terms and conditions. If the digital products are delivered at one time, it is easy to determine the amount of revenue or expense. However, the digital products may cover a bundle of digital goods or services, contracts or online services over a specific period of time or for specific level of information. The expense and revenue associated with these types of digital products or services should be recognized depending on their use or delivery.

## Measurement and Valuation

A second issue associated with the accounting of digital products is measurement and valuation. If digital products are capitalized, their initial value and subsequent valuation basis should be determined. A digital product recognized as an asset should initially be measured at cost. The cost of the acquired digital product is the purchase price (DOR, 2009) or if this cannot be reliably determined the price of similar products on the market. If a digital product is obtained by means other than a purchase, then the price will be determined using the retail price of a similar digital product. However, as Peitz and Waelbroeck (2006) indicated, while the cost of the first copy or right of digital products is high, subsequent copies are cheap to produce. Therefore, as a result of low marginal cost, low distribution costs, and ease of reproduction, the traditional pricing policies cannot be applied for digital products (Chen, 1997).

For subsequent measurement it would be appropriate to use the fair value for digital products. A fair value may be the market value or the net cash inflow into an entity expected from the use of the digital product. However, as net cash inflow is a subjective measure, it would be more appropriate to use market value for the subsequent valuation of digital products. If the price of the digital product cannot be determined on the market, the price of a similar product may be used as a measure of fair value.

Where digital products are capitalized, depending on the categorization of digital products as inventory or a long-lived intangible asset, an amortization problem arises. In accounting, depending on their characteristics, long-lived assets are subject to depreciation or amortization. For the amortization of digital products, first the useful life of the digital products should be determined. If these have an infinite life they should not be amortized. In contrast, digital products with a finite life should be amortized over the useful life of the digital product. Similar to intangible assets, the amortization period should reflect the pattern of benefits expected from digital products. In addition, in the case of capitalizing, another concern is how to control the products. According to Mullins (2005), adequate internal control standards are required for accurate management especially in recording and controlling of revenues.

If the digital product is classified as inventory, for subsequent measurements, the lower of cost or net realisable value (NRV) principle, where NRV is the estimated selling price less the estimated cost of completion and selling costs (IAS 2) or the lower of cost or market rule (FASB 151) should be applied.

## Reporting

A third issue associated with the accounting of digital products is reporting in financial statements.

Digital products that meet the definition of assets are reported in the balance sheet and other digital products are recognized as expense or revenue in the income statement. In addition to financial statements, the valuation base, amortization and the classification of the digital products should be clearly reported in the notes of the financial reports. For companies providing both digital and traditionally-traded products, a segment reporting should be required to report any profit or loss resulting from digital products' operations. In addition, entities trading only digital products should also prepare segment reporting by regions, because trading of digital products is across borders.

## Taxation

Entities pay tax to governments as a result of commercial activities. Of this tax, sales taxes and custom duties compose a great portion of tax revenues of governments (Mullins, 2005). As with other products, commercially traded digital products are subject to sales taxes and custom duties. However, in contrast to physically traded goods and services, it is difficult to measure the volume of trade or define the amount of tax associated with digital products, which are traded internationally, and are thus subject to different tax systems and legal procedures.

Therefore, in addition to valuation, revenue and reporting, taxation is a serious concern. Digital products, which are digitally traded with no physical carrier medium, are exempt from indirect taxes. This causes two major problems; uncontrollable trading volume of digital products and unfair trade terms for digital products as compared to other physically traded products.

Besides, digital products that are internally created and used for internal operations are not subject to taxation. However, in the case of the commercial trade in digital products, digital products accessed through any means (downloaded, streamed, subscription service, networking, etc.) or created internally for commercial sale are subject to sales and use tax (DOR, 2009). In a report published by DOR (2009) it is indicated that the taxation process may applied through a digital code specifically identifying the purchased digital product and at the same time giving purchaser a right to obtain or access the product on a single occasion or over a period of time.

## FUTURE RESEARCH DIRECTIONS

The accounting of digital products is an emergent issue. Both in the literature, and in regulations and standards, there is a lack of framework for the accounting treatment of digital products. In this chapter, current accounting standards have been interpreted for the accounting of digital products. However, the need for the clear definition of digital products, especially for those internally created ones, and the accounting standards to be applied should be clearly defined by standard setters and to accounting profession. For further research, the accounting treatment of digital products should be analyzed through financial reports of companies trading digital products. In addition, another important research topic is the free downloading of digital products. If these kind of digital products are under the control of entities and a future benefit is expected from their use, what would be the accounting for such digital products? Are they assets or not? This issue might be investigated in further studies.

## CONCLUSION

In recent years, there has been a rapid growth in digital products traded on the market. The digital economy has led to a change in the method of production and delivering of goods and services. There is now no need for a physical carrier medium for the trade in goods or services such as software programs and online newspapers, music, books, almost all kind of banking services, consultancy services, films, ticket sales, education, which are delivered electronically over the Internet. This new form of business has raised important problems on how to manage these digital products.

In this chapter, first, the challenges arising as a result of trading digital products has been explained briefly and then suggestions for overcoming these challenges have been proposed in terms of accounting. As digital products have similarities to intangible assets, this chapter presents a framework for the accounting treatment of digital products based on the accounting standards for intangible assets and the nature of digital products. Overall, it is important to say that, without a proper classification, it would be impractical to make suggestions and interpretation for the accounting treatment of digital products.

## REFERENCES

Chen, P. (1997). *Pricing Strategies for digital information goods and online services on the Internet*. Retrieved September 4, 2009, from http://www.mba.ntu.edu.tw/~jtchiang/StrategyEC/eec/report1/report1.htm

David, P. A. (2002). Understanding the digital economy's evolution and the path of measured productivity growth: present and future in the mirror of the past. In Brynolfsson, E., & Kahin, B. (Eds.), *Understanding the Digital Economy* (pp. 49–95). Boston, MA: MIT Press.

DeMark, E. F. (2004). Revenue recognition issues in a digital economy. *CPA Journal, 74*(5). Retrieved August 14, 2009, from http://www.nysscpa.org/cpajournal/2004/504/perspectives/nv3.htm

Financial Accounting Standards Board (FASB). (2006). Joint Conceptual Framework Project. Retrieved September 5, 2009, from http://www.fasb.org/fasac/Conceptual_Framework_12-06.pdf

Financial Accounting Standards Board (FASB). (2009). Position paper on IASB / FASB revenue recognition project. Retrieved September 5, 2009, from http://www.fasb.org/cs/BlobServer?blobcol=urldata&blobtable=MungoBlobs&blobkey=id&blobwhere=1175819078334&blobheader=application%2Fpdf

Gayer, A., & Shy, O. (2003). Internet and peer-to-peer distributions in markets for digital products. *Economics Letters, 81*(2), 197–203. doi:10.1016/S0165-1765(03)00170-8

Hui, K. L., & Chau, P. Y. K. (2002). Classifying digital products. *Communications of the ACM, 45*(6), 73–79. doi:10.1145/508448.508451

International Accounting Standard Board (IASB). (2009). *IAS 2- Inventories*. Retrieved December 5, 2009, from http://www.iasplus.com/standard/ias02.htm

International Accounting Standard Board (IASB). (2009). *IAS 18- Revenue*. Retrieved December 5, 2009, from http://www.iasplus.com/standard/ias18.htm

International Accounting Standard Board (IASB). (2009). *Framework for the Preparation and Presentation of Financial Statements*. Retrieved December 5, 2009, from http://www.iasplus.com/standard/framewk.htm

Kuan, A. H. Y. (2006). *Planning Intellectual Property for Marketing Strategies in the Digital Content Industry*. Unpublished masters dissertation, National Chengchi University, Taiwan.

Luxem, R. (2000). The impact of trading digital products on retail information systems. Paper presented at the meeting of the *33rd Hawaii International Conference on System Sciences (HICSS)*, Hawaii.

Mullins, L. J. (2005). Managing intellectual property in the digital product market. *Journal of Digital Asset Management, 1*(1), 59–66. doi:10.1057/palgrave.dam.3640010

Park, Y., & Scotchmer, S. (2005). Digital rights management and the pricing of digital products. National Bureau of Economic Research, NBER Working Paper No. 11532. Retrieved September 4, 2009, from http://www.nber.org/papers/w11532.pdf?new_window=1

Peitz, M., & Waelbroeck, P. (2006). Piracy of digital products: A critical review of the theoretical literature. *Information Economics and Policy, 18*(4), 449–476. doi:10.1016/j.infoecopol.2006.06.005

Poon, S., & Joseph, M. (2000). Product characteristics and Internet commerce benefit among small businesses. *Journal of Product and Brand Management, 9*(1), 21–34. doi:10.1108/10610420010316311

Rayport, J. F., & Sviokla, J. J. (1995). Exploiting the virtual value chain. *Harvard Business Review, 75*(6), 75–85.

Seddon, P. (1998). Digital products and processes: A critique of Whinston, Stahl, and Choi's Chapter 2. Retrieved August 14, 2009, from http://disweb.dis.unimelb.edu.au/staff/peterbs/research/DigitalProductsAndProcesses.doc

Shaw, M. J., Gardner, D. M., & Thomas, H. (1997). Research opportunities in electronic commerce. *Decision Support Systems, 21*(3), 149–156. doi:10.1016/S0167-9236(97)00025-0

State of Washington, Department of Revenue (DOR). (2009). *Digital Products Bill (ESHB 2075)*. Retrieved September 5, 2009, from http://dor.wa.gov/Content/GetAFormOrPublication/PublicationBySubject/TaxTopics/DigitalProductsQA.aspx

Statement No, F. A. S. B. 151, Inventory Costs. Retrieved December 5, 2009, from http://asc.fasb.org/

Tao, G., Yi-jun, L., Jing, G., & Long, G. (2006). Research on the economic features and pricing of digital products. In Y. Ji-rong (Ed.) Proceedings of the 2006 International Conference on Management Science & Engineering (13th) (pp.152-156). Harbin: Institute of Technology Press.

Vincent, S. W. (2004). WTO, e-commerce, and information technologies: from the Uruguay Round through the Doha Development Agenda. In J. McIntosh (Ed.) *A Report for the UN ICT Task Force*. Retrieved September 14, 2009, from http://www.iie.com/publications/papers/wunsch1104.pdf

Wang, Y., Tang, T., & Tang, J. E. (2001). An instrument for measuring customer satisfaction toward Web sites that market digital products and services. *Journal of Electronic Commerce Research, 2*(3), 89–102.

Whinston, A. B., Stahl, D. O., & Choi, S. Y. (1997). *The Economics of Electronic Commerce*. Indianapolis, IN: McMillan Technical Publishing.

## ADDITIONAL READING

Bakos, J. Y., Brynjolfsson, E., & Lichtman, G. (1999). Shared information goods. *The Journal of Law & Economics, 48*, 117–156. doi:10.1086/467420

Bhimani, A. (2004). *Management Accounting in the Digital Economy*. New York: Oxford University Press.

Bonsón, E., & Escobar, T. (2006). Digital reporting in Eastern Europe: An empirical study. *International Journal of Accounting Information Systems, 7*(4), 299–318. doi:10.1016/j.accinf.2006.09.001

Bostan, I., Mates, D., Grosu, V., & Iancu, E. (2009). Alternate methods of evaluation for Web sites concordant to IAS/IFRS Standards. *Journal of Computing, 1*(1), 141–148.

Cunard, J., Banlas, C. & Hall, K. (2003, November). Report on current developments in the field of digital rights management. World Intellectual Property Organization, Geneva.

Deshmukh, A. (2006). *Digital Accounting*. Hershey, PA: IRM Press.

Fetscherin, M., & Knolmayer, G. (2004). Business models for content delivery: An empirical analysis of the newspaper and magazine industry. *International Journal on Media Management*, *6*(1), 4–11. doi:10.1207/s14241250ijmm0601&2_3

Fisher, I. E. (2004). On the structure of financial accounting standards to support digital representation, storage, and retrieval. *Journal of Emerging Technologies in Accounting*, *1*, 23–40. doi:10.2308/jeta.2004.1.1.23

Lenk, M. M. (2007). Digital accounting: The effects of the Internet and ERP on accounting. *Issues in Accounting Education*, *22*(1), 122–123.

Meyer, M. H., & Zack, M. H. (1996). The design and development of information products. *Sloan Management Review*, *37*(3), 43–59.

Pearson, T. A., & Singleton, T. W. (2008). Fraud and forensic accounting in the digital environment. *Issues in Accounting Education*, *23*(4), 545–559. doi:10.2308/iace.2008.23.4.545

Rappa, M. (2000). Business models on the Web. Retrieved October 15, 2009, from http://digital-enterprise.org/models/models.pdf

Tackett, J. A. (2007). Digital analysis: A better way to detect fraud. *Journal of Corporate Accounting & Finance*, *18*(4), 27–36. doi:10.1002/jcaf.20305

Timmers, P. (1998). Business models for electronic markets. *Electronic Markets*, *8*(2), 3–8. doi:10.1080/10196789800000016

Varian, H. R. (2000). Buying, sharing and renting information goods. *The Journal of Industrial Economics*, *48*, 473–488. doi:10.1111/1467-6451.00133

## KEY TERMS AND DEFINITIONS

**Intangible Asset:** A non-monetary asset lack of physical substance.

**Electronic Commerce:** A business method where goods and services are traded electronically over Internet or other networks.

**Accounting (System):** An information system concerned with recording, classifying, summarizing and reporting the transactions of economic events in monetary terms.

**Recognition:** The process of recording and reporting an item in the financial statements.

**Measurement:** Determination the value of the items to be recognized in monetary terms.

**Valuation:** The process of determining and recording items' initial and subsequent market value.

**Taxation:** The imposition of tax by legal authorities.

# Section 3
# Issues and Strategies

# Chapter 7
# It's All about the Relationship:
## Interviews with the Experts on How Digital Product Companies Can Use Social Media

**Delaney J. Kirk**
*University of South Florida, Sarasota-Manatee, USA*

## ABSTRACT

*The rapid growth of social media sites has caught the attention of individuals and organizations hoping to use these to market their products and services. As digital products are sold online, it makes sense to tap into these networking communities to sell products as well as to share information and gather feedback. In this chapter, experts who are currently using social media in a variety of ways are asked to share their experiences. Tips are given on what to do as well as what not to do in order to participate successfully on social media sites. Future managerial and research implications are then discussed.*

## INTRODUCTION

Social media is defined as "the activities, practices, and behaviors among communities of people who gather online to share information, knowledge, and opinions using conversational media" (Safko & Brake, 2009, p. 5). With social media sites we are able to move beyond the one-way communication methods used in the past to promote products, to interactive Web tools that give people and organizations the ability to conduct on-line discussions and obtain immediate feedback. Given that digital products such as online music, magazines, and newspapers are stored, produced, and disseminated through the Internet, it makes sense to tap into social media networks to market these products.

To get an idea of just how big the phenomenon of social media is, consider this: It took 38 years for radio to gain 50 million users and 13 years for television to do the same. However, the Internet had 50 million users in four years, and Facebook-just one of many social networking sites-counted 100 million users in less than nine months (Qualman, 2009).

This chapter gives a brief history of some of the most popular social media tools being used today. It then presents insights gained from per-

DOI: 10.4018/978-1-61692-877-3.ch007

sonal interviews with a variety of experts who are currently using social media to market their own digital products. Strategies are given as to how these experts have been able to create their online communities and what companies can do to tap into this fast growing resource.

## BACKGROUND

There are hundreds of social networking sites on the Internet today. Some of the most widely used venues include blogs, forums, Twitter, Facebook, MySpace, and YouTube.

### Blogs

The term, weblog (commonly called blog), is defined as a "frequently updated website, normally with dated entries and usually with the newest entries at the top" (Merriam-Webster Online Dictionary, 2008). It is estimated that there were only 150 blogs in the late 1990s (Trammell & Ferdig, 2004). However, this number increased to 10.3 million blogs in 2004 (Quibble, 2005), 70 million in 2007 (Sifry, 2007), and to over 112.8 million by July 2008, with 175,000 new blogs being added each day ("Technorati website statistics," 2008).

Blogs differ from traditional websites in several distinct ways (Quibble, 2005). Websites tend to be static and do not change often; blogs are dynamic and usually added to frequently by the author (or authors). Also, the readers of the blog can respond to the writer's text by making comments that can then be read by other readers. In addition, blogs are much easier to add content to because they do not require the expertise and special programming software to start or update as websites do. In fact, the major reason for the growth in blogs is that "software companies created the database-driven content management tools needed to run blogs so that non-coders could start their own blogs" (Trammell & Ferdig, 2004, p. 61). In other words, blogs have become easier to create than a webpage. As noted by one website designer, "adding a new post [on a weblog] is as easy as sending an e-mail" (Demopoulos, 2007, p. 4).

### Forums

Internet forums were one of the first online communication tools and are still popular for people wanting to engage in an ongoing, interactive conversation on a specific topic. Forums are similar to chat rooms except that forums do not have to occur in real time. People in forums tend to form very strong bonds as they care about the subject matter and trust each other. Forums are seen as predecessors to the blog although the strong growth of blogs has not in any way led to a demise of the forum format (Safko & Brake, 2009).

### Twitter

Twitter was founded by Jack Dorsey, Biz Stone, and Evan Williams in 2006 as a way for users to post updates on what they were doing. These updates are limited to a maximum of 140 characters and can be posted using a computer, text message, or instant message ("Twitter stats," 2009). It is estimated that there were six million Twitter users in 2008 in the United States, or about 3.8 percent of all Internet users.

Twitter released its most recent stats in April of 2010, stating that there are now 106 million users of Twitter and these users post 55 million tweets per day. Twitter has been called the "water cooler of the Internet" and can be used for sharing information and links, networking, and marketing. Research by Professor Jim Jansen at Penn State reveals that "20 percent of all tweets…[on Twitter] are directly related to products or services" (McDermott, 2009).

### Facebook

Facebook was launched by Mark Zuckerberg on February 4, 2004, as a social networking site for students at Harvard University and quickly expanded to 30 more colleges within four months ("Facebook," 2009). The site has now evolved past targeting students to include people of all ages as well as organizations. As of September of 2010, over 500 million people all over the world have a Facebook account ("Facebook stats," 2010) and 58.6 percent of all the social networking site visits are to Facebook ("Facebook visits increased," 2009). "People spend over 700 billion minutes per month on Facebook" ("Facebook stats," 2010). The majority of users are over 25 years old due to the rapid increase in the number of older users, although young people between 18 to 25 still make up the largest age group (Walsh, 2009). However, visits to Facebook from persons 55 and older have increased by 77 percent in the past year ("Facebook visits increased," 2009).

## MySpace

MySpace was first launched in 2003 by several eUniverse employees. The site dominated the social networking scene in 2006 (Cashmore, 2006) but was overtaken by Facebook in 2008 ("Facebook: Largest, fastest growing social network," 2008). MySpace tends to be geared towards "friends" while Facebook includes colleagues and organizations and the network is seen as older and more professional (Salazar, 2007).

## YouTube

YouTube was launched in December 2005 as a way to let people share personal videos. Today, over two billion people watch videos each day ("YouTube Facts and Figures, 2010) and "every minute, ten hours of video is uploaded to YouTube" ("YouTube fact sheet," 2009). These videos allow others to discover new artists, watch first-hand accounts of current events, and share tape on their kids, pets, and hobbies. Users span in age from 18 to 55 and include both males and females equally.

## SOCIAL MEDIA STRATEGIES

Using social media tools such as the ones discussed above to market a company's products takes a different approach from that used with traditional media sources such as radio, television, film, and print ads. Experts in the area of social media would say it requires the acquisition of "whuffie." The term, whuffie, was first used by Cory Doctorow, author and creator of the blog Boing Boing, to describe social capital (Doctorow, 2003). Social capital isn't bought; it's earned through building relationships with others in your online community (Hunt, 2008). Individuals and organizations gain this social capital based on their reputation and whether others believe they are trustworthy. As noted by Brogan & Smith (2009), people with social capital today are the pioneers who understand these new technologies and are mastering the how-to's of one-to-many communications. They go on to say that we are currently living in a society where people have no confidence in advertising, where they are suspicious and even hostile to those appearing to have ulterior motives. This has led to the formation of loose networks of people with common interests who share information and are thus able to connect with and influence those within their circles. Although members of these social networks may never meet in person, there is a sense that one knows and trusts these "friends" or influencers on the Internet and thus would turn to them when making decisions about products and services to buy.

A good example of someone who has been able to successfully leverage social media to sell his digital products is singer-songwriter, Jonathan Coulton. Coulton left his job as a software programmer in 2005 to pursue a full-time music career. He decided to make a commitment to recording a new song a week and releasing it for free on his website as a podcast. As he notes, "It was an attempt to keep the creative juices flowing as freely as possible, and a way for me to push myself to take risks, work quickly and trust in the creative process. It was also a stunt designed to get people to notice me" (J. Coulton, personal communication, October 9, 2009). Coulton has a blog, hosts discussion forums, is on Twitter and MySpace, and urges his fans to use a website called Eventful that is aimed at creating demand for musicians. According to Coulton, his strategy was to put his songs out there in a format that was easy for the consumer to link to, download, and email to friends and to be as contagious and open as possible. In that first year, he used his blog to describe the process of writing his songs and as a result began having conversations with the people that came to his blog. Early on someone sent him a drawing illustrating one of his songs which made him realize that he wanted people to have fun with his music. He put the drawing on his blog and started encouraging people to expand on his songs to create new music and art work. When he saw people making videos using his music, he would link to them on his blog and give them a little piece of the spotlight. There are now hundreds of music videos on YouTube that people have made using his music. Coulton's website took off when several blogs with high readership numbers linked to one of his songs, pushing his audience from 100 visitors a day to 100,000 (www.jonathancoulton.com).

Online newspapers are also beginning to see the possibilities of social media. Greg Auman, a sports reporter for the *St Pete Times* online newspaper, has experimented with various formats for several years. Two years ago he began blogging about the University of South Florida football games and in 2009 this evolved into a live interactive chat with 200 people participating during the first two games of the season. According to Auman, "the blog really took off during the third game which was not televised. When one of the players hurt his knee during the game, fans at the stadium instant messaged their friends and within 20 minutes, there were over 1200 tweets on Twitter about the incident" (G. Auman, personal communication, September 28, 2009). This led others to log on to Auman's blog to get the updates. Auman sees social media as an amazing resource for a reporter to interact with the public.

Another type of digital product successfully using social media are companies producing webisodes, defined as "a serialized story that is broken into short installments, each just a few minutes long, and each of which often ends in a cliffhanger" (Miller, 2008, p. 261). For example, *Something to Be Desired* was created in 2003 and is the Internet's longest-running web sitcom. The story line revolves around life after college for the characters and it is produced in Pittsburgh, PA, with a local cast and crew. Writer/producer, Justin Kownacki uses Twitter and discussion forums to communicate when new webisodes are available and to talk with fans. In addition, fans can rate each webisode online.

James M. Maher of Tampa, Florida, shares his experience with authors Sharon Lee and Steve Miller who tapped into the Internet community in order to offer their science fiction books online. Subscribers were able to access their book, *Fledging*, a chapter at a time for free as it was written from January to October 2007. Fans were encouraged to make suggestions and give feedback and those who contributed money during the process were promised a hard copy of the final book if it ever sold. Fans of Lee and Miller set up "friends of Liad" groups on Facebook and on various discussion forums, creating a community that proved to publishers that there was indeed a market for the book. Lee and Miller were able to revise and sell *Fledging* to Baen Books which released the book in September 2009. Maher, along with 1199 other supportive fans then received their signed, first edition book for supporting the writing duo during the process (J. Maher, personal communication, October 22, 2009).

While the previous examples show digital product companies embracing all forms of social media, others have focused on just one type they felt was most suited to their audience. *Winding Road* is an online magazine in the automotive market that was first released in December 2004. According to the editorial director, Tom Martin, the magazine is targeted at automotive influencers (both passionate enthusiasts and people who follow new cars regularly but somewhat casually). The magazine is accessed by four million people all over the world although the magazine's primary focus is the U.S. market. Martin believes that offering the magazine online has advantages of faster timing, lower costs, and the ability to provide links and videos of new cars. The company's editors and their readers both initiate discussions in forums online to continue the conversation beyond the magazine. Martin states that forums work best for them in getting user participation and feedback and more page views. According to Martin, other sites such as Twitter, Facebook, and blogs are used at the title level but not the company level, meaning that individual employees may choose to use them but the company does not view these as effective methods to interact with customers (T. Martin, personal communication, September 28, 2009).

## ADVICE FROM THE EXPERTS

According to Safko & Brake, there are three guidelines to using social media to market products and services: (1) It's all about "enabling conversations; (2) You cannot control conversations, but you can influence them; and (3) Influence is the bedrock upon which all economically viable relationships are built" (2009, p. 5). As noted by Jonathan Coulton, each person or company has to approach social media by keeping in mind their product, image, and who their audience is. As he says, his were mostly "nerds who liked geeky songs and were already involved in social networking." His advice is to be authentic and honest, something echoed by others working with this new medium. As he observes, "there is a fine line between social media and marketing" and the public is educated to tell the difference. Social media is all about word of mouth or rather, word of Internet." (J. Coulton, personal communication, October 9, 2009). Appendix A has additional suggestions from Chris Brogan, co-author of the book *Trust Agents*, for businesses on how to get started using social media.

Microsoft found out what not to do in their recent attempt to convince the online community to promote house parties for their Windows 7 launch. A six minute instructional video was placed on YouTube illustrating how people could throw their own parties from October 22-29, 2009, to demonstrate the new product ("Hosting your party," 2009). While a number of people did sign up to get their free Signature Edition of Windows 7, along with complimentary party favors, many more got on Twitter and blogs to make fun of Microsoft's "forcing" the social media relationship. In fact, Lukas Neville, Ph.D. Candidate in Organizational Behaviour at Queen's University, saw this push by Microsoft to get the online community talking about their new product as a potential threat to the trust forged between the company and its brand community. "If the video makes volunteer 'hosts' feel like they're being made to shill versus participating in a community-building event," the trusting relationship between the firm and its community could be eroded, a true violation of social netiquette. The key to recovering this trust, Neville suggests, is to invest in an enduring relationship with the event hosts. "Involve them with new products, seek their feedback on ideas, etc. Attributions about the party would then be made in context of this broader relationship by those who have a positive, ongoing relationship with the firm." And as for the damage to Microsoft's branding? "If I were Microsoft, I'd just try to have some (self-effacing) fun with it at this point. Perhaps record a 'director's commentary' or a video of the cast biting their nails, waiting for the Oscar nominations" (L. Neville, personal communication, October 12, 2009).

Greg Auman agrees that the process cannot be hurried or manipulated. He believes that you want people to feel that they can come to you for accurate information. Although his newspaper is not sure how much time they want to invest on Twitter and other social networking sites if they don't see a direct return on investment, his thought is that any traffic you can get the better and thus you gain more currency for advertising which will, in time, pay off. As noted earlier, he uses both Twitter and Facebook to drive traffic to his sports blog and believes the blog attracts customers to the online newspaper. He does say that many of his colleagues fear that Twitter will be obsolete in 18 months and thus the newspaper will have wasted its time. He sees it differently and states that if this should prove to be true that companies should just jump to the new tool or be at risk of being seen as a dinosaur. And while his live blog audience who live in Iraq, Thailand, and Bangkok may not frequent restaurant owners in Tampa, he does see how an Internet site selling University of South Florida merchandise could sell products if they had a link on his site.

## THE REALITIES OF TODAY'S ONLINE WORLD

Jumping onto the social media bandwagon can be daunting for digital product companies due to the amount of time and expertise involved without any guarantee that this will be beneficial. However, one given in today's digital world is that everyone has a voice and can use it to talk positively or negatively about your product or company on various social networking sites. This is going to happen anyway so companies have to decide if they want to link in to see what is being said and decide how they want to respond. One issue to consider is whether any negative comment is an isolated case or a problem that is much bigger. If it's one customer, knowing the issue allows the company to address it and satisfy the customer and even get positive feedback at online sites from the customer afterwards. For example, the author of this chapter received an order from Zappos, an online shoe company. The new pair of shoes arrived scuffed and she tweeted on her disappointing experience on Twitter. Within 15 minutes, she received a tweet from the CEO of Zappos, Tony Hsieh, apologizing for the problem and giving a phone number to call to make it right. While shoes are not a digital product, this illustrates the type of response that makes for good customer relations.

If it's a rant from a group of people online, it's best to address the root problem that you can control and then communicate to the community what you did. Sean Reid of *Reid Reviews*, an online magazine on photography, recently had this experience. Reid writes reviews and essays about various cameras. On a blog discussing the Leica M9, a comment was made that Reid (and others) were "very carefully selected by Leica, presumably on the assumption that they would write 'friendly' dispatches about the new camera" ("The ethics of reviewing," 2009). Reid responded to this attack on his integrity by saying he would consult with his attorney to see if this was actionable slander. Others jumped in to comment, supporting both sides of the issue. After several days of this, the blogger where the original comment was made posted that, [Reid and I] "both regret the many misunderstandings and miscommunications that have characterized this disagreement from the start. I still read and recommend *Reid Reviews*. Please note that at no time did I ever mean to impugn Sean's professional integrity, directly or indirectly" ("Blog note update," 2009). Needless to say, the camera world watched this all unfold in close fascination. Reid then sent out an email to all subscribers urging them to email him with any questions they might have on the controversy. Obviously, taking a proactive approach and keeping the conversation transparent has helped Reid to retain the trust his fans have in him.

Another reality of the social media world is that few companies are able to make money right away. When Jonathan Coulton was asked if he was afraid that by giving his music away for free and not controlling who accesses it, that he would lose money, he stated that this is happening anyway, that it's easy to make a copy of a song and "if you think you can put the genie back in the bottle, you're crazy." But he says people understand the nature of the relationship:

*Things have really changed in the music industry over the last few years…I give away music because I want to make music, and I can't make music unless I make money, and I won't make any money unless I get heard, and I won't get heard unless I give away music. This is all part of the experiment. I believe it can work, but we all need to adjust our thinking about the relationship between artists and fans – the RIAA thinks that music listeners are criminals and that music should be locked up and protected. I disagree. I think there are times when free music and file sharing can greatly benefit an artist. Believe me, I spent many years making music and not sharing it with anyone, and that didn't get me anywhere.*

(J. Coulton, personal communication, October 9, 2009).

Thus, Coulton has made it very clear to fans that "we" create a situation where he, as a professional musician, makes music for them and they pay for it. Lee and Miller also adopted this strategy when they offered their book chapter drafts for free but made it clear that they needed to make money in order to continue. Their readers were encouraged to support their efforts by contributing to the process, a request that 1200 fans responded to gladly.

Ted Demopoulos, consultant and author of books such as *What No One Ever Tells You About Blogging and Podcasting* and *Blogging for Business*, states that social media is a natural process of getting to know people, developing trust and credibility, and only then to market your products. "If you're looking for hard measures such as we have with direct mail involving dollars or hours spent in comparison to sales, it's difficult to find. However, you can measure things such as increase in traffic to your website or rankings in search engines." He cites his co-author of *Blogging for Business* as saying, "It's like looking for return on investment on pants. There's no direct dollars earned but you need pants in order to do business" (T. Demopoulos, personal communication, October 14, 2009).

Relying on advertising is a possibility for digital product companies but as Tom Martin notes, advertisers really don't understand the digital product market. For that matter, many editors don't understand how radically different digital can be from print. For this reason Martin offers a physical copy of other magazines they publish and they have considered doing so with *Winding Road*. He says the biggest topics of discussion among digital magazine publishers are format-related software and advertising standards. However, any advertising has to fit in with your strategy and target markets. Jonathon Coulton tried using Google ads on his website but says he wasn't making that much money from these and that he thought they were distracting for his fans.

Figuring out how to approach the social media world can be challenging. According to Aaron Kahlow, "We're in the middle of an unfolding human experiment, as social networking tools become a primary means of communicating and marketing. It's all about understanding the human element of these new tools…and the human element really comes down to social networking etiquette" or what we are calling social netiquette (Kahlow, 2009). In order to use these new tools, we have to reevaluate our company brands in light of our overall relationship with our customers. This is not a bad thing as noted by Mike Wagner, CEO, White Rabbit Group:

*Social media is the technology solution brand managers and marketers have been waiting for. With a full social media toolbox a business can create both high tech delivery and a high touch experience. Every social media interaction creates a brand-building personal reaction. And branding is about multiplying the positive reactions with the right customers in the right ways. Smart companies are learning how to "live" their brands in this emerging and powerful social media marketplace. That's why today's brands must be social media savvy! (M. Wagner, personal communication, September 30, 2009).*

Wagner goes on to say that companies must use the technology that appeals to their customers. His company puts out a monthly newsletter for clients that is distributed by email rather than on the company's blog or on Twitter as his readers, mostly CEOs, prefer this channel.

Bruce Hendrickson, author of the popular marketing blog, "Show, Don't Tell," says that companies are struggling with how or even why to translate their business models to adjust to this new media paradigm shift. He agrees with Wagner that it's important to use a variety of social media tools in the proper contexts, in order to achieve the results you want. As he notes, "I've had clients say, 'Get us on Twitter!' or, 'We need more followers!' I've had others say, 'Social media is a waste of time.' Both of these are wrong-headed. People need to understand these tools and where they fit within the overall strategy, but they can't do that without evolving their point of view" (Hendrickson, 2009).

"It's all about building your network before you need it," states Liz Strauss, one of the leading experts in creating online communities. Strauss has hosted Open Mic night on Tuesdays at her blog, www.successful_blog, since May of 2006. While this has been widely successful for her and others in building community, she is even now thinking of moving to new venues. As she says, "part of this new world of media is being open to new possibilities, to know when to let go and move on."

Strauss says the whole point of social media is to make connections. "When people have a sense of ownership from the beginning, they will support you as a friend. They'll buy your product, show information about you online, and recommend you to others. And that's what it's all about" (L. Strauss, personal communication, October 14, 2009).

Ted Demopoulos agrees but advises digital product companies not to rely only on social media outlets:

*That might work if you're Chris Brogan [www. chrisbrogan.com] or Darren Rouse [www.problogger.net] as they already have a huge and loyal following. For the rest of us, social media is a way to get your name out there. It's important to gain this name recognition but most of your sales will come through focused search traffic-people who come to your website because they are looking for a particular answer or product. Facebook and Twitter are excellent ways to get feedback, to get your name out there, and to see what the community is saying about you. But you can't rush the process (T. Demopoulos, personal communication, October 14, 2009).*

## FUTURE OF SOCIAL MEDIA

It is anticipated that social networking sites will continue to grow worldwide over the next few years. Members are moving towards forming subgroups with others with like interests with whom they feel a connection. For example, people on Twitter can now form "lists" which allows them to group together and follow people such as others interested in photography or local music (Catone, 2009). Digital product companies will be attracted to those communities that have been formed around an interest in their product lines. Advertising can be specifically aimed at these target markets. However, as mentioned earlier, companies must concentrate on building a relationship with their customers in order to successfully tap into these social networks. Users are more likely to believe information that comes from a source they think they can trust.

The social networking sites themselves are experimenting with how companies can advertise their products to their users. Facebook recently introduced "social ads," where Facebook users can display products they like on their webpage, thus "recommending" these to their "friends" (Klassen, 2007).

Researchers will design both qualitative and quantitative studies in order to expand the current limited understanding of social media and how online networking sites can be better used for business purposes. As noted by Boyd and Ellison (2007), "researchers' ability to make causal claims is limited by a lack of experimental or longitudinal studies." This is changing but there is still much to understand as to who is using these sites (and who is not) and why. The next few years as more is understood about the long-term implications of social media will help managers and organizations in developing relationships with employees, customers, and other organizations. Companies will "become more serious about trying to measure social media's impact on sales," especially in terms of "how much a large fan base translates into sales or brand loyalty" ("2010 predictions round-up," 2009). CEO Geoff Ramsey predicts that companies will spend more ad dollars on digital advertising, will embrace the use of online video advertising, and will continue to attempt to use social media to build brand loyalty ("2010 predictions round-up," 2009).

Deciding not to engage in social media may even result in a negative impact on a company's bottom line. As noted by Jeremiah Owyang, "Consumers will rely on their peers as they make online decisions, whether or not brands choose to participate. Socially connected consumers will strengthen communities and shift power away from brands...eventually this will result in empowered communities defining the next generation of products" (2009). At any rate, companies should be prepared for the fact that their products and Web pages can and will be reviewed by those on the Internet, regardless of whether they plan to actively participate on social media sites. Appendix B presents details from a personal interview with David Henshaw, owner/editor of *A to B Magazine*, who has decided not to engage in social media in marketing his online magazine at this time.

## CONCLUSION

So what have we learned from the experts on using social media? Scott Monty, head of social media for Ford Motor Company, puts it this way:

*People still trust people like themselves; but the ones they know best are the ones they're most likely to trust. Therefore, it will be the people in their close networks - particularly from a geographic perspective - that will remain the closest. Brands will also realize that they can't be all things to all people, and will focus on those influencers who are the best fit for them (Monty, 2010).*

It makes sense that products that are produced and disseminated via the Internet would want to tap into online social networking sites to market these products. However, doing so is not something that occurs overnight. It takes time to create a community of loyal followers. However, the possibilities are there. Paul Chaney, a social media expert who is an author at MarketingProfs.com tells a story that is applicable here.

A shoe salesman is assigned a new territory, some islands in the south Pacific. He gets to one such island and finds that everyone is barefoot. He calls his boss and says, "We can never be successful here. No one wears shoes."

Needless to say, he is quickly replaced by another, more optimistic salesman. He shows up at the same island, calls his boss and says, "Send me 1,000 pair today! We're going to do great. No one here wears shoes!"

As Chaney notes, "No one here is using social media? Send me 1,000 blogs! We're gonna do great!"

## REFERENCES

2010 predictions round-up. (2009, December 31). Retrieved on January 18, 2010, from http://www.emarketer.com/Articles/Print.aspx?1007446

Blog note update. (2009, September 14). Retrieved September 16, 2009, from http://theonlinephotographer.typepad.com/the_online_photographer/2009/09/blog-note update.html

Boyd, D. M., & Ellison, N. B. (2007). Social network sites: Definition, history, and scholarship. *Journal of Computer-Mediated Communication, 13*(1), article 11. Retrieved on December 20, 2009, from http://jcmc.indiana.edu/vol13/issue1/boyd.ellison.html

Brogan, C., & Smith, J. (2009). *Trust Agents: Using the Web to Build Influence, Improve Reputation, and Earn Trust*. New York: John Wiley & Sons, Inc.

Cashmore, P. (2006, July 11). Myspace: America's number one. Retrieved on October 11, 2009, from http://mashable.com/2006/07/11/myspace-americas-number-one/

Catone, J. (2009, November 2). How to use Twitter lists. Retrieved January 13, 2010, from http://mashable.com/2009/11/02/twitter-lists-guide/

Chaney, P. (2009, October 13). Social media works for small business. I have proof. Retrieved on December 19, 2009, from http://www.mpdailyfix.com/2009/10/social_media_works_for_small_b.html

Demopoulos, T. (2007). *What No One Ever Tells You About Blogging and Podcasting: Real-Life Advice from 101 People Who Successfully Leverage the Power of the Blogosphere*. Kaplan Publishing.

Doctorow, C. (2003). *Down and Out in the Magic Kingdom*. New York: Tor Books.

Emarketer, Twitter Tally. (2009, April 28). Retrieved on October 10, 2009, from http://www.emarketer.com/Article.aspx?R=1007059

Facebook. (2009, May 27). Retrieved October 2, 2009, from http://topics.nytimes.com/topics/news/business/companies/facebook_inc/index.html

Facebook: Largest, fastest growing social network. (2008, August 13). Retrieved September 30, 2009, from http://www.techtree.com/India/News/Facebook_Largest_Fastest_Growing_Social_Network/551-92134-643.html

Facebook visits increased 194 percent in past year. (2009). *Experian Hitwise*. Retrieved on October 13, 2009, from http://www.hitwise.com/us/press-center/press-releases/social-networking-sept-09/

Hendrickson, B. (2009, November 20). Evolution in the media revolution. Retrieved on December 20, 2009, from http://theshortestdistance.biz/?p=476

Hosting your party. (2009, August 29). Video retrieved September 30, 2009, from http://www.youtube.com/watch?v=1cX4t5-YpHQ.

Hunt, T. (2009). *The Whuffie Factor: Using the Power of Social Networks to Build Your Business*. New York: Crown Publishing Group.

Kahlow, A. (2009, July 23). Marketing to people. Retrieved October 5, 2009, from http://www.clickz.com/3634482

Klaassen, A. (2007, November 26). Facebook's bid ad plan: if users like you, they'll be your campaign. *Advertising Age*, retrieved on September 30, 2009, from http://adage.com/digital/article?article_id=121806&search_phrase=%22social+ads%22

McDermott, C. (2009, September 30). Tweeting provides benefits for businesses. *The Daily Collegian Online*. Retrieved on October 6, 2009, from http://www.collegian.psu.edu/archive/2009/09/30/tweeting_provides_benefits_for.aspx

Merriam-Webster Online Dictionary. (2008). Retrieved June 25, 2008, from http://www.merriam-webster.com/dictionary/blog

Miller, C. H. (2008). *Digital Storytelling: A Creator's Guide to Interactive Entertainment* (2nd ed.). Focal Press.

Monty, S. (2010, January 2). Social media predictions for 2010. Retrieved on January 18, 2010, from http://www.scottmonty.com/2010/01/social-media-predictions-for-2010.html#ixzz0czJJdsg1

Owyang, J. (2009, April 27). The future of the social Web: in five eras. Retrieved January 16, 2010, from http://www.web-strategist.com/blog/2009/04/27/future-of-the-social-web/

Qualman, E. (2009, August 11). Statistics show social media is bigger than you think. Retrieved from http://socialnomics.net/2009/08/11/statistics-show-social-media-is-bigger-than-you-think/

Quibble, Z. K. (2005). Blogs and written business communication courses: A perfect union. *Journal of Education for Business, 80*, 327–331. doi:10.3200/JOEB.80.6.327-332

Safko, L., & Brake, D. K. (2009). *The Social Media Bible: Tactics, Tools, and Strategies for Business Success*. New York: John Wiley & Sons, Inc.

Salazar, C. (2007, May 23). Can Facebook win the battle over MySpace? Retrieved October 9, 2009, from http://ebizz.wordpress.com/2007/05/23/can-facebook-win-the-battle-over-myspace/

Sifry, D. (2007). The state of the live Web. Retrieved June 25, 2008, from www.sifry.com/alerts/archives/000493.html

Stats, Facebook. 2010. Retrieved September 26, 2010, from http://www.facebook.com/press/info.php?statistics

Stats, Facebook. Retrieved October 10, 2009, from http://blog.facebook.com/blog.php?post=136782277130

Stats, Twitter. Retrieved October 13, 2009, from http://www.crunchbase.com/company/twitter

Technorati Website Statistics. Retrieved July 30, 2008, from http://www.technorati.com/about/

The ethics of reviewing. (2009, September 14). Retrieved September 16, 2009, from http://theonlinephotographer.typepad.com/the_online_photographer/2009/09/the-ethics-of-reviewing.html

Trammell, K. D., & Ferdig, R. E. (2004). Pedagogical implications of classroom blogging. *Academic Exchange Quarterly, 8*(4), 60–64.

Walsh, M. (2009, March 26). Facebook users growing up fast. Retrieved September 29, 2009, from http://www.mediapost.com/publications/?fa=Articles.showArticle&art_aid=102973

Yarrow, J. (2010, April 14). Twitter finally reveals all its secret stats. Retrieved on September 26, 2010, from http://www.businessinsider.com/twitter-stats-2010-4#ixzz10fR5TZAu

YouTube Fact Sheet. Retrieved October 10, 2009, from http://www.youtube.com/t/fact_sheet

YouTube Facts and Figures. (May 2010). Retrieved September 26, 2010, from http://www.website-monitoring.com/blog/2010/05/17/youtube-facts-and-figures-history-statistics/

## ADDITIONAL READING

Algesheimer, R., & Dholakia, P. M. (2006). Do customer communities pay off? *Harvard Business Review, 84*(11), 26–30.

Barone, L. (2010, January 5). 100+ SMB blogging ideas to kick start 2010. Retrieved January 13, 2010, from http://smallbiztrends.com/2010/01/100-smb-blogging-ideas.html

Barone, L. (2010, January 12). 80 ways to use Twitter as a SMB owner. Retrieved January 13, 2010, from http://smallbiztrends.com/2010/01/how-to-use-twitter-as-a-smb-owner.html

Beal, V. (2009, December 4) Retweet tips for maximum Twitter exposure. Retrieved January 12, 2010, from http://www.webopedia.com/DidYouKnow/Internet/2009/twitter_retweet_tips.asp

Blossom, J. (2009). *Content Nation: Surviving and Thriving as Social Media Changes our Work, our Llives, and our Future*. New York: John Wiley & Sons, Inc.

Cassidy, J. (2006, May 15). Me media: How hanging out on the Internet became big business. *The New Yorker*. Retrieved on December 20, 2009, from http://www.newyorker.com/archive/2006/05/15/060515fa_fact_cassidy?printable=true

Chaney, P. (2009, December 1). Ten commandments for effective online social networking. Retrieved January 5, 2010, from http://www.marketingprofs.com/articles/2009/3213/ten-commandments-for-effective-online-social-networking

Denton, I. (2010, January). Tweeting for dollars: Social media sites can boost your business if you do it right. *Biz941 Magazine*, 11. Retrieved January 12, 2010, from http://www.biz941.com/Articles/2010/01/Tweeting-for-Dollars.asp

Digital marketing and media fast pack. (2007). *Advertising Age*. Retrieved January 12, 2010, from http://adage.com/images/random/digital-factpack2007.pdf

Galeotti, A., & Goyal, S. (2009). Influencing the influencers: A theory of strategic diffusion. *The Rand Journal of Economics*, *40*(3), 509–532. doi:10.1111/j.1756-2171.2009.00075.x

Gangadharbatla, H. (2008). Facebook me: Collective self-esteem, need to belong, and Internet self-efficacy as predictors of the iGeneration's attitudes toward social networking sites. *Journal of Interactive Marketing, 9*(2), retrieved January 12, 2010, from http://www.jiad.org/article100

Hagel, J., & Armstrong, A. (1997). *Net Gain: Expanding Markets through Virtual Communities*. Boston, MA: Harvard Business School Press.

Halligan, B., & Shah, D. (2010). *Inbound Marketing: Get Found Using Google, Social Media, and Blogs*. New York: John Wiley & Sons, Inc.

Hassan, S. S. (2008). Bringing lead-user innovations to the market: Research and management implications. *S.A.M. Advanced Management Journal*, *73*(4), 51–58.

Hosterman, A. R. (2009, December). Tools of the trades: Getting technical about using Twitter. *Intercom*, 12–14.

How to use Twitter for Business. Retrieved January 12, 2010, from http://www.hubspot.com/Default.aspx?app=LeadgenDownload&shortpath=docs%2FHowToUseTwitterForBusiness.pdf

Joel, M. (2009). *Six Pixels of Separation: Everyone is Connected. Connect your Business to Everyone*. New York: Hachette Book Group.

Kennett, J., & Matthews, S. (2008). What's the buzz: Undercover marketing and the corruption of friendship. *Journal of Applied Philosophy*, *25*(1), 2–18. doi:10.1111/j.1468-5930.2008.00391.x

Lin, C., & Yu, S. (2006). Consumer adoption of the Internet as a channel: The influence of driving and inhibiting factors. *Journal of Academy of Business*, *9*(2), 112–117.

Marketing News' Digital Handbook. (2009).. . *Marketing News.*, *43*(7), 9.

Nutley, M. (2007, May 3). It's the influencers, not the social media, that brands need to target. *Marketing Week*, 19-21.

Ridings, C., Gefen, D., & Arinze, B. (2003). Some antecedents and effects of trust in virtual communities. *The Journal of Strategic Information Systems*, *11*(3-4), 271–295. doi:10.1016/S0963-8687(02)00021-5

Shirky, C. (2008). *Here Comes Everybody: The Power of Organizing Without Organizations.* New York: Penguin.

Sledgianowski, D., & Kulviwat, S. (2009). Using social network sites: the effects of playfulness, critical mass and trust in a hedonic context. *Journal of Computer Information Systems*. Retrieved January 12, 2010, from http://www.allbusiness.com/marketing-advertising/marketing-advertising-overview/12723438-1.html

Snell, S. (2008, August 13). 35 must-read articles for social media markets. Retrieved January 12, 2010, from http://traffikd.com/resources/35-must-read-articles-for-social-media-marketers/

Social Media Revolution. http://www.youtube.com/watch?v=sIFYPQjYhv8&feature=player_embedded

Social networking online boosts bottom line. (2006). Retrieved January 12, 2010, from http://www.emarketer.com/Results.aspx?N=0&Ntk=basic&Ntt=social%20media%20boost%20bottom%20line

The Ten Ways Twitter Will Permanently Change American Business. (2009, May 26). Retrieved January 12, 2010, from http://247wallst.com/2009/05/26/the-ten-ways-twitter-will-permanently-change-american-business/

Thelwall, M. (2008). No place for news in social network Web sites? *Online Information Review*, *32*(6), 726–744. doi:10.1108/14684520810923908

Varey, R. J. (2002). *Relationship Marketing: Dialogue and Networks in the E-Commerce Era.* Chichester: Wiley.

What's working for social media marketers? (2010, January 5). Retrieved January 12, 2010, from http://www.emarketer.com/Article.aspx?R=1007449

Zeng, F., Huang, L., & Dou, W. (2009). Social factors in user perceptions and responses to advertising in online social networking communities. *Journal of Interactive Advertising*, *10*(1).

Zhang, J., & Daugherty, T. (2009). Third-person effect and social networking: Implications for online marketing and word-of-mouth communication. *American Journal of Business*, *24*(2), 53–63.

## KEY TERMS AND DEFINITIONS

**Blog (short for "web log"):** Technically a website that is usually frequently updated with dated entries and with the most current blog entry at the top of the webpage. Readers can leave online comments to blogposts.

**Forums:** Discussion boards on websites where people can start online conversations and ask questions on specific topics of their choice.

**Influencers:** People who are trusted by others in the online community who are able to persuade others based on this trust. Influencers can be bloggers, academics, politicians, celebrities, or persons with specific expertise.

**Social Media:** Refers to the communities of people developed online to share information and opinions on platforms such as blogs, forums, podcasts, Facebook, and Twitter. Also referred to as social networking.

*It's All about the Relationship*

**Social Netiquette:** A set of social conventions and protocols to help facilitate communication over social media sites. For example, typing in all caps is considered rude or the equivalent of yelling at someone.

**Whuffie:** Another name for social capital or the value an individual accumulates from being trusted by others in a social network.

## APPENDIX A

**5 Things Small Business Owners Should Do Today Online**

1. Start a blog – I can't think of any simpler website technology to start and master, and there are cheap and free platforms readily available. Why a blog? Because they're easy to create, because they're easy to update, because they encourage repeat visits, and because you can use them in many flexible ways. Need a good website address (URL)? Pick a name out at Ajaxwhois.com, which lets you search many variations at the same time. Then, click through to buy the domain at GoDaddy.com, and then decide if you want to buy hosting there, or from another site. The company Bloghost.me, offers $10/year hosting for WordPress blogs.
2. Start listening – People are talking about you. Find out where they are and who they are. When you're done with that, start finding new business opportunities. People tweeting or blogging about being in your neck of the woods? Reach out, if it makes sense.
3. Try Twitter OR Facebook – Let's not rush things. Facebook has many more users, but it's a bit harder to find customers, prospects, partners and colleagues. Twitter is easier to use and faster to connect with people, but there are far fewer users on there today. I'll let you choose. If you go with Facebook, make a personal account under your own name, and then start a fan page for your business.
4. Get the word out – If you're going to spend time building these social sites, let's presume that you want more people to contact you and interact with you through them. Print business cards with the company name, and/or the request for people to join your fan page or follow you on Twitter. Extra points if you give them a social-media-tool-only discount of some kind.
5. Try moving the needle – now let's *really* get crazy. See if you can fill the place up with social-media minded folks. Okay, this won't work for *every* business, but don't be too quick to count out the idea. Let's try inviting them to a store-only special event, or let's give them a discount code. You know, the stuff you already know how to do. Any difference in the results? See if you can do some kind of really special one-day-only push, and what that brings to you.

There's obviously much more to it than just starting and doing, or is there? One way that small businesses get all confused and thrown for a loop is by feeling a strong sense of analysis paralysis, or that notion that they should be using all the tools right out of the gate.

The reason I started with a blog is that it will give instant search juice to the organization. It doesn't mean your company will rise to the very top listing right away, and there's so much more to it than that, but starting with *any* presence is better than having none.

Source: Chris Brogan, (2009, September 14). http://www.chrisbrogan.com/5-things-small-business-owners-should-do-today-online/.

## APPENDIX B

## Case Study Based on Interview with David Henshaw, Owner/Editor of A to B Magazine

### Could Social Media be Used to Market *A to B* Magazine?

*A to B* is a digital magazine that started out as a newsletter for the Folding Society in England in August 1993 and focused on reviews of folding and separable bicycles. By August 1997, the organization had 1200 subscribers worldwide, the magazine had a new title, *A to B,* and a new tag line-to explore new ways of traveling from point A to point B. The magazine continues to feature folding bikes but has expanded to provide coverage on electric bikes, bike trailers, and public transport. Subscribers include bike manufacturers and distributors, public utilities, and train operating companies. The UK makes up 90 percent of their subscribers and the rest come from the U.S., Holland, Japan, and Germany with a total subscriber list of around 5,000.

In August 2006, the company signed up with a third party agent to offer the magazine digitally. The owner/editor of the magazine, David Henshaw, indicates that his digital experience has been "rather mixed...and not entirely positive." The magazine started with about 350 subscribers, and while the numbers shot up at first, they then dropped back sharply. Lately these seem to have rallied but as he noted, many people said they were tempted by digital, but have changed their minds. His biggest surprise was in the location of the subscribers. With their paper issues, about 90 percent come from the United Kingdom and with the digital magazine, it's about 80 percent. He had expected a lot more digital readers from the United States and Australia but states that the number of physical magazines that they send airmail hasn't really dropped.

When asked about his experience in offering the magazine online, Henshaw replied that there was "a lot of confusion over whether to provide digital as a free perk for paper subscribers. We dropped this as it was costing us $1,000 a year to provide it, with no clear benefits for us. We now charge about $3 a year for digital if people want it as well as the paper subscription, which more or less covers costs."

When asked if the magazine participated on any bicycle-related online discussion forums, his response was strongly against:

*Not at all, never, ever, ever! I find them appalling and the information that goes round and round there is often very inaccurate. Again, we did get enthusiastic in the early days of the Internet, but began to realize that the sort of - slightly nerdy - people who participate are not the sort who buy magazines, so we were wasting our time in terms of promoting what we do. I would participate in these often ill-informed discussions if I had the time, but I don't.*

Henshaw feels the same about using social networking sites such as Facebook or Twitter in marketing their online magazine. As he notes, "I think the marketing element has been much exaggerated." They have also not considered a company blog. As he says, "I don't think we need one. People email every day with thoughts and queries. If we think these emails warrant further discussion, we put them in one of the magazines. Running an online blog would take a lot of my time, and I'm not sure it would yield anything useful." He typically doesn't comment on other people's blogs on bicycles either. "Only on the rare occasions that I have been misquoted on one and someone has warned me."

When asked about his digital magazine experience overall, his comment was his experience was "very mixed, and a bit disappointing really. Digital is most useful as a marketing tool for back numbers. To buy all our back numbers would cost a great deal, but we can offer the whole set (back to Aug 2006) for $3 a year, if people subscribe to the magazine, and I hope this yields a few new subscribers. We lose out selling back numbers, of course, but the administration and packing was a real hassle, so it's a service for us too. I now print fewer magazines, and hold very few in stock in our back number catalogue. That's the biggest change."

**Questions to Ask?**

✓ Who are the core customers that *A to B* magazine wants to reach?
✓ Which social networking sites would be most likely to reach them?
✓ Does the company have the time and resources to put towards marketing their magazine on social networking sites?
✓ What would be the best way for Henshaw to ease into using social media sites?
✓ What advice would you give Henshaw as to whether or not to use social media to market his digital magazine?

# Chapter 8
# Digital Convergence and Horizontal Integration Strategies

Troy J. Strader
*Drake University, USA*

## ABSTRACT

*Digital technology is unique in that it enables convergence of access devices and content as well as convergence of industry participant operations and strategy. This digital convergence creates opportunities, and threats, for developing new business models and unique growth strategies for digital product companies. The purpose of this chapter is to discuss examples where digital product companies have taken advantage of digital convergence through horizontal integration strategies that enable them to create unique mixes of products/services and reach larger markets. Actual horizontal integration strategy examples are discussed for several industries where products are in a digital form. In addition, potential cross-industry integration strategies and online intermediary (cybermediary) strategies are identified along with tactical level strategies for mass customization and use of interactivity tools and social networking. Strategic alternatives for introduction, growth, and maturity lifecycle stages are also discussed. The rationale for these strategies and implications for managers of digital product companies are discussed along with directions for future research.*

## INTRODUCTION

Digital technology has been around since the invention of the electronic digital computer about seventy years ago. Over time the capabilities of digital technology have continually improved, but there has been an acceleration of activity associated with the commercialization of the Internet and the development of new methods for development and delivery of digital multimedia content. Examples of digital content and product industries include newspapers, magazines, electronic books (e-books), music, movies, games, and software. Digital technology and the Internet have enabled companies in each of these industries to independently transform their traditional product into a

DOI: 10.4018/978-1-61692-877-3.ch008

digital format. This led to some cost savings and access to larger markets, but the most successful companies have often taken this a step further to pursue strategies that take advantage of the opportunities created by digital technology, content, and industry convergence through integration strategies within and across industry boundaries. Operationally, the world of electronic business and electronic commerce has enabled the entire digital product order fulfillment process – information search, order placement, payment, product delivery, and associated customer service – to be converged which has created opportunities for cost reduction and creation of entirely new products and services. Strategically it has also created opportunities for focus on core competencies, access to larger markets, and creation of entirely new products and services.

The primary purpose for this chapter is to analyze one aspect of this overall phenomenon, the opportunity created by digital convergence and why this has led many digital product companies to choose a horizontal integration strategy as their path toward achieving competitive advantage. The focus is on digital product companies and not digital services. Digital products are informational, multimedia, and software products that are in a digital form and possession of the content/product is transferred to the consumer for either a period of time or forever. Examples of digital services that are outside the scope of this chapter would include online financial services, career services, travel services, auctions, and virtual communities.

This chapter is organized as follows. First, the concept of digital convergence is discussed along with an explanation of the goal and process used in strategic management as well as the range of alternative strategies available to companies. The main section then combines these issues to illustrate some of the types of horizontal integration strategies currently utilized by real-world companies to take advantage of the opportunities presented by digital convergence. The rationales for digital product managers choosing these strategies are discussed. Next, other potential cross-industry partnerships, intermediary strategies, mass customization strategies, interactivity and social networking strategies, and life cycle stage strategies are identified to illustrate the range of possible growth strategies that could be utilized in the future. The chapter concludes with a discussion of implications for digital product managers and directions for future research associated with future digital product strategy issues.

## BACKGROUND

Two issues are presented in this section to provide the background for understanding the rationale behind the strategies described in the next sections. These two issues include three dimensions of digital convergence (technological, content, and industry) and a description of the goal for strategic management, process used for strategic analysis, how technology changes are incorporated into this analysis, and the range of strategic alternatives available to companies that wish to pursue the opportunities presented by digital convergence.

### Digital Convergence

Technological convergence is the result of combining two or more media platforms into a single hybrid device (Laudon & Traver, 2010). Numerous examples illustrate the concept of technological convergence. Many of today's cell phones can be used as a phone and also can be used to access Web content and send e-mail messages or text messages. The device with the widest range of capabilities is probably the personal computer. Today's PCs enable users to access the Web, send e-mail, run software programs, listen to music, play games, make payments, and the list continues to grow. Essentially there is no limit to the capabilities of hybrid devices as long as the product/service can be digitized. The benefit to the user is a reduction in the number of devices dedicated to only one or a small number of tasks.

Content convergence is made possible by the unique ability of content to be digitized during its design, production, and distribution (Laudon & Traver, 2010). Content from older media is initially migrated into a new media without significant change, but later the content using new media will evolve to utilize the full capabilities of the new media. Online catalogs in the 1990s often looked similar to their printed counterparts, but later they began to incorporate video technology, sound, automated order placement, and interactive discussion and customer service support. The convergence of content with multimedia and supporting processes has become the norm. Digital content creation processes have also been impacted. The cost and time involved in developing digital content has been reduced, and it is somewhat easier to deliver digital content to different platforms, although there are still hurdles to overcome given the proprietary nature of many devices and operating systems. The downside for content convergence is that easier creation, copying, and delivery of digital products makes unauthorized possession and distribution easier.

Industry convergence involves the partnership or merger of separate digital product companies to create synergistic offerings across multiple markets and platforms (Laudon & Traver, 2010). The appropriate convergence of industry participants is the result of decisions made based on a strategic analysis process. To understand why many digital product companies have been involved with industry convergence, the following section describes the goal of strategic management, the strategic analysis process, how changes in digital technology are incorporated into the analysis process, and the range of strategic alternatives available.

## Strategic Management and Integration Strategies

The goal of strategic management is to identify an overall direction for an organization that will lead to some form of competitive advantage. Competitive advantage can be achieved in a variety of ways including developing better products/services, reducing costs, increasing market share and revenue, entering into unique partnerships and alliances, and maintaining or gaining control over various forms of resources (human, informational, financial, brand equity, customer service perceptions, materials), among others. The four elements of the strategic management process include environmental scanning, strategy formulation, strategy implementation, and evaluation and control (Hunger & Wheelen, 2007). In this section the focus will be on the first two steps.

Environmental scanning is the collection of information about the current and near future external and internal environment an organization is working within. External issues can be categorized as either societal issues or industry issues (Hunger & Wheelen, 2007). Societal issues impact a wide range of issues and include broad-based economic, technological, political-legal, and sociocultural issues. External issues can also be issues that are directly related to the industry in which an organization operates. All of these external issues are analyzed to identify factors that could positively impact an organization (opportunities), or those that may negatively impact an organization (threats). For example, the commercialization of the Internet is a societal level external issue that provided an opportunity for creation of a new business (eBay, Priceline, etc.), but also created a potential threat to existing print-based media companies (magazines, newspapers, and books).

The unique characteristics of the Web are the source of many of the opportunities, and threats, discussed throughout this chapter. It is a societal external issue because the same system is available to any company worldwide. The unique current characteristics of the Web include ubiquity, global reach, universal standards, richness, interactivity, information density and personalization/customization (Laudon & Traver, 2010). The Web is ubiquitous because it can be accessed any time and from any connected device whether it's a personal computer or wireless device. This means that answers to questions can be available any time, but it also means that the line between work and home may be blurred because managers know that you can do work anywhere. The Web has a global reach because it reaches across geographic boundaries. The same Internet and Web are used worldwide which provides an opportunity to reach a very large market place, but the threat is that competitors can reach the same people in the same way. The Internet and Web use many standard protocols and digital information formats. Text, images, audio, and video can be shared worldwide because of common standards. The threat from this is that hackers may attack websites anywhere in the world using the same system. Also, viruses can have a greater negative impact because so many people use common operating systems. The Web is a rich environment because it supports many forms of multimedia. It has many advantages over traditional print technologies. Increased use of multimedia has led to demand for greater bandwidth, but the information provided to people can be a much more sensory rich experience combining sight and sound. This may be a particularly important opportunity for advertising products that require more than just text. The Web is an interactive tool that enables two-way communication between people and other people or organizations. Questions can be asked, and answers given, to support sale of any digital product. The Web is also a dense information space with massive amounts of information available in a digital format through many different service providers including sites such as Google and Yahoo. Almost any search produces numerous possible answers, and the results are available very quickly. Finally, the Web is an interface that can be personalized or customized based on user requirements. Websites can offer a unique experience for any user if people are willing to provide their criteria. The potential threat is a reduction in privacy as more and more personal information is given to service providers and digital product companies. Improvements in digital technologies will have to be considered as they become available because they may create new opportunities as well as new threats.

Internal issues are specific to the organization doing the analysis. Many internal issues are associated with the business function such as finance, marketing, human resources, information technology, production, and logistics. Internal positives are described as strengths and internal negatives are weaknesses. Based on the information gathered in the environmental scanning process, strategic decision makers then work to generate strategic alternatives that attempt to take advantage of opportunities and strengths, or reduce problems associated with threats and weaknesses.

Alternative strategies can be classified into one of three categories – growth strategies, stability strategies, and retrenchment strategies. Growth strategies, the focus of this chapter, include vertical integration, horizontal integration, and diversification. Vertical growth can be achieved by taking over a function previously provided by a supplier or by a distributor (Hunger & Wheelen, 2007). An e-commerce vertical integration strategy example would be if an online book seller would buy a book publisher (supplier) or buy a delivery company (distributor). Horizontal growth can be achieved by expanding into new complementary product or service areas, or by offering products in new geographic markets (Hunger & Wheelen, 2007). If an online company's initial business was to only sell books in the US, then horizontal growth could include adding other products that book buyers may want, or beginning to sell books outside the US. Diversification is a less common strategy for digital product companies. Diversification involves growth into industries that are quite different from the current business. If an online book company bought a restaurant chain, this would be considered a diversification strategy. Stability strategies imply that the organization will continue with their current product offerings and markets, but will make tactical or operational changes to pursue competitive advantage. One example of a stability strategy for an online company would be to improve information search capabilities on their current website to improve their customer's experience and satisfaction which may lead to a current or future sale. Retrenchment strategies involve reducing or eliminating some or all aspects of an existing companies products/services or markets.

Changes in information technology capabilities impact strategy because it is part of companies' external environment. Early research in this area focused on the impact of information technology on the costs associated with various forms of electronic markets and hierarchies for governing transactions, and how these cost differences impact a firm's choice of one governance mechanism over another (Bakos, 1991; Clemons et al., 1993; Gurbaxani & Whang, 1991; Malone et al., 1987). The common finding across these studies was that increased information technology use lowers coordination costs without significantly increasing risk and production costs. This cost reduction led to the expectation that over time an increasing number of firms would choose markets instead of hierarchies to organize their activities. Firms were expected to focus on their core competencies and then form partnerships or alliances with other firms that complemented their current focus. For digital product companies this is even more pronounced because the digitized product also has production and distribution cost advantages when compared with physical goods (Strader & Shaw, 1997). The results theorized by these early studies appear to have come to fruition and can be seen in the strategic choices made by today's digital product companies.

In the first five years after the introduction of the Web, most companies pursued growth strategies, but in the next decade a wider range of strategies was seen as some companies grew, others transformed within their current business, and others became smaller niche firms or completely disappeared. The overall conclusion from this section, as shown in Figure 1, is that digital convergence is both an opportunity and a threat for digital product companies and one way to deal with this external issue is by choosing an appropriate strategy from a long list of alternatives with the goal of achieving competitive advantage.

## Example Digital Product Company Horizontal Integration Strategies

Given the opportunities presented by digital convergence, in this section a number of real-world horizontal integration strategies are presented including a brief discussion of benefits and risks that may result from these strategic choices. The examples are summarized in Table 1. The examples are not intended to illustrate all possible forms of horizontal integration, but they are sufficient to identify patterns across industries.

*Figure 1. Digital convergence and digital product company strategy*

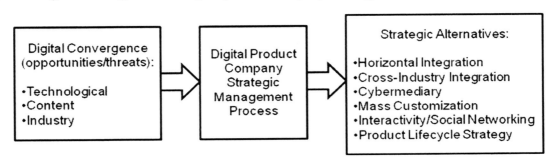

## Newspapers

Printed newspapers have been around for centuries, but online newspapers are relatively new. In 2009 the newspaper industry continued to go through a transformation. There were about 10,000 online newspapers in the world while readership for traditional printed newspapers continued to decline. Many of the largest newspapers had several million visitors to their websites each month and online newspaper readership had been growing at about 17% per year (Laudon & Traver, 2010). Many of the largest online newspapers had readers, but identifying a business model that generates significant revenue online has been elusive. Given that many of these sites offer similar news content, it is apparent that differentiation through the offering of unique content or value-added services is a necessary strategy, but is difficult to achieve.

In the mid-1990s Yahoo! Inc. provided an online directory and over the next decade they had expanded their site to encompass a number of other complementary businesses. In 2006 they announced a strategic partnership with more than 150 daily US newspapers to provide search, content, and advertising across their website and the newspapers websites (Yahoo, 2006). Their stated objective was to secure a leading position in an area where they see the greatest potential for growth by leveraging the complementary strengths of Yahoo and its partners. The newspapers receive access to technology and audience while Yahoo receives access to local content and an expanded advertising network to differentiate itself from its competitors. This strategy is an example of content and industry convergence across the online directory and newspaper industries.

In another digital integration strategy example that may impact the future of online news, the Associated Press (AP) and Yahoo are working on a deal to impose tighter restrictions, and potentially a higher price tag, for AP stories distributed on Yahoo's news site (Adams, 2010a). Major Internet portals utilize AP news articles to offer news content to their users, but the threat to AP has been that they would like more control over how and where the articles appear and they would like a larger portion of the profits realized from the news content. This is an example of a common threat to digital content, it is easy to store and distribute but difficult to track and control. Also, if AP news content is available on many different sites, then readers may view it as a commodity and may not be willing to pay for it.

*Table 1. Digital product company horizontal integration example summary*

| Digital Product Industry | Company | Strategic Partner |
|---|---|---|
| Newspaper | Yahoo | 150 daily US newspapers |
| Newspaper | Associated Press | Yahoo |
| Magazine | Sports Illustrated | CNN |
| Electronic book | Amazon | E-book hardware + publishers |
| Electronic book | Texas Instruments | E-reader developers |
| Electronic book | Amazon | Macmillan |
| Music | Apple | Lala Media |
| Music | Hewlett-Packard | Omnifone |
| Movies | Youtube | MGM |
| Games | Electronic Arts | Playfish |
| Games | MTV Games | User-generated content |
| Software | Adobe | Research in Motion |

Profit sharing between news agencies and online news distributors is an important strategic issue now and it will continue to be an important issue in the future. This is another example of content and industry convergence in the online directory and news content industries.

Pricing is also a major issue for online newspapers. What is a reader willing to pay for a print subscription? And what are they willing to pay for an online subscription? For the past decade the answer has been that they are willing to pay to have a paper delivered to their home, but unwilling to pay for anything they see online. Online newspaper revenues have been on the decline, especially as the economy has faltered and advertising revenues have dropped. In January 2010, the New York Times decided to make a tactical move to charge readers for unlimited access to the Web version of their newspaper (Adams, 2010b). This was seen as a risky move that could push away an already declining number of subscribers. The plan will be launched in 2011 with people who read more than a certain number of articles in a month being charged a flat monthly or annual fee for additional access. Print subscribers will have full access. Few papers have been willing to try this strategy. The hope is that the newspaper's core readers will be willing to pay for this service. It is a strategy that requires a large established user base that feels that the product is worth paying for. Over time, if enough large digital content sites move toward this strategy, it may begin to change the consumer's mindset that material viewed online should be free.

## Magazines

Online magazines have been around since the mid-1990s, but it has been a rough road. Most online magazines have failed, and many of these failures were at least to some extent due to their inability to differentiate their content to attract paying subscribers. Many of the issues facing online magazines were similar to those faced by online newspapers. Digital convergence enabled cost reduction and extended their market reach, but it was a hypercompetitive environment because all magazines had access to the same Web technology capabilities and Internet content distribution network.

Sports Illustrated (SI) is an American sports magazine that published its first issue in 1954. SI became a successful magazine during its first forty plus years of publication, but one example of their entrance into the online magazine world began just after the turn of the most recent century when Sports Illustrated and CNN partnered to offer sportsillustrated.cnn.com (Schreiber, 2001). Sports Illustrated has a sports-related core competency and CNN is a news network, but together they can reach a larger market with a merging of their complementary content-related strengths. They can continue their primary business while offering this co-branded website. The result is access to the partner's viewers, reduction of overall costs, and content extensions using Web multimedia technology, all without the need to move away from their individual strengths. This is one example of a horizontal integration strategy where a unique content product is offered to a broader market. Content and industry convergence enabled this partnership that would have been very difficult and much more costly in the offline environment.

## Electronic Books

Books have been either hand-copied or printed for centuries, but electronic books are a relatively new phenomenon. Digital technology offers many advantages to book publishers including lower production and distribution costs, but there are also a number of disadvantages that need to be overcome. It is easier to lose control over digital content because it is so easy to copy and distribute stolen content, and what may ultimately be the hardest issue to overcome – readers still seem to prefer the traditional printed book. Despite some hesitance by readers to embrace e-books, revenues are estimated to surpass $200 million in 2011 making this a viable and growing market (Laudon & Traver, 2010). E-book content can be accessed through the Web, downloaded to a PC, or read using a dedicated e-book reader. Despite a wide range of content access methods, there is still no clear winning technology which provides a potential opportunity for hardware, software, and publishing companies.

Amazon.com began business as an online book seller, but they have expanded into a wide range of other product areas over the past fifteen years (Amazon, 2009). One of their most recent ventures is the development of the e-book reader named Kindle. The Kindle is a departure from their traditional growth strategies where they continually added new products and extended their website sales to new markets. The rationale for this new strategy is described by CEO Jeff Bezos as one where they identified a customer need and then worked backward (Salter, 2009). Electronic ink display and improved wireless connectivity provided two technological opportunities. Kindle sales have grown quickly, but the project is still a small part of their overall annual revenue. The Kindle dedicated e-book reader hardware with associated book and other content is an example of technology and content convergence. The Kindle also offers the potential for industry convergence if publishers wish to provide their books on this new e-book platform.

One outcome of the introduction of the Kindle has been a fight between publishers and Amazon over pricing. Whenever there is a disruption in a supply chain where one firm gains an advantage and tries to increase their revenue share, there is the potential for a fall out between partner firms. In early 2010 this happened when Amazon stopped selling Macmillan books after a dispute over how to price e-books (Fowler, 2010c). The fight created a lot of debate about the right price for e-books. Publishers, retailers, and customers all know that the marginal cost of selling an addition e-book unit is near zero, but total revenues from all sales must also cover all of the fixed costs involved in developing the content. One perspective would be that the right price is the one that covers costs plus some profit margin. Another would be to charge what the market would allow based on price differences when compared with printed book alternatives. Hard back books cost more than e-books which cost more than paperback books. If readers are willing to pay a price, then that is the price. By the end of January 2010 the situation seemed to have been resolved when Amazon began selling Macmillan books again, but this is just one of many potential fights between publishers and e-book retailers that may arise as the industry searches for appropriate prices for a digital product.

Competition in this relatively new market is already heating up as new entrants develop their own e-book readers to compete with the Kindle. New e-book readers were introduced in 2009 by several companies including Spring Design, Samsung, Hearst Corp., Plastic Logic, Entourage Systems, and DMC Worldwide, and each of these products hopes to grab market share with its e-book reader and unique user services (Fowler, 2010a).

Another entrant into the e-reader market is Texas Instruments (TI). Texas Instruments designs and sells semiconductors and related electronics technologies to electronics designers and manufacturers worldwide (Texas Instruments, 2010). They are developing new e-book technology which will enable manufacturers to produce e-readers that are more energy efficient and cost less (Fowler, 2010b). They have simplified the components used for content display and improved the e-book reader's battery life. TI's improved technology will potentially reduce the product development time for new e-book hardware that is produced by their partners. This is an example of technological and industry convergence where an experienced hardware manufacturer grows their company through a horizontal integration strategy to serve new partners in an electronic book industry where there is the potential for massive growth in the future.

## Music

Music has been played for centuries and recorded music has gone through many media formats during the past century. Music files stored in a digital format have a number of advantages including higher quality sound representation, fast and simple distribution, reduced storage and distribution costs, and they can be played on a wide range of devices. The primary problem in the early days of the online music industry was to identify a business model where users were willing to pay for access to digital music files, but this seems to at least some extent been solved because it is estimated that online music revenues in 2011 will be nearly $4 billion (Laudon & Traver, 2010).

In the late 1970s Apple Inc. developed, manufactured, and sold personal computers and software. In the past decade the company has expanded into a much broader range of hardware and software products and services (Apple, 2009). Continuing this horizontal expansion, in 2009 they acquired Lala Media to expand their presence in the digital music industry (Smith & Kane, 2009). It is not clear how their current strategies will be merged. Lala lets users pay ten cents for songs that are streamed via a Web browser, while Apple has focused on selling downloaded songs for $1 and albums for $10. One possibility is to utilized Lala's distribution model with an iPhone app. The online music industry is highly competitive so control over content and proprietary player technology is critical. This is primarily an example of industry convergence, but also to some extent technology and content convergence.

Hewlett-Packard (HP) develops a wide range of technology products and services and these products and services are provided to companies in many different industries (Hewlett-Packard, 2010). Recently they have decided to adopt a horizontal integration strategy to enter the digital music market (Vitorovich, 2010). They are launching a subscription music service in ten European countries by partnering with Omniphone, a UK mobile music company. HP will offer a subscription based digital music service, using Omniphone's service, and it will be called H-P MusicStation. The strategy will enable HP to deliver music to millions of their PC users in these countries and revenues will be split between HP, Omnifone, and the music companies. This is an example of content and industry convergence between an experienced technology company, a start-up digital music company, and existing music labels.

## Movies

Movies and movie theaters have been a popular form of entertainment since early in the twentieth century. Television brought movies to people's homes in the last half of the twentieth century and beginning in the 1970s the industry further evolved due to the invention of the VCR which enabled people to choose a movie and watch it at home when it fit their schedule. More recently movies have been digitized and distributed on DVDs or through various forms of Internet file transfer. During each of these transformations the movie industry has had to develop new business models to deal with the new threats and opportunities enabled by technology. One disadvantage of digital movies is that the file sizes are very large. As more users gain access to broadband Internet connections this has become less of a problem. It is estimated that online video revenues in 2011 will be more than $3 billion (Laudon & Traver, 2010).

YouTube debuted in 2005 as a site that provided access to a huge array of online video content. In 2006 they were purchased by Google (Stone & Barnes, 2008). User created clips are interesting and draw a lot of traffic to the site, but advertising revenues can be enhanced by offering access to commercial TV shows and movies. Control over unique content is a differentiation strategy for companies in the online video industry. In 2008 YouTube announced that they would partner with MGM to show some full-length television shows and films (Stone & Barnes, 2008). This partnership is an important step toward adding content that will enhance their advertising revenues. Competitors like Hulu are also pursuing similar strategies with other television and film companies. The YouTube MGM partnership is primarily an example of industry convergence. Secondarily it could be considered content convergence because large amounts of content can be distributed to viewers through a single website.

## Games

Video games have come a long way from the arcades of the 1970s and 80s. Today, video game hardware and a wide range of games are available to people who want to play games at home. Video games have all the advantages of any digital product, but like software products there is a significant development process and associated fixed costs. This increases the barrier to entry for new video game companies to enter this market. Revenue for all digital entertainment products continues to grow and video games are no exception. Overall video game revenues are growing and online video game revenue is expected to exceed $1.5 billion in 2011 (Laudon & Traver, 2010).

Electronic Arts, Inc. is the world's biggest publisher of video games (Electronic Arts, 2009). In 2009 they announced an increased focus on a digital strategy that includes downloaded games and content. This is a transition from their traditional business that has had an impact on their profitability and has led to job reductions. According to EA's CEO John Riccitieloo, the impact digital technology had on the print, television, and music industries should also be expected in the digital game industry. To extend their products they acquired Playfish, a maker of social games (Kane, 2009). EA's COO John Pleasants described their strategy as a test of new distribution channels and business models that is intended to enhance their competitive position in the industry and increase their profitability in the future (Kee, 2009). This is an example of horizontal integration enabled by content and industry convergence.

Companies may also grow by introducing partnerships with their users for developing and selling video game content. In a unique arrangement, MTV Games is hoping to improve sales for their 'Rock Band' game with a new service that lets users create and sell video game versions of their own music (Smith, 2010). The service will be called Rock Band Network Store. There is great potential for this service, but this is not as simple as creating a text document describing a news event. Users will have to be rather technologically savvy to create and upload their own songs. Cost will also be a deterrent. The service may be most popular with existing bands that want to increase their exposure. They have the music and money to develop video game versions of their songs which may enable them to reach new audiences. A new industry has also been created where companies offer to help individuals and unknown bands develop video game versions of their songs. This is an example of content and industry convergence within the overall music industry.

## Software

Software is used by individuals and businesses to support a wide variety of applications such as documents, spreadsheets, databases, presentations, multimedia, electronic communications, and support for nearly all business functions and order fulfillment sub-processes. Annual revenue for worldwide software sales is many billions of dollars and continues to grow. Because software is in a digital format, it can take advantage of all of the opportunities presented by technology, content, and industry convergence.

Adobe Systems Incorporated is one example of a software company. One of their main software products is the Flash multimedia player. In 2009 they announced that they are working with Research in Motion (RIM) to bring Flash support to the BlackBerry platform (Adobe, 2009). The objective is to remove barriers to publishing content and applications seamlessly across a wider range of devices. This alliance is one piece of the Open Screen Project that includes close to 50 companies working together to provide consistent runtime environments across mobile phones, desktops, and other devices. This is an example of technology and industry convergence trying to overcome current content convergence limitations.

## Managerial Implications

The following managerial questions can be answered based on an analysis of the digital product company strategies described in the previous section: What are the overall implications associated with digital convergence for managers of companies in these industries? What are the expected benefits from horizontally integrating a digital product company? And what are the potential risks? The expected benefits and risks are summarized in Table 2.

The first conclusion that can be made from these examples is that every company has access to the same digital technology and Internet. This makes it very difficult to differentiate products based on technology or product distribution channels alone. The differentiators are more likely to be unique content, unique partnerships and alliances, or novel business models or methods for charging users for digital product sales. Strong brands are important, but they are typically not enough to guarantee success.

The second conclusion is that digital product managers decide to expand their company through horizontal integration strategies for a number of reasons. Operational improvements may result, but most of the motivations are at the strategic level. Multiple sources of unique content can be combined to complement each other, extend an existing digital product, provide enhanced value to the customer, and reduce overall costs. Partnerships with content companies can increase control over the distribution of that content and increase entry barriers for other firms that may result in a competitive advantage beyond the short term. Digital product company partnerships also allow separate firms to reach new and larger markets at a reasonable cost. Partnerships also enable firms to adopt the partner's unique business model or unique revenue model (pay-per-download, etc.). And horizontal integration strategies also may create new distribution channels either through the creation of new hardware devices or distribution of content through large digital content distribution service providers.

Digital product company horizontal integration strategies are enabled by the unique characteristics of digital technology and the ability of information technology to reduce overall coordination costs across multiple firms. But working with other companies also introduces some new risks. Control over a firm's content is at least to some extent lost or reduced when it is being shared with partners. Partners must be chosen carefully because any fiscal or ethical problems a partner firm has will be seen as a problem that may tarnish all partners.

*Table 2. Digital product horizontal integration strategy benefits and risks*

| Benefits | Risks |
|---|---|
| • Revenue growth<br>• Complementary content synergy<br>• Control over unique content<br>• Reach larger market with minimal additional cost<br>• Each partner can focus on their core competency<br>• May adopt partner's business model and revenue sources<br>• Product extensions may involve less additional cost<br>• May raise entry barriers for potential competitors | • Problems can arise over the issue of revenue sharing<br>• Growth involving development of hardware devices can involve significant fixed costs<br>• May pull focus away from core competencies<br>• Potential for loss of control over content<br>• Partner ethical and financial issues impact all horizontally integrated partners<br>• Mixing business models and revenue sources may confuse customers |

## POTENTIAL CROSS-INDUSTRY HORIZONTAL INTEGRATION STRATEGIES

In this section, additional potential horizontal integration strategies will be identified. To organize the discussion, the digital product industries will each be combined into the following categories based on their product characteristics: informational content (newspapers and magazines), entertainment and educational content (e-books), multimedia (music and movies), and software entertainment (games). As seen in previous discussion, digital products also share some common attributes with digital services in that they are available in a digital format and often distributed through the Internet and Web. In addition to the digital product categories, online services will be considered as potential horizontal integration partners in the following examples. The goal in this discussion is to illustrate the wide array of potential combinations that could occur in the future as companies cross over into new complementary industries to provide unique combinations of products and services to their customers by fully utilizing the opportunities presented by digital convergence.

### Newspaper/Magazine + E-Books

Online newspapers, magazines, and electronic books each provide content to their users. A cross-industry integration example in these industries could involve the promotion and delivery of e-book content through a magazine Web site. An online magazine could provide reviews for newly released books written either by the magazine staff or by readers. The online magazine could be integrated into an e-book online retailer to provide efficient ordering of books by the magazine's readers. This would simplify the ordering process for the users, but would also create a beneficial partnership between the separate companies. They could each expand their online exposure and share any profits resulting from their online sales. A concern would be to make sure that book reviews are perceived as independent and unbiased so that the book buyers see the recommendations as useful in addition to the value offered by the efficient ordering process.

### Newspaper/Magazine + Music/Movies

A similar arrangement could be made between an online magazine and an online music or movie retailer. New music is sold online either through CDs or through download. Movies are sold online through DVDs, and as bandwidth improves, they will also be able to be distributed online to more users. New music and movies are reviewed on many online sites and digital convergence enables the integration of the magazine Web sites with the online retailer's ordering process. They each benefit by expanding their market and providing useful information and an efficient ordering process for their users. The full potential will be realized if they can provide a download of the new product which saves time and distribution costs. An online magazine related to these industries can be a stand-alone success, but integrating with other industry participants is probably beneficial to the industry and its customers.

### Newspaper/Magazine + Games

An online magazine focused on computer games could provide another potential cross-industry partnership. This time, instead of selling the entire product, the magazine's Web site could provide game samples to advertise new game products. The online game companies could provide the software as a means of advertising their new products and gamers would benefit by reading about news in the industry and also having a hands-on experience with new products. The online magazine and game companies will benefit from expanding their market and also providing a multimedia advertising experience for their new

*Digital Convergence and Horizontal Integration Strategies*

and returning customers. The unique nature of online game products may benefit the most from enabling customers to experience the product first hand. As with all of these integration examples, revenues received through sale of the online games can be shared between the online magazine and the game company.

## Newspaper/Magazine + Online Services

Up to this point, this chapter has focused entirely on digital products. An array of transactional services is also available online (financial services, career services, travel services, online auctions, etc.) and these companies provide additional potential partnerships for digital product companies. One service provided by traditional newspapers is job listings. An online newspaper and an online career services company could partner to provide online job ads viewable through the newspaper's Web site. Each company could focus on their core competence, but the readers would benefit from having access to up-to-date job information in addition to the newspaper content. The companies could reach new customers and any revenues received through the service could be shared between the online career service company and the newspaper.

## E-Books + Music/Movies

Electronic books and movies are often intended to entertain. Content convergence enables book content and images to potentially be integrated with movie video content. An integration strategy across these industries could involve an electronic book written about the movie industry. Any time a specific movie is discussed in the book, a link could be made to a clip from the movie or to a retailer who sells the movie online. Customers could benefit from fully utilizing the multimedia capabilities of the Web and improving the overall value received from the e-book. The companies benefit by being able to offer a better e-book product with little additional cost. The movie clips can be stored and shared online at a very low marginal cost. E-book readers may be willing to pay more for these multimedia books and the additional revenues can be shared between the e-book publisher, retailer, and movie companies.

The same idea could be utilized by integrating e-books with music clips. The technology is similar, but it would be even easier in the music industry because the file sizes are much smaller when compared with video content.

## E-Books + Games

Similar to the example above, e-books could be written about online games. The e-book could link to a site that provides either clips from new games or ways to order the games. The reader/gamer receives a more valuable e-book product which they may be willing to pay more for. The additional revenues can be shared between the e-book companies, game companies, and retailers. As with movies, bandwidth issues will be a concern because book content involves fairly small files while complex multimedia games can involve very large files and the need for streaming download.

## E-Books + Online Services

Another integration strategy possibility would be between an e-book company and an online travel service. If you go to a book store there is probably a section that includes books about travel destinations. An e-book could be written about various locations around the world and the book could provide links to an online travel site that sells trips to these places. The online travel company could provide text, images, and multimedia content about different travel destinations along with information about the prices they would charge for a travel package. The e-book and service site can be linked allowing the reader to go back and forth between reading about the destination and finding up-to-date information about travel opportunities. Again, the traveler may pay more for an e-book that links with this service and they are able to get the information they need without having to go to several separate Web sites.

## Music/Movies + Games

Movies and games are entertainment. New movies typically have a promotional Web site and one feature they could add would be to provide access to a game that is related to the movie. This cross-promotion could create more interest in the movie so people see it in the theater, but it may also interest them enough to purchase the game. A movie company and a game company could co-produce the two products and then introduce them at the same time. Additional revenues may be realized at the theater and at an online game retailer. Digital convergence enables this strategy and the companies and their customers can all benefit.

## Games + Online Services

A final cross-industry integration strategy could combine an online game company with an online service. There are numerous examples of multi-player fantasy games online where users create their own characters and over time the value of these characters can increase either by possessing objects or through their conquests. An online game could be linked to an online auction site that would allow players to trade valuable items, or entire characters, that are part of the game. Digital convergence enables the game software to be linked to an auction service Web site and players may be willing to exchange items and money through these transactions. Gamers pay money for updated versions of the games, but also may pay for the additional auction service. The companies can benefit by providing this additional feature for the game without much additional cost. This could be an interesting product extension that is very difficult to offer in a traditional off-line game environment.

## Cross-Industry Horizontal Integration Strategy Conclusions

The conclusion from this section is that the number of possible partnerships is infinite. Digital convergence offers incredible opportunities for combining products and services to offer greater user value, with minimal additional costs, and greater market reach and increased revenues.

# POTENTIAL DIGITAL PRODUCT INTERMEDIARY STRATEGIES

Some of the most successful and powerful e-commerce companies are online intermediaries (cybermediaries). In the early days of e-commerce activity in the mid-1990s, there was a theory that intermediaries would disappear. This phenomenon is called disintermediation. The idea was that Web technology would enable producers and customers to directly interact to cut out the middleman. If this was possible, then both the producer and customer would benefit by splitting the profit that was previous paid to the middleman. The interesting result from the first decade of e-commerce was that the opposite has occurred in many industries. This is a counter-intuitive result so further analysis is need to identify how online intermediaries are providing value to buyers and sellers.

The overall finding of a study by Jin and Robey (1999) was that cybermediaries often provide value that is not available in disintermediated channels. Some of the theories used in the study include transaction cost economics, social exchange theory, and knowledge creation theory. Transaction cost economics would explain consumer behavior as a search for the lowest cost channel. Some of the common costs used to compare intermediated versus disintermediated channels would be information search costs, risk costs, and product price. The channel that includes an intermediary provides value because it would be able to reduce information search costs by more efficiently organizing information from multiple sources, it would reduce prices by aggregating buyer demand to negotiate better prices from sellers, and it would potentially reduce risk costs by protecting buyers from opportunistic behavior on the part of the seller. Intermediated channels also had some non-cost related advantages. Social exchange theory would tell us that buyers feel a sense of reciprocity toward trusted channels that provide places for interaction with other channel participants. If a person reads product reviews on a Web site, they feel that they should also provide reviews even if they are not compensated. Intermediaries are able to provide tools for facilitating communications with experts in a particular industry, or between customers who want to discuss their feelings about products and sellers. Intermediaries can also provide value to consumers by providing industry knowledge that is not available to individual sellers through a disintermediated channel. They can gather information about every transaction in their site, between all buyers and sellers, and provide customers with not only information about their own purchases, but also recommend new products based on other people's transactions. The recommended products may be from several different sellers.

Given the value that intermediaries can provide in an online market, the question is whether digital product companies can succeed by choosing this strategy. The following provides some insights into potential intermediary strategies for the digital product industries.

## Newspaper/Magazine Intermediary Strategy

News content comes from many sources. Stories can be written by journalists who are employed at a newspaper or who are freelance writers. Readers get their news from a variety of sources ranging from local newspapers, to regional or national newspapers, or through other media such as television. One strategy for a newspaper would be to provide an intermediary service between news agencies and the public. It could consolidate news from many sources into one website. They could provide a valuable service by reducing the time and energy needed to find stories about many different topics. Everything would be on one site, and it would be searchable. They could also provide opportunities for interaction with experts or other readers through message boards or blogs. This would be valuable if a critical mass of readers was included in one site because any topic could be discussed at any time through one site. And news trends could be identified based on reader access activity to provide highlights for other readers. This could be a form of knowledge creation that is available through online retail product intermediaries. Readers may value a site that offers this consolidation service and they may be willing to pay for it because it will save them time and provide news about a wide range of topics. They will also be able to track issues important to all of the sites readers through the various interaction spaces and recommendations. The news cybermediary itself will be able to provide value through existing Web technologies which may enable them to charge a fee that is sufficient to produce a profit. News contributors can receive compensation either through a per-article fee or based on the number of hits their stories receive online.

Magazines would be in a similar situation to newspapers. There are thousands and thousands of magazines that are focused on various niche markets. An online magazine content intermediary could provide the same valuable services described above, but they could do it for a particular topic that does not require daily content updates. The same benefits would be available to readers, but discussions with experts could be more focused on a particular topic – for example, a hobby, sport, educational topic, or social issue. The intermediary could compensate article writers and still potentially make a profit because the cost to run an online intermediary site would be relatively low.

## Electronic Book Intermediary Strategy

Millions of books have been written, and in the past they would all have to be printed before they were sold. The electronic book technology available today enables writers to provide their product to a reader in a very different format, but it may still be difficult for people to find a book they are interested in. An online e-book intermediary could provide sales and supporting services for all e-books from many writers including writers who wish to self publish their work.

The e-book cybermediary would provide a common marketplace for everyone to interact. Placing information about many e-books into one site will reduce information search costs. Readers can search for the books they are looking for and receive results from many different sources. They will also have some protections resulting from going through an intermediary rather than buying directly from an author. Since authors have little marginal cost for selling each additional copy of their e-book, they can offer the product for relatively low prices. This may be a particularly important scenario for new authors who are trying to become known in the market and have not been able to get a publishing deal. The e-book cybermediary will also be able to provide information about past purchases and recommend new products on a particular topic. They would be viewed as relatively un-biased because they are not the author, but only the intermediary service provider. Revenues can be shared between the site and the authors, and removing some middleman costs will provide an opportunity to sell books at a low price. The potential deterrent to this strategy is if publishers control much of the book content and feel that they are being pushed aside in this new online e-book intermediary site. The intermediary site will only be a success if it can reach a critical mass of authors and readers to produce a large volume of transactions.

## Online Music/Movie Intermediary Strategy

Music all begins with the artists and ultimately some of songs are owned by record companies and some are not. An online music intermediary could provide a site for interaction between musical artists and the people who listen to the music. The site can include songs from many different sources which would allow customers to search for information and select any combination of songs they wish. This reduces their information search costs. They will also have some protection against problems with the bands because the music cybermediary can assist them. As with e-books, the marginal cost for a copy of a digital format song is near zero so prices on these sites may be low. There will also be opportunities for fans to interact with music industry experts, with the bands themselves, or with other fans through message boards and blogs. This will create a lot of traffic and if people are coming to the site for music news, they may also stay and purchase music through the same site. If there is a lot of activity on the site, then transaction information can be provided to buyers to indicate what they purchased in the past and to recommend new bands that other people have liked. This is very difficult to do in a direct sales channel with one record company. Popular bands may get some benefit from the site, but new acts would be particularly interested because this would give them a way to become known. Ultimately they may almost give away copies of some of their music in hopes that people will buy other songs or buy tickets to their concerts. The online music intermediary can share revenues with the artists and still cover their costs because running an online music site selling digital music will be low.

An online movie intermediary will operate much the same as an online music intermediary. One difference would be that most movies will come from movie production companies, but there would still be opportunities for independent movies to be included. Another difference would be that movies are stored in much larger digital files which will require higher bandwidth to make the service usable for direct download. A movie cybermediary can reduce information search costs, risk, and prices for movie buyers. They can also provide interaction with movie experts or other movie buyers as well as provide knowledge about trends in movie purchases seen through the transactions on their site. Independent movie producers may particularly like this site because it will give them a channel to provide their work that will generate some revenue and help them gain popularity.

## Online Game Intermediary Strategy

Most online games are created by a fairly small number of large game developers, but there are opportunities for individuals to create entirely new games on their own. An online game intermediary could bring all of the game titles and the people who play the games into one site. Games can be played on computers, game machines, or online. Relative to newspapers, magazines, e-books, music, and movies, they are a much more complex digital product which may make a game cybermediary more valuable to buyers. A large number of games in all genres can be sold through the site. This will reduce the time and effort needed by game players to find and purchase several games from different producers. The cybermediary will also be able to protect the buyer if there are technical problems and also negotiate better prices because of the high volume of sales. Games are an experience good so there are numerous opportunities for interaction with game developers and with other game players. The site can glean a great deal of knowledge from these discussions. The site can also tell people what they have purchased in the past and recommend new products as they become available. As with the movie intermediary, there will be a need for high bandwidth connections for direct download. Individuals who wish to produce their own games can sell their products through this site to get some revenue and also to become known in the market without having to expend large amounts of money for marketing.

## Digital Product Intermediary Strategy Conclusions

A number of conclusions can be made based on an assessment of the potential for cybermediaries in the newspaper, magazine, e-book, music, movie, and game industries. The value provided to buyers is the same in every industry. The cybermediary potentially reduces information search costs, risk costs, and product prices. It also enables interaction between all industry participants in one site. And it becomes a trusted source for industry knowledge based on its ability to analyze all of the transactions that take place on the site as well as customer comments on message boards. The fixed costs to get the site started will be fairly low because there is no need to build stores or hire large numbers of employees. Identifying product suppliers will be critical, especially at the start. Given the low marginal costs for producing product units, the revenue flowing through the site does not have to be extremely large to survive. A digital product cybermediary can be a great success if it can do one thing – reach a critical mass of users to generate enough traffic to increase the value provided to both sellers and buyers and cover the fixed costs that were needed to get started.

## POTENTIAL DIGITAL PRODUCT MASS CUSTOMIZATION STRATEGIES

In this section the discussion of digital product company strategic alternatives continues by considering opportunities presented by the paradigm of mass customization where large amounts of personalized products are produced according to the specifications of individual consumers or small niche groups of consumers. Creation of physical products in the past century typically used a mass production approach (Pine et al., 1993). The focus was on efficiently producing large numbers of the same product with no variations in features. Products produced would be stored in inventory until they were needed. This would reduce the production cost per unit which would enable products to be sold at lower prices so more people could afford to buy them. This worked fairly well for markets that had large numbers of people who wanted the same products and products could continue to be sold for many years (Pine et al., 1993). This was also a rationale strategy during this time period given the technical limitations for information sharing between supply chain partners which made coordination of manufacturing, inventory control, and distribution difficult. In the past two decades, increased use of digital technologies and computer networks to support order processing, manufacturing, logistics, and distribution have created opportunities to break away from the old mass production approach in an attempt to provide greater value to individual customers.

Early research compared mass production and mass customization approaches in physical product industries (Kotha, 1995). Most early studies viewed mass customization as an idea that had great promise, but the reality of the past decades has been that it has had limited successes in the physical product industries. The cost involved in creating and distributing physical goods create limits for how successful mass customization can be. It does have a much greater potential as a strategy for digital products because content, multimedia, and software can be digitally reproduced quickly and cheaply (Lee et al., 2000). Integrating digital product production systems with online order entry and customer preferences provides a perfect infrastructure for providing fully customized products to a mass online marketplace. Digital product customers are also demanding more from providers so these companies must consider not only the horizontal integration strategies discussed earlier, but also new strategies like mass customization. The question is to what extent mass customization can be used in the digital product industries to provide customized and personalized content, multimedia, or software. The following provide some examples of potential mass customization strategies for newspapers/magazines, e-books, music/movies, online games, and software.

### Newspaper/Magazine Mass Customization

Traditional printed newspapers and magazines are mass produced. Each issue has the same content in the same sequence which allows many copies to be produced at lower cost. Online magazines and newspapers could conceivably use Web technology to provide a personalized copy of their content to each of their individual readers based on their requirements. The requirements could be specified by the reader, or they could be 'learned' by tracking the order in which someone views the content each day and the links that they follow. For example, people may each read different sections of a daily newspaper, and they may read them in a particular order that does not match the printed copy that is delivered. An online newspaper can create a unique front page for each user based on the reader's preferences. The non-linear nature of the Web allows readers to view the online content in any sequence they prefer. Content can be provided in a wide range of languages, and can be translated by the user if desired. They can also be alerted to breaking news in their main areas of interest through e-mail or text message at the time the news occurs rather than waiting until the next issue is published.

The reader may view personalization as a valuable feature because they can access the content that interests them the most, and do it more efficiently. Newspapers will still need to provide all of the different types of content that they did in the printed version, but they may realize increased revenues and brand loyalty by providing their readers with a personalized experience. The difficulty will be to provide a personalized experience without giving the reader the feeling that they are being monitored or losing their privacy.

## Electronic Book Mass Customization

Textbooks in the past were printed and included all of the chapters in a pre-determined order. These books could be mass produced to reduce their cost, but they were not easily customizable to fit unique classes or learning environments. Today, digital technology has allowed most textbook publishers to offer customizable electronic versions of all of their books. Professors can choose a subset of chapters and have their customized book printed for the students. Or they can take it a step further by offering textbooks as e-books that are downloadable to a PC or e-book reader with access for the semester paid through a subscription fee. The publishers benefit by reducing their marginal production, inventory, and distribution costs. The students benefit by having access to all of the class material at a lower price and in a more flexible format. There is also no need to return textbooks at the end of a semester. Eventually, course material may be developed as modules that can be combined as needed. The electronic format also allows for the incorporation of multimedia such as audio and video which will further enhance the value of the electronic textbook.

## Music/Movie Mass Customization

The music industry is now offering their product much like the electronic textbook publishers. In the past a set of songs would be on a CD that someone purchased. You either bought the whole CD, or nothing. Songs are now developed and stored digitally and this same digital technology allows individual songs to be purchased, distributed, and played on PCs or other devices. Users have a customized experience because they can purchase individual songs and listen to them anywhere, anytime, and in any order. Online services can make recommendations when new music becomes available. The music publishers were initially hesitant, but this business model has now become the norm. People are willing to pay a little money for each song, and ultimately may pay as much or more for all of their music because they have more choices. They can purchase an MP3 file, or they can subscribe to a service for a period of time. Established musicians will still receive large amounts of revenue through song sales and concerts. And new acts can reach a large number of potential fans at a low cost to try to increase their popularity.

## Online Game Mass Customization

Traditionally, online or computer games provided the same experience for every player. Each copy of the game software was the same for every person that purchased it. Games that were distributed on CDs were impossible to customize for each person because the cost was prohibitive. Digital technology and mass customization allow game developers to offer a customized and unique experience for each player based on their desires. Game software can include a set of customizable parameters that are set by the player. This can be done in real-time for online games accessed through the Internet. For example, in a driving game, a car's exterior and interior can be customized as well as its driving performance. Tracks can be customized and so can the number of players and their cards. Each person can have a unique game after they have customized it.

As with the previous examples, the game providers benefit by offering a more valuable customized product at a moderate extra cost. The people playing the games may be willing to pay more for a customized experience, and they may become more loyal to the game developer. A downside is that customer service and upgrading will be more difficult because the game software is more complex and there are more potential unique problems.

## Software Mass Customization

One of the most common software applications is office automation suites that include components that support development of documents, spreadsheets, presentations, and databases. Other components may include Internet browsers and e-mail readers. Users often have to purchase the entire package to get the parts they need. Mass customization would enable users to purchase, or subscribe to, just the minimal applications that they need at any given time. The software-as-a-service model may provide a paradigm that would make widespread customized software a viable business model. Customization would also allow any setting in any application to be changed as the user desires. This is available today to some extent, but it can be a tedious task to alter a large number of settings. Software should be able to be fully customized because it is an entirely digital product. The software company is already developing a working version of each software suite component, but what will change is the flexibility in how the product is offered.

As with all of the earlier examples, software companies could benefit by offering customized software because users may value the product more and have greater brand loyalty. It may be difficult for a user to switch to a different software product once they have customized their current applications which creates a switching cost. Users may be willing to pay more for a customized product. And offering software through a software-as-a-service model may make the process of maintaining and upgrading software easier.

## Digital Product Mass Customization Strategy Conclusions

Digital products must still be developed and this often involves a significant fixed cost. What has changed is that mass customization can be used to provide a customized and personalized experience on the user's end. Each person can have a unique experience when reading an online newspaper, reading a book, listening to music, or playing a game. Across each of these digital product industries, marginal customization costs are near zero, but the value produced can be significant. A potential issue that would reduce acceptance of mass customization would be perceptions that a person's privacy is being invaded. Individuals may be wary if they must provide personal information as part of the customization process. In a hypercompetitive online marketplace, mass customization may be a requirement in the future to provide value to users, increase brand loyalty, and ultimately survive.

## POTENTIAL TACTICAL INTERACTIVITY AND SOCIAL NETWORKING STRATEGIES

The next set of strategies discussed is intended to take advantage of the potential for online two-way interaction between everyone involved with digital product industries. One of the unique characteristics of the Web is that it is an interactive technology. The interface enables people to send and receive e-mail or other forms of electronic communications any time and any where they have access either through personal computers or wireless devices. This is a major advantage for consumers compared with some of the traditional ways they received product information through direct mail, catalogs, radio ads, or television ads. Social networking systems promote even broader interaction because of the sheer number of users and the tools available in a single site. The question is how these technology capabilities could be adopted into the tactical level strategies for digital product companies. In this section, potential strategies for interactivity and social networking in digital product industries are identified along with their potential benefits and risks.

## Online Newspaper/Magazine Interactivity and Social Networking Strategy

Online newspapers can publish content for any news stories or feature topics, and they can publish it as soon as the story is ready. Access to these stories is important to readers, but an even more valuable service is when the same site provides a way for them to discuss the story with other readers or with experts. Blogs provide a simple tool for commenting on an ongoing thread of related comments. E-mail and online forms can be used to submit questions to a news provider that will be replied to at a later time. It is also simple for a reader to send a link to a story to a friend who may be interested in the topic. Printed newspapers do not have this capability, so if this interactivity strategy is chosen by a news agency they may build more traffic to their site. The traffic may generate revenue directly from readers through subscriptions or per-article download fees, or it could be used as the basis for charging higher fees to advertisers. The cost to add this capability is minimal, but it may create some value for readers and for the news site. This strategy may be a way for a newspaper to differentiate itself from its competitors to gain at least a short term advantage. The risk is that the interaction space may be used by people who are unhappy with the site either because of the articles they publish or their customer service. Online newspapers and magazines should also be open to the idea of using stories written by readers. This can provide a large amount of potential material from which to select a few of the best stories that can be edited and published.

## Electronic Book Interactivity and Social Networking Strategy

Electronic books are another industry where interactive technology and social networking may provide value to readers. Consider linking an online e-book retailer to a social networking site so that people can comment on books they have read. Readers who use social networking sites are already linked to many other people, and this would give them a simple way to share their likes and dislikes regarding books they have read. This may generate more traffic on the e-book website, and also on the social networking site. If a company can be the first retailer to do this, then they main gain an advantage in the market and become the primary place where books are discussed online.

## Online Music/Movie Interactivity and Social Networking Strategy

Music and movies would be similar to e-books. People listen to new music, for example, and want to share their comments with other music listeners or their friends on a social networking site. If the services can be linked there may be more traffic on the retailer's site which may produce more sales or at least the potential for greater advertising fees. Because the music is in a digital format, samples can be automatically shared through the same system used to support the interactivity. New bands could also create social networking site profiles to add friends and promote their new music, perhaps for no fee at all. The risk is that messages in these sites can become viral meaning that good or bad comments will be available to many people very quickly which could be difficult to overcome for an unknown band.

## Online Game Interactivity and Social Networking Strategy

Games are unique because many of them require multiple players. The online environment seems to be a perfect environment for supporting large scale interaction between gamers. People can use simple interactive tools on the Web, or use social networking sites, to find game competition worldwide, or just to discuss their experiences with new games. Game sites build traffic which may increase sales or advertising. Negative or positive comments can be tracked as a free source of market research. If enough traffic occurs in these sites then they may be used for beta testing new games to strengthen the connection between game developers and the people who play their games.

## Potential Interactivity and Social Networking Strategy Conclusions

There is great potential for interactivity and social networking strategies for digital product companies. Digital convergence enables all aspects of the order fulfillment process to be digitized including pre-order information search, order placement, payment, distribution, and communication with customer service personnel or other customers. These activities can be done worldwide and the site can be available any time and accessed using any device. These strategies require a critical mass of users to make it cost effective, but if this can be achieved then it may raise an entry barrier for companies trying to duplicate the first-mover's strategy. These strategies may lead to an explosion of low-cost market research data. Successful interactivity strategies should also create brand equity. The primary risk is if the interaction surrounding a product becomes very negative it can reflect badly on the seller and the original content or product producer.

## STRATEGIES FOR DIGITAL PRODUCT LIFECYCLE STAGES

Many different strategies have been discussed up to this point, and most of them have applied to firms who have a digital product that has been developed and is now in its growth or maturity phase. In this section the concept of digital product lifecycle is discussed along with potential strategies to use at different points in a company's evolution. Products typically go through four stages – introduction, growth, maturity, and decline. The time involved in each stage could be days, or it could be years. The cost structure for a digital product makes lifecycle strategies very important because most costs are incurred in the introduction phase and significant revenues may not be realized until much later. The following provides some insights into strategic alternatives for digital products during the introduction, growth, and maturity stages. To illustrate how these strategies may be applied, an e-book retailer will be used as an example.

## Digital Product Introduction Phase Strategy

In the introductory phase for a new e-book retailer they will develop their website with all of the supporting services, identify e-book publishers to provide products to sell on the site, and develop initial marketing plans. The goal during the introductory phase is to increase traffic on the site and to begin generating some sales. E-book retailers will have high fixed costs in this stage, but their marginal costs will be fairly low. The cost to acquire new customers will be high and they must be innovative to differentiate themselves from other online e-book companies. One way to gain market share is by offering the e-books at very low prices, perhaps even losing money on each unit sold. Special pricing deals and other incentives can be used to bring people to the site which may increase traffic in the short term. When monster.com started they paid companies to advertise jobs on their site. They knew that they had to starting building traffic through any means necessary. Other services such as personalization or providing interaction with other readers may increase traffic, but there will be little value in these services if the number of other users is low. Eventually, if the company successfully moves to the growth phase, additional strategic alternatives may be available.

## Digital Product Growth Phase Strategy

In this phase the e-book company has an established and growing number of customers. They can now begin increasing prices for some of their products. Pre-ordered books can be sold at a discount. E-books that are just released can be priced higher, and prices can drop as e-books get older and there is less demand. They can also use the power of their customer base to get revenue from advertisers. In the industry there will be a growing number of competitors and one way to grow is through the various horizontal integration strategies discussed earlier. The e-book retailer can look for partners that sell complementary products or they can broaden their geographic market. They can also look for unique cross-industry partners to increase their customer base without dramatically increasing their marketing costs. Product description pages can include places for customers to provide comments which should make the site more valuable and continue to add traffic. At this point they should become profitable.

## Digital Product Maturity Phase Strategy

Once an e-book retailer has matured they will experience more competition and will need to continue searching for ways to differentiate their site. They can rely on their existing partnerships and offer additional services to their existing customer base such as website personalization for each customer. They may also provide additional product extensions for their basic e-book products. If they are large enough they will be able to survive fluctuations in demand, but they will need to continually search for ways to incorporate new digital technology capabilities into their site. Having covered their fixed introduction costs, their profit margin will increase even if their prices do not change. Their size may also impact their relationship with suppliers because they will be in a stronger position. They may receive new e-book products earlier or they may be able to retain a greater proportion of their revenues. The unique characteristics of digital product industries may mean that no company is ever truly in a maturity phase. They may all continue to search for growth opportunities and partnerships with other successful digital product firms.

## Digital Product Lifecycle Stage Strategy Conclusions

Different stages require different strategies. For many digital product companies it is an all or nothing struggle to grow and reach a critical mass of customers. Everyone has the same technology, and in many digital product industries they sell the same products. This makes it difficult to differentiate one company from its competitors, but if a company can continue growing through product sales or other services it can be highly profitable.

## FUTURE RESEARCH DIRECTIONS

Beyond the managerial implications noted above, there are also a number of directions for future research in this area. From the firm perspective, empirical studies are needed to identify the characteristics of successful digital product horizontal integration strategies. One example would be to test whether the relative size of firms produces better or worse results. Is a horizontal integration strategy better if there is one large company working with a smaller company? Or would it be better if the partners are of equal size so one partner does not take a disproportionate share of the benefits? Studies across multiple industries are also needed to identify the combinations of digital products that produce competitive advantage and those that do not. Is horizontal integration within an industry better, or is it better if the partners are currently in two or more existing industries? Strategies for improving trust between partner firms will also be an important research issue.

Another issue to address would be whether the current technology environment provides any situations where vertical integration is preferred. Models can be developed and tested to identify circumstances where vertical integration is better than other growth strategies. At what point do the costs of vertically integrating become less than the benefits gained through control over more steps in the value chain? These same issues can be studied for cross-industry integration strategies, cybermediary strategies, mass customization strategies, interactivity and social networking strategies, and product lifecycle strategies. Case studies would also be useful for providing a deeper understanding of digital product strategic decision making. These cases could also be used for class discussion in information systems, electronic commerce, and strategic management courses.

From the digital product user perspective, studies may address the factors that impact individual use of particular combinations of digital products. Research in this area is in its early stages, but some findings have already been published. Gill and Lei (2009) studied technological convergence in the electronics industry. They used an experimental methodology to address the issue of which electronics product brands may gain from the inclusion of some type of new functionality. Functionalities were classified based on their level of congruence (similarity or difference of functionalities added to a base product) with the existing capabilities of a device. New functionality did not imply that the function was entirely new to the market, but that it was a new capability added to an existing product. They found that lower quality brands gain from the addition of congruent functionality, while higher quality brands gain from the addition of technology that adds an incongruent function. This is an important finding because it indicates that perceptions of technological convergence are complex and that adding functions to an existing digital product may not bring added value to the consumer.

A second study of technological convergence consumer perceptions addressed the issue of when consumers prefer a converged product versus a dedicated product (Han, et al., 2009). The methodology used a series of four studies and products included in the studies were categorized as either low level technological performance or high level technological performance. They found that consumers of lower level performance technologies preferred converged products, while consumers of higher level performance technologies preferred dedicated products. Results also indicate that consumers do not always want the highest technology performance, but instead desire a product that fulfills a need. Again, this is an interesting finding that illustrates the complexities of technological convergence and its association with consumer preferences.

Several other questions could also be addressed by future research studies. Why do people read one online magazine and not another? Why do they buy e-books at one site and not another? And why do they view online videos at one site and not another? Each site has the same technology available, but what factors motivate the users. What are the unique benefits offered by particular sites that provide competitive advantage? From a technology acceptance theoretical perspective, how do issues such as perceived usefulness, perceived ease of use, and established market size impact adoption and use of a new digital product or combination of products? For the long term it is also useful to identify the contexts under which users will pay for digital products to enable digital product companies to increase revenue and stay in business.

## CONCLUSION

Digital convergence provides opportunities and motivation for digital product companies to adopt a wide range of horizontal integration strategies, cross-industry strategies, intermediary strategies, and tactical level mass customization and interactivity strategies, but once a direction is chosen it is often duplicated by competitors. This creates a situation where there is a continual search for short-term competitive advantage. The winner in all of this appears to be consumers (Crovitz, 2009). Digital product companies will continue to try to serve their needs using the unique capabilities of technology, content, and industry convergence. Information technology capabilities continue to expand, and businesses have incentives to continue searching for new content, multimedia, and software products that fill consumer needs. Those firms that are successful will tap into the growing global revenue in these industries. Many firms will fail in these endeavors, but new companies will continue to enter the market and existing organizations will continue to evolve in an attempt to survive. Digital product company managers know that they cannot stand still in this hypercompetitive environment. The digital nature of all of these products will blur the lines between industries. It is easy for companies in separate digital product industries to work together and we should expect an increase in cross-industry strategies leading to a smaller number of large online media and digital product organizations. These sites will reach a large market and spread their risk across several industries and product line.

## REFERENCES

Adams, R. (2010, January 14). AP, Yahoo near deal on content use. *The Wall Street Journal*. Retrieved from http://online.wsj.com/

Adams, R. (2010, January 21). New York Times to charge for Web. *The Wall Street Journal*. Retrieved from http://online.wsj.com/

Adobe. (2009). RIM joins Open Screen Project. Retrieved December 8, 2009, from http://www.adobe.com/aboutadobe/pressroom/pressreleases/pdfs/200910/100509RIMjoinsOSP.pdf

Amazon. (2009). Profile for Amazon.com Inc. *Yahoo! Finance*. Retrieved December 8, 2009, from http://finance.yahoo.com/q/pr?s=amzn

Apple. (2009). Profile for Apple Inc. *Yahoo! Finance*. Retrieved December 8, 2009, from http://finance.yahoo.com/q/pr?s=aapl

Bakos, J. Y. (1991). A strategic analysis of electronic marketplaces. *Management Information Systems Quarterly*, *15*(3), 295–310. doi:10.2307/249641

Clemons, E. K., Reddi, S. P., & Row, M. C. (1993). The impact of information technology on the organization of economic activity: The 'move to the middle' hypothesis. *Journal of Management Information Systems*, *10*(2), 9–35.

Crovitz, L. G. (2009, October 11). Media moguls and creative destruction. *The Wall Street Journal*. Retrieved from http://online.wsj.com/

Electronic Arts. (2009). Profile for Electronic Arts Inc. *Yahoo! Finance*. Retrieved December 8, 2009, from http://finance.yahoo.com/q/pr?s=erts

Fowler, G. A. (2010, January 7). More makers jump into the e-reader market. *The Wall Street Journal*. Retrieved from http://online.wsj.com/

Fowler, G. A. (2010, January 7). Texas Instruments to enter e-reader market. *The Wall Street Journal*. Retrieved from http://online.wsj.com/

Fowler, G. A. (2010, January 31). Amazon backs down from e-book publisher fight. *The Wall Street Journal*. Retrieved from http://online.wsj.com/

Gill, T., & Lei, J. (2009). Convergence in the high-technology consumer markets: Not all brands gain equally from adding new functionalities to products. *Marketing Letters*, *20*(1), 91–103. doi:10.1007/s11002-008-9050-5

Gurbaxani, V., & Whang, S. (1991). The impact of information systems on organizations and markets. *Communications of the ACM*, *34*(1), 59–73. doi:10.1145/99977.99990

Han, J. K., Chung, S. W., & Sohn, Y. S. (2009). Technology convergence: When do consumers prefer converged products to dedicated products? *Journal of Marketing*, *73*(4), 97–108. doi:10.1509/jmkg.73.4.97

Hewlett-Packard. (2010). Profile for Hewlett-Packard Company. *Yahoo! Finance*. Retrieved January 26, 2010, from http://finance.yahoo.com/q/pr?s=hpq

Hunger, J. D., & Wheelen, T. L. (2007). *Essentials of Strategic Management*. Upper Saddle River, NJ: Pearson Prentice Hall.

Jin, L., & Robey, D. (1999). Explaining cybermediation: An organizational analysis of electronic retailing. *International Journal of Electronic Commerce*, *3*(4), 47–65.

Kane, Y. I. (2009, November 25). EA chief wagers on digital future for games. *The Wall Street Journal*. Retrieved from http://online.wsj.com/

Kee, T. (2009, June 17). EA COO John Pleasants: 'going digital' is key to returning to profitability. *Paidcontent.org*. Retrieved December 4, 2009, from http://paidcontent.org/article/419-can-eas-new-digital-strategy-bring-back-its-glory-days/

Kotha, S. (1995). Mass customization: implementing the emerging paradigm for competitive advantage. *Strategic Management Journal*, *16*(7), 21–42. doi:10.1002/smj.4250160916

Laudon, K. C., & Traver, C. G. (2010). *E-Commerce: Business. Technology. Society*. Upper Saddle River, NJ: Pearson Prentice Hall.

Lee, C. H., Barua, A., & Whinston, A. (2000). The complementarity of mass customization and electronic commerce. *Economics of Innovation and New Technology*, *9*(2), 81–110. doi:10.1080/10438590000000005

Malone, T. W., Yates, J., & Benjamin, R. I. (1987). Electronic markets and electronic hierarchies. *Communications of the ACM*, *30*(6), 484–497. doi:10.1145/214762.214766

Pine, B. J. II, Victor, B., & Boyton, A. C. (1993). Making mass customization work. *Harvard Business Review*, *71*(5), 108–122.

Salter, C. (2009, February 11). The Fast Company 50 – 2009, #9 Amazon. *Fast Company*. Retrieved December 8, 2009 from http://www.fastcompany.com/list/amazon

Schreiber, J. (2001, March 7). Five reasons to co-brand your Web site. *MarketingProfs*. Retrieved December 4, 2009, from http://www.marketingprofs.com/2/cobrand5reasons.asp

Smith, E. (2010, January 19). MTV games seeks buzz for 'Rock Band'. *The Wall Street Journal*. Retrieved from http://online.wsj.com/

Smith, E., & Kane, Y. I. (2009, December 4). Apple acquires Lala Media. *The Wall Street Journal*. Retrieved from http://online.wsj.com/

Stone, B., & Barnes, B. (2008, November 10). MGM to post full films on YouTube. *The New York Times*. Retrieved from http://www.nytimes.com/

Strader, T. J., & Shaw, M. J. (1997). Characteristics of electronic markets. *Decision Support Systems*, *21*(3), 185–198. doi:10.1016/S0167-9236(97)00028-6

Texas Instruments. (2010). Profile for Texas Instruments Inc. *Yahoo! Finance.* Retrieved January 8, 2010, from http://finance.yahoo.com/q/pr?s=txn

Vitorovich, L. (2010, January 25). H-P enters Europe music venture. *The Wall Street Journal.* Retrieved from http://online.wsj.com/

Yahoo. (2006). Yahoo! forms strategic partnership with consortium of more than 150 newspapers across the U.S. Retrieved December 8, 2009, from http://yhoo.client.shareholder.com/press/releasedetail.cfm?ReleaseID=219204

## ADDITIONAL READING

Adobor, H., & McMullen, R. S. (2002). Strategic partnering in e-commerce: Guidelines for managing alliances. *Business Horizons*, *45*(2), 67–76. doi:10.1016/S0007-6813(02)00190-8

Amit, R., & Zott, C. (2001). Value creation in e-business. *Strategic Management Journal*, *22*(6-7), 493–520. doi:10.1002/smj.187

Ansari, A., & Mela, C. F. (2003). E-customization. *JMR, Journal of Marketing Research*, *40*, 131–145. doi:10.1509/jmkr.40.2.131.19224

Barney, J. B. (1991). Firm resources and sustained competitive advantage. *Journal of Management*, *17*(1), 99–120. doi:10.1177/014920639101700108

Benjamin, R., & Wigand, R. (1995). Electronic markets and virtual value chains on the information superhighway. *Sloan Management Journal*, *36*(2), 62–72.

Boczkowski, P. J. (2005). Multiple media, convergent processes, and divergent products: Organizational innovation in digital media production at a European firm. *The Annals of the American Academy of Political and Social Science*, *597*(1), 32–47. doi:10.1177/0002716204270067

Bucklin, L. P., & Sengupta, S. (1993). Organizing successful co-marketing alliances. *Journal of Marketing*, *57*(4), 32–46. doi:10.2307/1252025

Das, T. K., & Teng, B.-S. (2000). A resource-based theory of strategic alliances. *Journal of Management*, *26*(1), 31–61. doi:10.1016/S0149-2063(99)00037-9

Dellaert, B. G. C., & Stremersch, S. (2005). Marketing mass-customized products: Striking a balance between utility and complexity. *JMR, Journal of Marketing Research*, *42*, 219–227. doi:10.1509/jmkr.42.2.219.62293

Dewan, R., Jing, B., & Seidmann, A. (2000). Adoption of Internet-based product customization and pricing strategies. *Journal of Management Information Systems*, *17*(2), 9–28.

Dupagne, M., & Garrison, B. (2006). The meaning and influence of convergence: A qualitative case study of newsroom work at the Tampa News Center. *Journalism Studies*, *7*(2), 237–255. doi:10.1080/14616700500533569

Ernst, D., Halevy, T., Monier, J. H. J., & Sarrazin, H. (2001). A future for e-alliances. *The McKinsey Quarterly*, *2*, 92–102.

Goodman, S. E., Press, L. I., Ruth, S. R., & Rutkowski, A. M. (1994). The global diffusion of the Internet: Patterns and problems. *Communications of the ACM*, *37*(8), 27–31. doi:10.1145/179606.179733

Hagel, J., & Armstrong, A. G. (1997). *Net Gain.* Boston: Harvard Business School Press.

Huffman, C., & Kahn, B. E. (1998). Variety for sale: Mass customization or mass confusion? *Journal of Retailing*, *74*(4), 491–514. doi:10.1016/S0022-4359(99)80105-5

Jenkins, H. (2006). *Convergence Culture: Where Old and New Media Collide.* New York: NYU Press.

Kalakota, R., & Whinston, A. B. (1996). *Frontiers of Electronic Commerce*. Reading, MA: Addison-Wesley Publishing Company, Inc.

Kaplan, A. M., & Haenlein, M. (2006). Toward a parsimonious definition of traditional and electronic mass customization. *Journal of Product Innovation Management, 23*(2), 168–182. doi:10.1111/j.1540-5885.2006.00190.x

Klinenberg, E. (2005). Convergence: News production in a digital age. *The Annals of the American Academy of Political and Social Science, 597*(1), 48–64. doi:10.1177/0002716204270346

Liechty, J., Ramaswamy, V., & Cohen, S. H. (2001). Choice menus for mass customization: An experimental approach for analyzing customer demand with an application to a Web-based information service. *JMR, Journal of Marketing Research, 38*, 183–196. doi:10.1509/jmkr.38.2.183.18849

Lumpkin, G. T., & Dess, G. G. (2004). E-business strategies and Internet business models: How the Internet adds value. *Organizational Dynamics, 33*(2), 161–173. doi:10.1016/j.orgdyn.2004.01.004

Malone, T. W., & Rockart, J. F. (1991). Computers, networks and the corporation. *Scientific American, 265*(3), 128–136. doi:10.1038/scientificamerican0991-128

Malone, T. W., Yates, J., & Benjamin, R. I. (1989). The logic of electronic markets. *Harvard Business Review*, (May-June): 166–170.

Park, S. (2007). *Strategies and Policies in Digital Convergence*. Hershey, PA: Information Science Reference.

Perry, M. L., Sengupta, S., & Krapfel, R. (2004). Effectiveness of horizontal strategic alliances in technologically uncertain environments: Are trust and commitment enough? *Journal of Business Research, 57*(9), 951–956. doi:10.1016/S0148-2963(02)00501-5

Porter, M. E. (1985). *Competitive Advantage: Creating and Sustaining Superior Performance*. London: Free Press.

Porter, M. E. (2001). Strategy and the Internet. *Harvard Business Review, 79*(3), 62–78.

Randall, T., Terwiesch, C., & Ulrich, K. T. (2007). User design of customized products. *Marketing Science, 26*(2), 268–280. doi:10.1287/mksc.1050.0116

Rayport, J. F., & Sviokla, J. J. (1994). Managing in the marketspace. *Harvard Business Review, 72*(6), 141–150.

Rubinstein, H., & Griffiths, C. (2001). Branding matters more on the Internet. *Brand Management, 8*(6), 394–404. doi:10.1057/palgrave.bm.2540039

Schoder, D., Sick, S., Putzke, J., & Kaplan, A. M. (2006). Mass customization in the newspaper industry: Consumers' attitudes toward individualized media innovations. *International Journal on Media Management, 8*(1), 9–18. doi:10.1207/s14241250ijmm0801_3

Stigler, G. (1961). The economics of information. *The Journal of Political Economy, 69*, 213–225. doi:10.1086/258464

Tryon, C. (2009). *Reinventing Cinema: Movies in the Age of Media Convergence*. Piscataway, NJ: Rutgers University Press.

Tseng, M. M., & Piller, F. T. (2003). *The Customer Centric Enterprise: Advances in Mass Customization and Personalization*. New York: Springer.

Wallin, J. (2006). *Business Orchestration: Strategic Leadership in the Era of Digital Convergence*. Chichester, England: Wiley.

Weiber, R., & Kollmann, T. (1998). Competitive advantages in virtual markets – perspectives of 'information-based marketing' in the cyberspace. *European Journal of Marketing, 32*(7–8), 603–615. doi:10.1108/03090569810224010

Williamson, O. E. (1985). *The Economic Insitutions of Capitalism*. New York: Free Press.

Yoffie, D. B. (1997). *Competing in the Age of Digital Convergence*. Boston: Harvard Business Press.

Zipkin, P. (2001). The limits of mass customization. *Sloan Management Review*, *42*, 81–87.

## KEY TERMS AND DEFINITIONS

**Competitive Advantage:** A situation where customers in a marketplace perceive that an organization has some form of advantage (cost, product, service, resources, etc.) over its competitors.

**Content Convergence:** Convergence of digital content for distribution through a single delivery mechanism.

**Digital Product:** Informational, multimedia, or software products that are in a digital format where, unlike a digital service, possession of the product is passed to the user permanently or for a period of time.

**Horizontal Integration Strategy:** A growth strategy where an organization expands to offer complementary products to their existing product/service or expands into a new geographic market.

**Industry Convergence:** Partnership or merger of separate digital product companies to create synergistic offerings across multiple markets or platforms.

**Mass Customization:** Provide a customized product to a large number of individual customers. Digital products can be customized more easily than traditional physical goods so there has been a growth of customized digital products.

**Strategic Management:** Choosing an overall strategic direction for an organization based on an analysis of external and internal factors with the goal of achieving competitive advantage.

**Technological Convergence:** Convergence of two or more media platforms into a single hybrid device.

# Chapter 9
# The Role of the Internet in the Decline and Future of Regional Newspapers

**Gary Graham**
*University of Manchester, UK*

## ABSTRACT

*Digital technology has had a significant impact on the newspaper industry in many different areas of the world. The Internet and digital content technologies enable online newspapers to reach a wide audience and to reduce many of the costs associated with print newspapers, but there have also been some negative impacts including a loss of readers and advertising revenue for traditional printed newspapers. In this chapter, focus groups and interviews are used to investigate the following issues: (1) the role of the Internet in the decline of the social/business influence of regional newspapers, and (2) the impact of developments such as Web 2.0 on the future of regional news supply. The chapter concludes with a discussion of managerial implications for the future.*

## INTRODUCTION

The Internet is contributing to a loss in newspaper readership and advertising revenue. Doom laden warnings about the future of the industry and public service[1] journalism are prevalent (Meyer, 2008; Pincus, 2009). Beam et al. (2009) notes that: "society should care what is happening to newspapers as without them democracy will be diminished". Meanwhile Byrne (2009, pp. 13-15) believes that the social influence and business model of newspapers is under grave threat from: (1) an ever shrinking audience for local/regional news products,[2] (2) a reduction in staff and public service journalism, and (3) the increased trivialization (sensationalizing) of editorial content. In an interview[3] with the editor/publisher of the Spring Hope newspaper in North Carolina, the business/social challenge facing newspapers was apparent: "there are fewer workers and fewer news organizations around to do professional journalism of any kind, public service or otherwise". While newsrooms have been rationalized, paradoxically at the same time the work demands on news

DOI: 10.4018/978-1-61692-877-3.ch009

journalists have increased. Journalists now have to submit multi-media digital copy material in addition to their traditional story/feature writing (Meyer, 2008).

These changes in the way newspapers report the news were expanded on by a social media specialist[4] we interviewed at the San Diego Tribune: "newspapers are evolving into news sites. They will continue to create must-read, must-have content, delivered in a medium that suits readers. In essence there is a need to transfer their editorial impetus online". Therefore in this chapter we set forth: (1) to investigate the role of the Internet in the decline of the social/business influence of regional newspapers and (2) to assess the impact of developments such as Web 2.0[5] on the future of regional news supply.

## LITERATURE REVIEW: HISTORY OF THE LOCAL PRESS

Newspapers have a long industrial history and are a phenomenon of the late nineteenth century with the development of the industrial printing press in the 1850s. For most of their history they had a simple business plan and that was to have large readerships to attract high amounts of advertising revenues. Stories were community sourced from the general public, police, courts and local government (though this view has been revised to include the increasing amount of local news content sourcing from interest groups and public relations professionals (Picard, 2004)).

Regional newspapers typically had high profitability with figures between '25% and 30%' commonplace for many local newspaper companies (Dear, 2006, p. 8). Meyer (2004) believes their high profitability was due to the following factors: (1) They had a monopoly in the production/distribution of printed local news content. High barriers to entry restricted the entry of competitors (e.g. production facilities, training, distribution outlets). (2) High circulations in the local area enabled them to obtain economies of scale of production and keep variable costs (e.g. journalist wages) low. (3) And high circulations attracted advertisers (in particular classified (property and recruitment) and display advertisements).

While the national newspapers were, in the main, generalized daily and weekly newspapers with headquarters in London[6] published in the morning and aiming for sales across the whole of the UK, the regional newspaper was distributed in the evenings and read almost entirely in the area of production (Meyer, 2004, p. 55). They each look towards each other as a source of news. For instance, the Iraq and Afghanistan wars are national stories capable of being made regional by particular local connections (e.g. soldiers, specialist medical teams).

The 'golden age' of the regional press, particularly the dailies in the large urban conurbations (such as Manchester) peaked in the period of post-World War II prosperity (Freer, 2007). Ownership was often in the hands of small family businesses and they often "rubbed shoulders with the journalists" (Freer, 2007, p. 93). But from the early 1960s the industry began to contract. For instance, Manchester had two evening papers until, in 1964, the Manchester Evening Chronicle closed despite sales of around 250,000, leaving the Manchester Evening News with a near monopoly in the city (Franklin & Murphy, 1991, p. 7).

The 1990s witnessed the national press (Trinity Mirror, DMGT and the Guardian Media Group) entering the regional newspaper market as they acquired some of the leading regional newspapers including the Manchester Evening News, Liverpool Daily Post, Liverpool Echo and Bristol Evening News. Today the four biggest regional publishers are Trinity Mirror, the Daily Mail and General Trust (DMGT) (owner of Associated Newspapers and Northcliffe Media), Johnston Press and Newsquest Media Group (a wholly owned subsidiary of Gannett plc (US)). These four now have almost 70% market share across the UK. In addition to their acquisitions of regional newspapers, all but one of these groups (and most of the smaller groups) has significant cross-media interests (Mintel, 2007).

## LITERATURE REVIEW: MANAGING DECLINE, FEWER PAPERS AND FEWER READERS

Using Newspaper Society (NS) data we calculated that over the period 2001-08, the 50 largest UK regional/local newspapers experienced a 26% decline in their circulations. For instance, the Sheffield Star (– 40%), the Liverpool Daily Post (-33%), the Nottingham Evening Post (– 33%), the Birmingham Post (-25%), and the Bristol Evening Post (-24%). Only two newspapers have experienced growth during the same period, but this is modest: the Rochdale Observer (5%) and the Stockport Express (4%). At an aggregate level, in Figure 1, Mintel illustrates that the decline of regional newspaper copy sales will continue from 1.7bn copies in 2005 to 1.4bn in 2010.

In 2007, the National Readership Survey (NRS) found in its study[7] of newspaper readership that 48% read online news services. The decline in national daily readership of printed newspapers has been steepest among young adults. With the number of 15 to 24 year olds reading newspapers falling by 37% and the number of 25 to 34 year olds by 40% (from 2002 to 2006). Table 1 reveals that there has been an increase in the number of 11 to 19 year olds using the Internet to read online news: 1) in the 11 to 14 cohort from 9.2 to 17.2%, and 2) 15 to 19 cohort from 18.1 to 26.9%. In its most recent study (2009)[8] the NRS reported that 33% of 15 to 19 year olds used the Internet to read online news.

The regional press is the largest print advertising medium in the UK, taking £2.3 billion a year and accounting for 14.3% of all advertising revenue. Online recruitment advertising in the regional press grew by 17.1% in 2008 (from £76.1m to £89.2m). Total advertising spending in the regional press is over five times the total spend on radio, and a little more than the combined total for radio, outdoor and direct mail (Advertising Association, 2009). Equally important is advertising's contribution to overall revenue which stands at a massive 80% of the overall income of local newspapers compared to approximately 46% for national papers (Mintel, 2007).

*Figure 1. Forecast of the UK regional newspaper sales (by volume), 2005-10. Source: Mintel, 2007, p. 7*

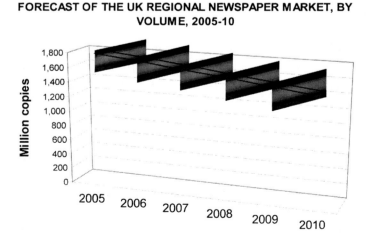

Table 1. Percentage of 11 to 19 year olds – Using the Internet for newspapers/Nnews (2002/06). Source: National Readership Survey, 2009, p. 43

| Age Group Category | 11 | To | 14 | 15 | To | 19 |
|---|---|---|---|---|---|---|
| Newpapers/news Any | 2002 9.2 | 2004 9.8 | 2006 17.2 | 2002 18.1 | 2004 22.2 | 2006 26.9 |
| At home | 3.9 | 5.3 | 8.2 | 11.3 | 15.4 | 18.4 |
| At school/college/ work | 6.0 | 7.0 | 12.2 | 10.1 | 12.5 | 16.3 |
| Elsewhere | 0.5 | 0.6 | 0.9 | 0.8 | 1.6 | 2.0 |

## The Internet Earthquake

Advertisers have followed consumers into the online space, gradually shifting their attention to online or non-news channels (Domingo et al., 2008). The Internet has taken an important chunk of classified advertising away from regional news media organizations (Pincus, 2009). For instance, in the case of Trinity Mirror the largest regional newspaper group, advertising revenue across the group's regional division fell 34.5% (in 2008) to £104m. Recruitment and property advertising, which account for 33% of total advertising revenues, fell 50.4% and 52.5% respectively (Trinity Mirror, 2009).

## Web 2.0 and the Rise of the Citizen Journalist

Web 2.0[9] allows users, as individuals or social networks, to produce and distribute news items on the basis of their observations or opinions, and computer-based selection and management systems support collective work processes to gather the news that is spread across the whole network. The economic logic behind such developments is that news firms now have unprecedented opportunities for user participation and (re)engagement with their audiences (Li & Bernhoff, 2008; Tapscott, 2008). They can also cut costs by 'crowdsourcing' tasks that were formerly performed by paid professionals (Howe, 2008).

Web 2.0 facilitated the development and evolution of citizen journalism[10] and blogging and this has inevitably caused tension within existing newsrooms (Tapscott, 2008). News organizations operate in tightly controlled command and control (production) workplaces while 'citizen' journalists work more flexibly (to their own deadlines and work schedules). Howe (2008) proposes that amateur or citizen journalists will continue to supplement the activities of paid correspondents rather than replace them. He draws examples of this from the Canadian-based NowPublic.com where the 130,000 users upload photos and videos and report when they witness a newsworthy event. This content is then sold on by NowPublic.com to Associated Press (AP) to be used in their reports. He cites AP's director of strategic planning, Jim Kennedy, who talks about looking to crowdsurfing in order to find the 'golden nugget' that can enhance a report. Gannett (US), one of the largest global news media organizations is embedding crowdsourcing into their overall business model. This is referred to as the 'Gannett crowdsourcing strategy' (GCS) and Michael Maress, the architect of GCS, explains the need to mix professional journalists with amateur contributors saying that at Gannett they: "...had to train our newsroom to accept they didn't have monopoly on ideas and opinions" (Howe, 2006, p. 104). With newspapers cutting back on staff we were interested to investigate the current and future role of citizen journalists and blogger given the dramatic cuts currently occurring in editorial staff.[11]

## The Future Strategic Focus of Regional Newspapers

Within this context of greater access to worldwide news through multiple channels, an increased emphasis on 'hyper-local' news has emerged. 'Hyper-local' news is the news that relates specifically to a narrow geographical region or local community (Thurman & Myllylahti, 2009, p. 691). For instance, in a city area such as San Diego, there are six distinct community zones targeted for 'hyper-local' content by the Tribune [12] (e.g. La Jolla, South Bay). Mersey (2009, p. 357) believes that: "the challenge to regional newspapers in light of dwindling circulation figures nationwide is to stay geographically relevant". The desire for hyper-local news has provided a niche that regional news organizations are in a strong position to occupy (Meyer, 2008, p. 35).

We investigated, using a focus group methodology, three key topics (raised in the literature regarding the decline and future of regional newspapers): 1) the long term decline in the social influence of newspapers and the recent fracturing of their business model (economic recession, online news); 2) emerging online value streams, and 3) the future role of citizen journalists/bloggers in content co-creation[13] and news production. To supplement the focus group work, three in-depth interviews were undertaken with senior managerial staff at News Chronicle (a regional newspaper in the Manchester area).

## METHODOLOGY: EXPERT FOCUS GROUP (LONDON, FEBRUARY 2009)

The expert focus group technique was selected as the most appropriate method of gathering the data for a number of reasons; for example, it is seen to be ideal for obtaining data about feelings and opinions (Basch, 1987). We followed Krueger and Casey (2000) in design. Three groups of news media suppliers and one group of bloggers took part in the research (see Table 2) in London in February 2009. The focus group discussion was recorded and later transcribed. From the transcripts, key themes were pulled together to allow comparisons to be made.

## KEY RESULTS: LONDON/MANCHESTER NEWS MEDIA ORGANIZATIONS

Three key questions were posed during the expert focus group: (1) We questioned participants about the decline of regional newspapers. (2) We then probed them about the potential of the Internet for online wealth/value creation and (3) We finished our questioning by asking them about the future role of citizen journalists in news media production. The results are presented on a question by question basis.

*Figure 2. Focus group participants*

| |
|---|
| The 'radio' group - two participants involved in the production and distribution of offline/online content. |
| The 'television' group – one participant who develops online content to support evening news programmes. |
| The 'newspaper' group - two participants who are developing online social (community) eco-systems. |
| The 'blogging' group – two participants who are professional bloggers providing training programmes to news media organisations. |

## Decline in Readership, Advertising and Internet Disruption

Some clear consensus emerged from the group. Firstly, they acknowledged the long term decline in printed newspaper readership and the recent acceleration in the decline of printed advertising revenue (since 2007). It was generally felt that newspapers had been slow to respond to and utilise the full potential of the Internet. They explained that the strength, character and direction of this resistance to the Internet reflects a host of factors which might be: workplace-specific (journalists' attitudes to change, for example, along with the strength of trade-union membership); title-specific (a newspaper's market position, circulation strength, profitability); company- or group-specific (the relative willingness of different newspaper companies to commit resources to change); or, finally, resistance may be reader- or consumer-specific (readers' preferences for certain types of news, how it is presented and on which platform, as well as their willingness/ability to pay).

Though the group did point out that the decline in newspaper circulation was not only a consequence of the growth in the Internet. They stated two other reasons: (1) because people are no longer reading habitually, fewer newspapers are being bought and (2) as other forms of media are replacing the reading habits of people, the time they spend reading newspapers is reduced. Both newspaper participants (of the focus group) acknowledged that the Internet was accelerating the circulation decline of their main titles.

We asked the same question about the decline of newspapers to the Deputy Sales Manager[14] at the News Chronicle and received the following answer: "the emergence of the Internet has reduced the number of readers of our printed newspapers. This led News Chronicle (in 2006) to launch the Chronicle 'hybrid distribution model' ('half paid-for'/'half free'). There were two main objectives in launching the model: (1) to increase the total circulation of the News Chronicle, and (2) to counter the steep decline in the number of younger readers. For instance, we are offering a free newspaper in the centre of City A where, during the midweek daytime period, there is a population of approximately 180,000 people".

## Emerging Online Value Streams

In respect to potential online revenue streams, consensus appeared around the idea of news organizations developing highly specialised commercially valuable 'news' on the Internet. This would be likely focused on specific areas such as science or technology and offered to consumers for a fee alongside freely available conventional news sources. The notion of 'paywalls'[15] restricting access to paid subscribers on news content sites was felt to be only workable for more specialised areas of news or for access to reports/stories of leading 'branded' journalists. Another idea was to make readers pay for accessing online video footage.

The Internet was identified to have a number of economic advantages for advertisers over the printed format: (1) It provides opportunities for collecting behavioural data on consumers and building profiles of them. (2) It can be utilized to identify potential advertisers' criteria. (3) Through its ability to match consumer profiles to advertisers' criteria, the Internet reduces waste. Levels of online advertising revenues were reported to be rising for each of the respective news media organizations. But the ratios to print remained stubbornly low. One example of a local newspaper was quoted in Bristol which had £39m printed advertising revenue/£2m online advertising revenue.

In respect to cutting costs, participants identified that the Internet was cutting out many stages of the production process. For example, many news organizations had no photography department. Freelance photographers now perform work using advanced digital technology and email their photographs to the organization. The Deputy Managing Director[16] at News Chronicle explained when interviewed that: "…. by launching an online version of the newspaper the total costs of production are reduced. We have been able to cut our staff budget (from (£15m), cut printing costs (from £3.5m), eliminate press and distribution costs. Since the cost of online newspapers in comparison to the printing of newspapers is inherently low (almost £700,000 per year), online versions of newspapers can potentially create us some value".

## The Role of Citizen Journalists, Bloggers and Consumers

Web 2.0/social media was seen to be challenging news organizations to extend the level of their direct engagement with audiences as participants in the processes of gathering, selecting, editing, producing and communicating news. Theoretically newspapers are able to lower their cost base by having editorial content outsourced to a growing body of bloggers/citizen journalists (e.g. photographs are now increasingly sourced from citizen-journalist sites such as scoopt.com).

Our findings support Domingo et al. (2008) by identifying the general reluctance of regional newspapers to open up their high-value operations to citizen journalists. We were told that citizen journalists/bloggers were offered only limited opportunities for engagement with the editorial process and were largely restricted to 'debating current events', while being firmly excluded from other aspects of news production, such as reporting and sub-editing. Finally, the issues of the uncontrollable consumer came to the fore. A consumers' willingness to participate and contribute content is likely to differ throughout the day and the week with most participation occurring at the weekend.

The Group Regional Managing Director[17] at News Chronicle confirmed: "co-creation in the context of the Chronicle's printed newspapers is tightly controlled. Even if the company obtains information from consumers it is not normally utilized unless it has been carefully checked for quality, authenticity and it is legally sound". The newspaper utilizes the Internet as a means by which it can interact with consumers; it exchanges information, recognizes its potential consumers' interests, and addresses these interests through a variety of media. They also run several online surveys and it uses the information collected to customize its news content and advertisements. Through its online website discussion forums, News Chronicle actively encourages citizen journalists, bloggers and consumers to become involved with local news issues.

## CONCLUSION, THE FUTURE OF REGIONAL NEWSPAPERS AND MANAGERIAL IMPLICATIONS

The early discussion in the chapter tended to focus on the negative impact that the Internet was having on the regional newspaper industry. But our findings indicate that Web 2.0 is facilitating the development of new electronic commerce value-creation activities (through the growth of co-creation news (editorial) content and consumer-driven moves towards a multi-media platform of production and distribution (including television, online, mobile and printed forms)). Though there is a strategic gap between the reality of news producer/consumer interactions and the theory (Howe, 2008; Tapscott, 2008). The newspaper industry is moving towards co-creating value with its consumers through Internet mechanisms such as blogs and discussion forums, but this is still very much at the customization stage with news content personalized to user preferences. Further, it is the consumers of news who have instigated the moves into online modes of distribution, rather than organizations looking to save costs or improve the quality of their interactive services (through incorporating Internet technologies). There are concerns that news-media organizations are moving online into social media spaces such as Facebook, Twitter, Linkedin and MySpace rather than creating their own space. This raises questions over whether consumers will be willing to interact with newspaper in these spaces or whether they will wish to create/find their own news spaces (Mersey, 2009).

With the advent of the Web 2.0, the advertising model that has long served newspapers will not be sufficient to sustain previous levels of social influence and a profitable business model (Meyer, 2008). So what are the alternatives? On August 6, 2009, Rupert Murdoch announced that his London newspapers had plans to start charging for access to their websites. The logic is that readers should pay to access a particular story, a specific journalists work or video footage. Murdoch envisages newspapers setting up their own individual online accounts and having readers pay through subscriptions or a mechanism of micro-payments. Though there are many who doubt whether this will be successful strategy. Particularly vociferous in doubting Murdoch's ability to succeed with this strategy is Gospill (2009, p. 1), who emphasizes that in the UK only the Financial Times (FT) has succeeded in building a strong pay wall round its content. They have over 100,000 subscribers paying £155 a year for full access; readers can get up to ten stories a month free of charge. It is evident that the FT is still profitable, but it has specialized and a valuable service on offer – not to mention a wealthy and corporate readership. Likewise, in the US, a number of business magazines also charge for their web content, such as the Wall Street Journal.

A second alternative is for the complete transformation of the business model into a news service experience operation whereby newspapers, through Web 2.0, supply a discursive, commentary-based and participative style of news. Unless consumers are given much greater opportunities to interact, contribute and supply user generated content, newspapers we believe in the digital age will struggle to retain their social influence and hence their commercial influence (Meyer, 2008). Moreover Web 2.0 requires the development of new business models and value chains rather than their protection. Interestingly this is a key conclusion of the UK government in its recent Digital Britain report[18] which tentatively outlined its plans for the future of regional news provision.

## ACKNOWLEDGMENT

We would like to acknowledge the contribution of the participants in the focus group workshop and other meetings. These are Zoe Smith (ITN), Anthony Munnelly (Blogger), Sarah Hartley (Manchester Evening News), Kirsten Schlyder (Journalist), Jennifer Tracey (BBC Radio 4, iPM), Chris Vallance (BBC Radio 4, iPM), Graham Holliday (Blogger), Cagri Yalkin (King's College London) and Peter Kawalek (Manchester Business School).

## REFERENCES

Advertising Association. (2009). UK media advertising expenditure 2008. Retrieved December 1, 2009, from http://www.adassoc.org.uk/aa/index.cfm/adstats/

Atton, C. (2008). Alternative media theory and journalism practice. In Boler, M. (Ed.), *Digital Media and Democracy: Tactics in Hard Times* (pp. 213–226). Cambridge, MA: MIT Press.

Basch, C. E. (1987). Focus group interview: An underutilized research technique for improving theory and practice in health education. *Health Education Quarterly*, *14*(4), 411–448.

Beam, R. A., Brownlee, B. J., Weaver, D. H., & DiCicco, D. T. (2009). Journalism and public service in troubled times. *Journalism Studies*, *10*(6), 734–753. doi:10.1080/14616700903274084

Byrne, C. (2009). Funding the regional news of tomorrow. *Press Gazette*, 13-15.

Dear, J. (2006, January 2). Put people before profits. *The Guardian*, 8.

Department of Innovation and Skills (and Department of Culture). (2009). *Digital Britain*. Retrieved November 27, 2009, from http://www.culture.gov.uk/images/publications/digitalbritain-finalreport-jun09.pdf

Domingo, D., Quandt, T., Heinonen, A., Paulussen, S., Singer, J. B., & Vujnovic, M. (2008). Participatory journalism practices in the media and beyond: An international comparative study of initiatives in online newspapers. *Journalism Practice*, 2(3), 326–341. doi:10.1080/17512780802281065

Franklin, B., & Murphy, D. (1991). *What News? The Market, Politics and the Local Press*. London: Routledge.

Freer, J. (2007). UK regional and local newspapers. In Anderson, P., & Wood, G. (Eds.), *The Future of Journalism in the Advanced Democracies* (pp. 89–103). London: Ashgate.

Gospill, T. (2009). Look after the pennies. *Free Press*, 1, 8.

Howe, J. (2006). The rise of crowdsourcing. *Wired*. Retrieved May 21, 2009, from http://www.wired.com/archive

Howe, J. (2008). *Crowdsourcing: Why the Power of the Crowd is Driving the Future of Business*. New York: Crown.

Krueger, R. A., & Casey, M. A. (2000). *Focus Groups. A Practical Guide for Applied Research*. Thousand Oaks, CA: Sage Publications.

Li, C., & Bernoff, J. (2008). *Groundswell: Winning in a World Transformed by Social Technologies*. Boston, MA: Harvard Business Press.

Mersey, R. (2009). Online news users' sense of community: Is geography dead? *Journalism Practice*, 3(3), 347–360. doi:10.1080/17512780902798687

Meyer, P. (2004). *The Vanishing Newspaper: Saving Journalism in the Information Age*. Columbia, MO: University of Missouri Press.

Meyer, P. (2008). The elite newspaper of the future. *American Journalism Review*, 32-35.

Mintel. (2007). Regional newspapers. *Mintel Report*, London: Mintel.

Murphy, D. (2008). Earthquake undermines structure of local press ownership: Many hurt. In Franklin, B. (Ed.), *Pulling Newspapers Apart: Analysing Print Journalism*. London: Routledge.

National Readership Survey. (2007). Readership of newspapers and online news. Retrieved December 1, 2009, from www.nrs.co.uk

National Readership Survey. (2009). Readership of newspapers and online news. Retrieved December 15, 2009, from www.nrs.co.uk

Newspaper Society. (2009). Local and regional newspaper historical data. Retrieved December 5, 2009, from http://jiab.jicreg.co.uk/JIAB.cfm?NoHeader=1

Nichols, J. (2009). It's crunch-time for journalism. *Free Press*, 1-2.

Office for National Statistics. (2009). Social trends, No. 39. Basingstoke, Hampshire: Palgrave Macmillan.

Picard, R. (2004). Commercialism and newspaper quality. *Newspaper Research Journal*, 25(1), 54–64.

Pincus, W. (2009). Newspaper narcissism: Our pursuit of glory led us away from readers. *Columbia Journalism Review*. Retrieved June 1, 2009, from http://www.cjr.org/archive

Retrieved May 21, 2009, from http://www.trinitymirror.com/documents/2009%20Interim%20Announcement%20final.pdf

Tapscott, D. (2008). *Grown Up Digital*. New York: McGraw Hill.

Thurman, N., & Myllylahti, M. (2009). Taking the paper out of news: A case study of Taloussanomat, Europe's first online-only newspaper. *Journalism Studies*, *10*(5), 691–708. doi:10.1080/14616700902812959

Trinity Mirror. (2009). Half yearly financial report for the 26 weeks ended 28 June 2009.

Vargo, S. L., & Lusch, L. S. (2006). Service-dominant logic: What it is, what it is not, what it might be. In Lusch, L. S., & Vargo, S. L. (Eds.), *The Service-Dominant Logic of Marketing: Dialog, Debate, and Directions* (pp. 43–56). New York: ME Sharpe.

Waugh, E. (1938). *Scoop*. Boston: Little, Brown and Co.

## ADDITIONAL READING

Adams, J. W. (2006). Striking it niche – extending the newspaper brand by capitalizing in new media niche markets: Suggested model for achieving consumer brand loyalty. *Journal of Website Promotion*, *2*(1/2), 163–184.

Brill, A. M. (2006). The print and online newspapers in Europe: A comparative content analysis in 18 countries in western and eastern Europe. *Journalism & Mass Communication Quarterly*, *83*(4), 953–955.

Bush, V., & Gilbert, F. (2002). The Web as a medium: An exploratory comparison of Internet users versus newspaper readers. *Journal of Marketing Theory and Practice*, *10*(1), 1–10.

Chyi, H., & Lasorsa, D. (2002). An explorative study on the market relation between online and print newspaper. *Journal of Media Economics*, *15*(2), 91–106. doi:10.1207/S15327736ME1502_2

Consoli, J. (1997). Online usage: 'More than a fad'. *Editor & Publisher*, *130*(32), 26.

D'Haenens, L., Jankowski, N., & Heuvelman, A. (2004). News in online and print newspapers: Differences in reader consumption and recall. *New Media & Society*, *6*(3), 363–382. doi:10.1177/1461444804042520

Dans, E. (2000). Internet newspapers: Are some more equal than others? *International Journal on Media Management*, *2*(1), 4–13.

Deleersnyder, B., Geyskens, I., Gielens, K., & Dekimpe, M. (2002). How cannibalistic is the Internet channel? A study of the newspaper industry in the United Kingdom and the Netherlands. *International Journal of Research in Marketing*, *19*(4), 337–348. doi:10.1016/S0167-8116(02)00099-X

Dimmick, J., Chen, Y., & Li, Z. (2004). Competition between the Internet and traditional news media: The gratification-opportunities niche dimension. *Journal of Media Economics*, *17*(1), 19–33. doi:10.1207/s15327736me1701_2

Ferguson, D. A., & Perse, E. M. (2000). The World Wide Web as a functional alternative to television. *Journal of Broadcasting & Electronic Media*, *44*(2), 155–174. doi:10.1207/s15506878jobem4402_1

Flavián, C., Guinalíu, M., & Gurrea, R. (2006). The influence of familiarity and usability on royalty to online journalistic services: The role of user experience. *Journal of Retailing and Consumer Services*, *14*(5), 363–375. doi:10.1016/j.jretconser.2005.11.003

Flavián, C., & Gurrea, R. (2006). The choice of digital newspapers: Influence of reader goals and user experience. *Internet Research*, *16*(3), 231–247. doi:10.1108/10662240610673673

Gentzkow, M. (2007). Valuing new goods in a model with complementarity: Online newspapers. *The American Economic Review, 97*(3), 713–744. doi:10.1257/aer.97.3.713

Gunter, B. (2003). *News and the Net*. Mahwah, NJ: Lawrence Erlbaum Associates.

Ihlström, C., & Palmer, J. (2002). Revenues for online newspapers: Owner and user perceptions. *Electronic Markets, 12*(4), 228–236. doi:10.1080/101967802762553486

Ihlström Eriksson, C., & Svensson, Jesper. (2009). A user centered innovation approach identifying key user values for the e-newspaper. *International Journal of E-Services and Mobile Applications, 1*(3), 38–78.

Kaye, B., & Johnson, T. (2004). A Web for all reasons: Uses and gratifications of Internet components for political information. *Telematics and Informatics, 21*(3), 197–223. doi:10.1016/S0736-5853(03)00037-6

Levins, H. (1998). Growing US audience reads news on Net. *Editor & Publisher, 131*(8), 14.

Li, X. (2006). *Internet Newspapers: The Making of a Mainstream Medium*. Mahwah, NJ: Lawrence Erlbaum Associates.

Lin, C. A. (2002). Perceived gratifications of online media services among potential users. *Telematics and Informatics, 19*(1), 3–19. doi:10.1016/S0736-5853(01)00014-4

Rathmann, T. (2002). Supplement or substitution? The relationship between reading a local print newspaper and the use of its online version. *Communications, 27*(4), 485–498. doi:10.1515/comm.2002.004

Salminen, A., & Hakaniemi, K. (2007). Facing the challenges of multi-channel publishing in a newspaper company. *Journal of Cases on Information Technology, 9*(1), 54–72.

Soderlund, G. (2006). Digitizing the news: Innovation in online newspapers. *American Journal of Sociology, 111*(4), 1221–1223. doi:10.1086/502988

Tucker, P. (2009). Newspapers face the final edition. *The Futurist, 43*(5), 8–9.

Van Oostendorp, H., & Van Nimwegen, C. (1998). Locating information in an online newspaper. *Journal of Computer-Mediated Communication, 4*(1), 1–14.

## KEY TERMS AND DEFINITIONS

**Crowdsourcing:** Using content from amateur contributors.

**Hyper-Local News:** News that relates specifically to a narrow geographical region or local community.

**Paywalls:** Restricting access to paid subscribers on news content sites.

**Public Service Journalism:** Journalism's role in educating the public or community, helping society function properly, and taking actions to benefit the public.

**Regional Newspapers:** Newspapers that focus on news for a geographic region rather than being a national newspaper.

## ENDNOTES

[1] Public service is often coupled with references to journalism's normative role in educating the public or community, in helping society function properly and in taking actions to benefit the public (Beam et al. 2009).

[2] The terms 'local' and 'regional' newspapers have no precise, agreed definition. Sometimes they are used almost interchangeably. Where a distinction is drawn it is usually by reference to an area of circulation or distri-

bution, local newspapers being regarded as those covering a single town or a small number of towns, and regional newspapers as having a wider geographical coverage.

3   Interview undertaken in the School of Journalism, University of North Carolina–Chapel Hill, November 11, 2008.

4   Interview undertaken at the San Diego Tribune, San Diego, February 27 2009. Research visit funded by EPSRC/Oxford University e-Research Centre, grant reference. EP/G001979/1.

5   Web 2.0 is defined in a later section.

6   With the exception of the Western Mail in Wales and the Scotsman and Herald in Scotland.

7   The NRS canvasses a panel of 36,000 randomly sampled people to find if they read newspapers/online news.

8   Cited in Social Trends (2009, p. 195). Note that readership data was only presented for the 15 to 19 age category.

9   The advent of Web 2.0 has led to the development and evolution of web-based communities, hosted services, and web applications. Examples include social-networking sites, video-sharing sites, wikis and blogs (Tapscott, 2008).

10  Chris Atton's definition of citizen journalism as: "a philosophy of journalism and a set of practices that are embedded within the everyday lives of citizens, and media content that is both driven and produced by those people.... (and whose) practices emphasise first person, eye witness accounts by the participants" (2008: 267) aptly captures the essence of the concept.

11  Nichols (2009) claims that 10,000 full-time journalists lost their jobs in 2008 on US-based newspapers.

12  Interview undertaken at the San Diego Tribune, San Diego, February 27 2009.

13  Vargo and Lusch (2006) define co-creation as "the participation of the consumer in the creation of the core product/service itself" (p. 48). The economic logic behind co-creation is that media businesses now have unprecedented opportunities for user participation and (re)-engagement with their audiences (Li & Bernhoff, 2008; Tapscott, 2008). Co-creation implies lower production costs through the outsourcing of news reporting to citizen journalists and bloggers (Thurman & Myllylahti, 2009; Meyer, 2008).

14  Our first interview was conducted at News Chronicle on July 4, 2009.

15  Recently introduced by Rupert Murdoch on the web additions of his News International titles.

16  A second interview at News Chronicle was conducted on July 10, 2009.

17  The final interview at News Chronicle was conducted on July 20, 2009.

18  Department of Innovation and skills/Department of Culture. (2009) Digital Britain, http://www.culture.gov.uk/images/publications/digitalbritain-finalreport-june09.pdf.

# Chapter 10
# Software as a Service and the Pricing Strategy for Vendors

**Nizar Abdat**
*Utrecht University, The Netherlands*

**Marco Spruit**
*Utrecht University, The Netherlands*

**Menne Bos**
*Accenture, The Netherlands*

## ABSTRACT

*Software as a Service (SaaS) has been a dominant information technology (IT) news topic over the last few years. It is a new phenomenon where software as a digital product, instead of being locally installed and delivered as a product, has been shifted to being installed in data centers and delivered as a service. The users do not need to worry about the installation and maintenance of their software since these tasks have now become the responsibility of the vendor. In reality, many people are still puzzled about SaaS with other new technologies. Next to that, there are numerous enterprise users who hesitate to adopt SaaS solutions because of the idea of storing data outside their company. This chapter elaborates on the state-of-the-art of SaaS from both scientific and business perspectives to help readers better understand this technology.*

## INTRODUCTION

This chapter describes the diverse aspects of Software as a Service, which makes it valuable for many different readers. It starts with the background of SaaS, followed by its definition and main characteristics. Next to that, we show how readers can see SaaS differently from traditional software and how it is interrelated to other technologies such as cloud computing, Web 2.0, Application Service Provider (ASP), and Software plus Services (S+S). The benefits and risks of adopting SaaS from different literatures will also be compared and explained in detail. In addition, we will present several scenarios for how SaaS is currently being delivered in the market and introduce our Pricing Strategy Guideline Framework (PSGF). In this chapter, several new deliverables that are produced during our studies and interviews with nineteen different SaaS companies will be presented. The

DOI: 10.4018/978-1-61692-877-3.ch010

main deliverable itself is the PSGF framework which aims at providing SaaS vendors with a set of guidelines to ensure that all the fundamental elements with respect to pricing are included in their pricing strategy. Our framework will be useful especially for small-to-medium SaaS vendors – particularly the startup vendors, which tend to have less experience in pricing their SaaS solutions and dealing with several issues in the aforementioned scenarios like low sales cycles, chaotic pricing, and entering a new market segment (Geisman & Nelson, 2008). The framework including all layers and elements will be described in detail with several examples as well. Our framework has been constructed from several existing theories and has been successfully validated by a number of experts in the field.

Several contributions from both scientific and business perspectives are conveyed from this book chapter. From the scientific perspective, the matrices of SaaS key principles and SaaS benefits and risks, which also specify to which groups they are applied, certainly add value to the existing SaaS literature. Furthermore, the six scenarios introduced in this chapter have never been examined in previous SaaS studies. However, this chapter's most significant contribution is the framework itself, as this research presents the first SaaS pricing framework available. In most of the available academic software pricing literature, authors are using mathematical formulas, whereas this book chapter does not, making it more favorable to readers. Finally, from a practical business point of view, our framework presents all fundamental elements related to the pricing of SaaS solutions which is very likely to be quite useful for many SaaS vendors.

## BACKGROUND

'Software' is a general term used to describe the computer programs, procedures, rules, and the associated documentation, in relation to the operation of a computer system which are stored in a read/write memory unit as part of the digital system (Langholz, et al., 1998; Wordreference. com, 2009). This term was first introduced by Tukey (1958), and has become an integral element of the English language and has been included in many other languages. At that time, when the computer era began, the manufacturer sold the computer as a physical machine, which included the operating system and rudimentary software at no additional cost. This situation has changed after IBM announced on June 23, 1969, that it would unbundle the hardware and software in the future. Later on, this has been seen as the birth date of the software industry as we know today (Kittlaus & Clough, 2009). Within a few decades the software market has dramatically grown.

From that time until today, in most situations, customers are required to buy the software license, install and run the software on their local computer or server before they can use it. The installation files can be stored on CDs or other storage devices like diskettes, USB sticks, etc. The licenses and CDs are usually sold by software vendors. Since this type of software is installed on the premises (in the building) of the users rather than on a remote facility, it also commonly known as *'on-premise software'*. Also, because this type of software has been used for ages compared to the new software phenomenon, it is also known as traditional software.

The Internet boom in the mid-1990s has shifted the way in which companies - including software vendors - are doing business. The Internet has become not only the medium for marketing their software, but also turned out to be the major requirement to deliver their software. In other words, without the existence of an Internet connection, this new type of software cannot be utilized by their customers. This new phenomenon began in the late 1990s with the concept of Application Service Provider (ASP), which has evolved into another software hype, most popularly known as 'Software as a Service' or SaaS (Hoch et al., 2001).

The trend towards SaaS has been predicted by Rust and Kannan (2003) as early as 2003 when they still called it 'e-service'. In SaaS, customers have the freedom to use the software as they require. Hence, SaaS is often referred to as *'on-demand'* software. SaaS applications are being installed in data centers and no longer delivered as a product (physical object), but as a service. This fact has become the reason why the term 'product' does not fit any longer in the world of SaaS. The term 'solution' is being used instead.

In rough economic times like those seen today, the SaaS adoption rate for 2009 has been increasing dramatically from 36% to 42% over 2008 (Mahowald, 2009). This is happening because the customers want to perform reasonable cost cutting without affecting the efficiency of their business operations. Adopting SaaS solutions has been the more favorable alternative rather than investing in completely new on-premise software or in outsourcing.

## What is SaaS?

At present, there are many definitions for SaaS from different sources and some of them even refer to SaaS using other terms. For example, the term 'E-services' was introduced by Rust and Kannan (2003) and 'Software Oriented Computing (SOC)' was brought by Papazoglou (2003). Besides these different terms, many people are still misunderstanding the concept of SaaS. These facts have motivated us to make a base-definition from the diverse available literature which could assist us in our research and to provide our readers with a more clear definition of SaaS.

Subsequently, in order to come up with this definition, we have selected twelve different published articles. The literature is comprised of seven scientific and five business studies. The base definition is set out from the similar elements stated in each article. The following table describes all important elements of SaaS definitions which are explicitly stated in these studies.

On Table 1, the most stated and important elements are highlighted in gray. And based on these results, the following base-definition is generated:

*"SaaS is a software delivery model that supports multi-tenancy in which the vendors host and operate their software on a data center (either independently or through third-party) and provide it to their customers over the Internet and typically on a subscription basis and/or pay-for-use basis."*

The term *'SaaS vendor'* is used for the company who builds and develops the software while *'SaaS provider'* is used for the third party company who provides the infrastructure in order to ensure that the software from the vendor is properly delivered as a service to end users.

Related to the definition above, researchers have categorized the two main scenarios regarding how the software and money cycle for SaaS is currently in the market. The first scenario is when the company develops, hosts and operates their software on their own data center or infrastructure. In this case, the term *SaaS vendor* is also applied to this company although they act both as vendor and provider. And for the second scenario, the company hosts and operates their software via SaaS provider (third party). The illustration of these two scenarios can be found in Figure 1.

## Key Principles of SaaS

Based on the matrix table in the previous section and the supporting literature, we have listed several key principles or main characteristics of SaaS in order to help readers understand its concept. The principles are:

- The architecture of SaaS has been designed to support *multi-tenancy* where one configuration code base can support multiple different customers, allowing them to share all resources including databases (Carraro

*Table 1. A matrix of SaaS definitions*

| Sources Elements | Accenture (2008a) | Blokdijk (2008) | Carraro and Chong (2006) | Chou and Chou (2008) | Greschler and Mangan (2002) | Hoch, et al. (2001) | IBM (2008) | Kaplan (2009) | Kittlaus and Clough (2009) | Rayner (2008) | Sääksjärvi, et al. (2005) | Sessions (2006) | Number of Elements | Base definition |
|---|---|---|---|---|---|---|---|---|---|---|---|---|---|---|
| Software model | x | x | - | x | x | x | x | - | x | x | - | - | 8 | x |
| Software approach | - | - | - | - | - | - | - | - | - | - | - | x | 1 | - |
| Software solution | - | - | - | - | - | - | - | x | - | - | - | - | 1 | - |
| Remotely access (Internet based) | x | x | x | x | - | x | x | x | x | x | x | - | 10 | x |
| Managed by providers | x | x | - | - | - | x | - | x | - | x | - | - | 5 | x |
| Hosted service | x | x | x | x | x | x | x | x | - | - | x | x | 10 | x |
| Subscription basis, recurring fee | | x | - | x | - | x | | - | x | x | - | - | 5 | x |
| Usage metric | - | - | - | - | - | - | - | - | x | x | - | - | 2 | - |
| Pay for use | x | x | - | x | - | - | - | - | x | x | - | x | 6 | x |
| Configurability | - | - | x | - | - | - | - | - | - | - | - | - | 1 | - |
| Multi-tenancy | - | - | x | - | - | - | - | - | - | x | - | - | 2 | x |
| Scalability | - | - | x | - | - | - | - | - | - | - | - | - | 1 | - |
| Developed by vendors | x | - | - | - | - | - | x | - | - | - | - | - | 2 | - |
| Data center/server | x | x | - | - | x | x | - | - | - | - | x | - | 5 | x |
| Internal or external data center/ third party | x | - | - | - | x | - | - | - | - | - | - | - | 2 | x |
| Time & location independent | - | - | - | - | - | - | - | - | - | - | x | - | 1 | - |
| Attractive payment | - | - | - | - | - | - | - | - | - | - | x | - | 1 | - |
| On-demand installation & maintenance (support) | - | - | - | - | x | - | - | x | x | - | x | x | 5 | x |
| No cost (advertisement model) | - | - | - | - | - | - | - | - | x | - | - | - | 1 | - |

& Chong, 2006; Tarzey et al., 2007). Many articles, especially from the business perspective, tend to skip this principle and focus more on the other benefits of SaaS. Many business people think that the multi-tenancy helps the economics but does not do anything for the customer. However, this multi-tenancy is the most significant key for SaaS since it makes SaaS different from other online applications such as ASP.

- The SaaS offerings are *scalable*; they are easily scaled up or down depending on the demand from customers (Carraro & Chong, 2006; Chou & Chou, 2008; Tarzey et al., 2007). Customers have the freedom to change their usage of the SaaS solutions within a certain period of time. For instance, customer X required 30 users last month to operate a certain SaaS offering, but because many of their employees are currently on summer vacation, this customer changed its usage to only 15 users for the next two months.
- The applications are hosted and installed at a site the vendor has chosen (internal or via third party provider) (Accenture, 2008a; Blokdijk, 2008; Carraro & Chong, 2006; IBM, 2008; Kaplan, 2009; Merchant & Geisman, 2006).
- Activities are managed in data centers rather than at the customer's site

*Figure 1. On-premise vs. software as a service*

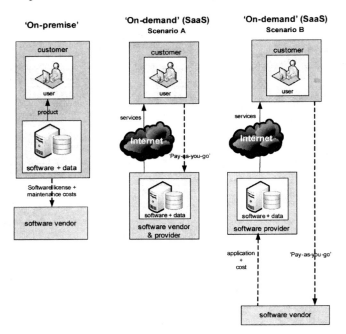

(Accenture, 2008a; Blokdijk, 2008; Greschler & Mangan, 2006; Hoch et al., 2001; Sääksjärvi et al.,2005).

- SaaS enables the customers to access the applications remotely via the Web at anytime and anywhere (Dym, 2009; Lassila, 2006; Rowell, 2009; Sääksjärvi et al.,2005). Thus, an Internet connection is a mandatory requirement to be able to use the applications.
- SaaS vendors control the upgrades of the application, which eradicates the need for customers to download patches and upgrades (Caldwell & Eid, 2007; Hoogvliet, 2008; Kaplan, 2009; Lassila, 2006).
- In most cases, the upgrades are done more frequently, and in smaller releases, rather than in one big release as it is normally done with on-premise applications (Geisman, 2008). Since the vendor has full control over the installed applications, they have the tendency to upgrade their applications and fix the bugs more often from their data centers.
- Customers are always working with the latest version of the application (Accenture, 2008a; Choudary, 2007; Hoogvliet, 2008; Mahowald, 2009; Rowell, 2009).
- The responsibility for technical infrastructure including their cost is allocated from customers to provider. Therefore, the provider controls the upgrades of the back end system (operating system, hardware, network, etc.) and is responsible for maintaining and fixing any failures of infrastructure, if it happens (Caldwell & Eid, 2007; Hoch et al., 2001; Lassila, 2006; Mahowald, 2009; Pring et al, 2007; Sääksjärvi et al.,2005; Sysmans, 2006; Tarzey et al., 2007).
- Customers are no longer the 'owner' of the application, but 'rent' the service of the applications (Chou & Chou, 2008; Hoch et al., 2001; Merchant & Geisman, 2006). The customers do not need to buy the licensing to use the applications any-

more. The applications are owned only by the vendors.
- Customers are charged using a subscription (fixed fee) model: monthly, quarterly, or annually; usage-based pricing based on metrics, and no costs with embedded advertisement like Google (Blokdijk, 2008; Chou & Chou, 2008; Hoch et al., 2001; Kaplan, 2009; Kittlaus & Clough, 2009; Rayner, 2008; Sysmans, 2006; Wang, 2009). Nevertheless, there are always possibilities that SaaS vendors might come up with new charging alternatives in the future.
- SaaS is frequently integrated into a larger network of communicating software - either as part of a mash up or as a plug-in to a platform as a service. Hence, additional skills and efforts are absolutely required to do it.

The aforementioned main characteristics of SaaS are also useful to differentiate between SaaS solutions and on-premise software in the market.

## On Premise vs. SaaS

On-premise software is installed on the customer side while for SaaS it is are installed in data centers. This is shown in the following figure where software and data are placed in a separate box for on-premise and not in the case of SaaS. Referring to the earlier base definition for SaaS, software and data can be installed either in the infrastructure of the SaaS vendor (scenario A) or through the third party or SaaS provider (scenario B). These two scenarios are chosen since they are the most common situations for SaaS.

In addition, Table 2 provides detailed information about the different essential aspects of on-premise and SaaS. The information is collected from different sources such as Accenture (2008b), Kittlaus and Clough (2009), Pring et al. (2007), Sysmans (2006), Tarzey et al. (2007), Wilson and Basiliere (2008), and lastly York (2008).

## SaaS and Related Technologies

As a new hype in the market, many people are still puzzled by how SaaS is related with other technologies. This section will provide a better explanation about its relationship with Cloud computing, Web 2.0, its predecessor – ASP, and Software + Service.

## Cloud Computing

Cloud Computing has been a dominant buzzword in the IT industry for some time. This term refers to any virtualized resources that are delivered as a service from data centers over the Internet and accessible from anywhere in the world (Lin, et al., 2009; Armbrust et al., 2009; Buyya et, al., 2008; Hayes, 2008; Vouk; 2008). All data and software applications that were originally located on desktop and corporate server rooms are being swept up and installed in "the cloud" or online resources. According to Lin et al. (2009), Cisco, one of the global leading suppliers of networking equipment and network management for the Internet, sees virtualization and automation as the key enabling technologies of cloud computing.

A cloud is categorized as *Private Cloud* or *Public Cloud* based on the location of the data center where the services are being virtualized. For private cloud, the data center is built internally behind a firewall and not shared outside the enterprise. Full control is retained by the organization. On the contrary, in public cloud, the service providers manage the infrastructure and offer public customers the ability to deploy and consume services over the Internet (Armbrust et al., 2008).

The cloud-based services are not only restricted to applications, or what is called "Software as a Service", but could also be the platform and the hardware (infrastructure). For example, Google

*Table 2. On-premise vs. software as a service*

| On-premise | SaaS |
|---|---|
| **Model** Software is delivered to customers for installation on the customers' computers. They are accessed on-site. Unless contracted separately, customers are typically responsible for installation, maintenance, access time, hardware performance, and applying any updates after receiving updated software from licensor via download or disk. The customers require in-house expertise in all technical aspects. Slow to deploy. The architecture model is more complex since there might be different code bases for different systems. E.g. different installation files for different OS. **Ownership** Ownership of the intellectual property assets belong to the vendors (licensors). Customers receive a physical copy of software (usually available with the code). **Pricing and Licensing Method** The customers need a license, which includes the pricing model in order to use the software. Mostly is sold with fixed one-time fee or perpetual license. **Warranty** Warranty of compliance with documentation or system specifications usually runs for a period of time (90-180 days), which expires afterwards. Any further problems are handled under separate maintenance contract. **Migration** It could be a compatibility issue of data migration on certain software and operating systems. The switching procedures require additional costs and efforts. **Marketing and Finance** Typically is sold for niche focus and used in low volume. On-premise software is typically counted as assets (capital expenses) and has bigger financial risk for the customers. | **Model** Software is accessed online (via Web) and installed in data centers chosen by the vendors. The vendors or providers are responsible for installation, maintenance, access time, performance, and updates. The customers do not require in-house expertise in the technical aspects. Faster and less expensive implementations. The architecture is designed to use one single code base for different OS. **Ownership** Ownership and possession of the intellectual property assets belong to the vendors. The customers rent the software. **Pricing and Licensing Method** The customers need a service contract, not license to use the software. Pricing model is also applied to SaaS offerings. The software is typically sold in subscription based (monthly, quarterly, annually, etc) and/or usage based on several metrics. **Warranty** All support, training, security risks are mostly included in SaaS fees. Customers will often have the option to discontinue using the software if service does not significantly comply with documentation. The refund of pre-paid fees is usually not available. **Migration** Typically not an issue because software is designed already to be compatible with many different operating systems. The switching procedure is relatively quicker and simpler. **Marketing and Finance** The software is sold for mass market and used in high volume. SaaS cannot be counted as assets on a balance sheet because it is considered more as operational expenses (opex), which of course has a smaller risk of investment. |

App Engine offers the users a complete development stack that uses familiar technologies to build and host Web-based applications; Amazon Elastic Compute Cloud (EC2) and Amazon Simple Storage Service (S3) provide the users with the ability to resize the capacity like bandwidth, processor and storage required for running their applications. Here, the users are mostly the software developers who implement their applications for, and deploy them in, the cloud (Youseff et al., 2008).

From the service type and its architectural point of view, Creeger (2009), Lenk et al. (2009), Lin et al., (2009), and Vaquero, et al. (2009) have distinguished three layers of cloud computing: SaaS, PaaS, and IaaS. The illustration of these groups can be found in Figure 2.

1. *Software as a Service (SaaS):* Since this is the subject of this research, a more detailed explanation about it can be found in other sections of this chapter. Some examples are Google Apps, Microsoft Office Live (office application), Salesforce.com (CRM application, Workday (HRM application), and NetSuite (accounting application).
2. *Platform as a Service (PaaS):* This is the application development platform that enables the runtime environment for cloud applications. The providers supply the users (developers) with a programming-language-level environment with a set of well-defined APIs to facilitate the interaction between the environment and the cloud applications, as

*Figure 2. Ontology of cloud computing*

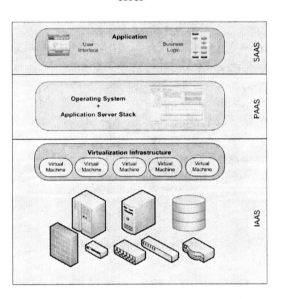

well as to accelerate the deployment and support the scalability needed for those cloud applications (Youseff et al., 2008). In order to enable this environment, the providers must ensure that the operating system and the supported application server stack are installed. These application server stacks might be a software bundle used for languages like Perl, Python or PHP, *Ruby on Rails* for Ruby, and *Tomcat* for Java (Zhen, 2008). Examples of PaaS are *Google App Engine* and *Force.com*. Google App Engine provides a Python and Java runtime environment together with the APIs for applications to interact with Google's cloud environment (Google, 2009). Force.com offers the Apex language that allows the developers to design, along with their applications' logic, page layout, workflow, and customer reports (Salesforce.com, 2008). To have a better visualization, we can think of PaaS that runs the software like MS Visual Studio to compile and to run a Web page. Instead of being used to deliver the cloud applications, PaaS can also be used as a way to integrate different SaaS applications (Giurata, 2008).

3. *Infrastructure as a Service (IaaS):* IaaS makes it easy for the developers to provision resources such as servers, connections, storage, and related tools (firewalls, routers, switches) in order to build a cloud application environment from scratch. It was previously known as Hardware as a Service (HaaS), which was first coined by Carr (2006). IaaS can be seen as having a data-center-in-the-cloud, which underlies the infrastructure of PaaS and SaaS.

Virtualization becomes the enabling technology for IaaS that allows the users unprecedented flexibility in configuring their settings while protecting the physical resources of the providers' data center (Youseff, et al., 2008). Some examples are Amazon Web Services S3 and EC2, Terremark, Flexiscale, and Rackspace hosting. Cisco and Microsoft have also stated that virtualization and automation are the key enabling technologies of cloud computing (Lin et al., 2009; Microsoft, 2009a).

Referring to the earlier definition of SaaS, a SaaS vendor may build its own platform and infrastructure to support its SaaS solution, or outsource it to the providers. In this case the PaaS companies are the providers. The same situation occurs for these PaaS companies. They can also build their own infrastructure or outsource it by making use of the available IaaS in the market.

## Web 2.0

According to Knol et al. (2008) and Barnatt (2008), Web 2.0 involves making new improved forms of online connection between two or more people (*interpersonal computing*), two or more online services (*Web services*), and between the individual user and software applications (*software as a service*).

Interpersonal computing is about using online technology to connect people with each other via social networking sites (Twitter, Facebook, LinkedIn, etc.), wikis, blogs, and online videos. Web services are components of online functionality which can be plugged together in order to create integrated online offering or *mash-up*. The developers are now using various Web APIs to create mash-ups to solve all types of problems from integrating the online payment service *PayPal* with other websites to the esoteric mash-ups that record the location of an object (Maximilien et al., 2008). As the final aspect of Web 2.0, SaaS is changing the way people use software and freeing users from the burden of installation and maintenance of the software. Web 2.0 also allows the creation of a new business models such as AdWords, AdSense, Analytics, Map for Google.

In conclusion, SaaS is just a part of the large picture of Web 2.0 and Cloud Computing, which are the continuing evolution of how we exploit the Internet and computer applications in general (Danielson, 2007; Obannon, 2009).

## ASP

The Application Service Provider (ASP) model became popular in the late 1990s with the emergence of the first wave Internet enabled application. Therefore, ASP is considered to be the predecessor of SaaS that came a few years later. Unfortunately, many people still think that SaaS is just a new name for ASP. Even worse, the CEO of Kleinsteen BV and the project coordinator at KC Beuningen BV also stated, *"there are several companies in the Netherlands who offer ASP functionalities, but market them as SaaS"*. This approach is actually a good marketing strategy for these companies, but consequently, many customers become more confused with these two terms.

ASP allows the computer applications to be hosted remotely from the user. The applications are simply hosted by the provider from a server farm located at its data center (Walsh, 2003) without fundamentally changing the architecture of the applications. Although the costs are spread over many customers, the hosting cost has still proved to be too much for ASP to survive in the market (Hoogvliet, 2008). Additionally, at the time where ASP was booming, there was only a small segment of the market that was willing to outsource their applications. They considered their applications as a strategic asset that needs to be kept safe under their own roof (Hoogvliet, 2008). This situation is different from the SaaS model where the applications are built from scratch to be Web based applications. A more detailed overview on the characteristics of ASP and SaaS can be found in Table 3. These characteristics are adapted from the article Accenture (2008d), Kittlaus and Clough (2009), Luit Infotech (2008), and Tarzey et al. (2007). We believe the aspects listed in this table describe the major differences between ASP and SaaS.

*Table 3. ASP vs. SaaS*

| ASP | SaaS |
|---|---|
| **Model** ASP applications are traditional single-tenant applications but hosted by a third party who usually do not have specific application expertise. The applications are not written as net-native (Web based) applications. Quicker to deploy than on-premise, but slower than SaaS. Access to client is provided via direct connection, or more recently the Internet, typically through client-server software or remote access like Citrix and Microsoft Terminal Services. The customer (organization) may still need in-house expertise in the application itself such as installation of the software, separate hardware, network and security configuration. The applications are single and usually a highly customized version for one customer. The upgrades are infrequent because the ASPs often depend on the commercial software suppliers. The upgrades are deployed when the supplier issued them, usually once a year or less. **Usability** Difficult. Customized version of an already complex application requires a great amount of training and orientation. **Ownership** Ownership and possession of the intellectual property assets belong to the vendors (licensor) or the ASP provider in some cases. **IT support** Exclusive, depending on the degree of the customization, additional maintenance might be required. | **Model** SaaS applications are multi-tenant applications, hosted by a vendor that has all the application expertise and they have been designed as net-native applications. Quicker to deploy than ASP. Access to client is provided via the Internet. Requires no in-house expertise in the technical aspects to use the application. SaaS offers standard software in a multi-user mode that the individual customer can only customize to a limited extent. There is only one code base for all users. The upgrades are relatively often with small releases since the applications are only deployed in data centers and made available to the all users. **Usability** Easier. The users can start using the applications immediately by utilizing the provided FAQ or demos on the vendor's site. **Ownership** Ownership and possession of the intellectual property assets belong to the vendors (licensor). **IT support** Inclusive, IT support becomes part of the service paid by the users. |

## Software + Services (S+S)

Software + Service (S+S) has been seen as a new way to combine the concept of on-premise and SaaS to reduce the limitations of both. So, on-premise users are no longer tied to a certain device and location where the software is installed since they can now access the software via cloud (Internet). In contrast, SaaS users that previously depended on the Internet connection to use software can also do it offline now. The data then will be automatically synchronized when SaaS users are connected to the Internet. The term *Software plus Service* was first outlined by the Microsoft Chief Architect, Razy Ozzie, in 2007 (Microsoft, 2009b). Especially due to the privacy concern, a vendor of S+S may provide their customers options to move non private to the cloud and sensitive data can reside locally without losing any software capability. This seems to bring a win-win situation for the vendors and their customers. However, we keep questioning the licensing issue for S+S. *Is this new model going to apply the licensing principle of on-premise or 'renting' SaaS within their market?* Unfortunately, by now we have not found any clear answers.

Additionally, S+S can also be seen as complimentary to SaaS, which increases the adoption of S+S. As an example, Google with its Gmail has adopted S+S by providing the offline capability through the existence of Google Gears (Clayton, 2009).

## Benefits: Risks of SaaS

Among the three layers listed in the cloud computing section, SaaS currently has the biggest number of vendors and has become the most widely used layer in the market (Ping & Stahlman, 2009). As a new software model that is now being increasingly adopted, SaaS also conveys several benefits and risks to the parties who are involved in the deployment and delivering process of the software.

## Benefits

Based on the SaaS definition and literature review, we have categorized the benefits and risks into three different groups: the vendors, the providers, and the customers (end users). Until now there have been more studies from the business perspective than the scientific perspective that discuss these issues. In fact, many of them talk more about the benefits of SaaS rather than its risks. Additionally, we have collected information from our interview respondents as well.

The benefits listed above are identified from twelve different studies from both the business and scientific perspectives. From Table 4 and additional input from our respondents, the most discussed benefits of adopting SaaS are described as follows:

- *SaaS enables rapid deployment cycles.* The SaaS architecture model that supports multi-tenancy, where one configuration code base supports multiple different users, allows for easy and rapid deployment (including: development, integration, implementation, and upgrades process) (Accenture, 2008c; Dym, 2009; Geisman, 2008; Hoogvliet, 2008; Kaplan, 2009; Mahowald, 2009; Merchant & Geisman, 2006; Pring et al., 2007; Rowell, 2009; Sääksjärvi et al., 2005).
- *SaaS provides access to the most recent software version.* In SaaS, maintaining the application becomes the responsibility of the vendors. The maintenance includes bug fixes and updates where the vendors install the patches or releases in their data centers. As a result, the users are assured that they all are always using the same and up-to-date version of the application (Accenture, 2008c; Choudary, 2007; Geisman, 2008; Hoogvliet, 2008; Kaplan, 2009; Lassila, 2006; Mahowald, 2009; Merchant & Geisman, 2006; Rowell, 2009; Sääksjärvi et al., 2005).
- *SaaS offers payment flexibility.* The 'pay-as-you-go' pricing model that comes with SaaS has brought flexibility to customers. This model allows them to scale their usage up and down and to pay for it as a recurring fee (Geisman, 2008; Kaplan, 2009; Mahowald, 2009; Merchant and Geisman, 2006; Pring et al., 2007; Rowell, 2009; Sääksjärvi et al., 2005).
- *SaaS application is accessible anytime and from anyplace via any devices with an Internet connection.* Because SaaS applications are Web based and installed in data centers (not in the users' site), the users can utilize them from wherever and whenever they want to. They only need to ensure that they have an Internet connection to access the applications from their devices such as desktop, notebook, PDA, or mobile phone (Accenture, 2008c; Dym, 2009; Hoogvliet, 2008; Lassila, 2006; Rowell, 2009; Sääksjärvi et al., 2005).
- *SaaS requires lower up-front investment for infrastructure (hardware) and operational management (maintenance staff).* Customers are not required to invest in hardware or additional IT staff in order to use SaaS applications. It is now the responsibility of the vendors to do so. And for an exchange, customers need to pay a recurring fee for using the application including some additional benefits such as maintenance and support (Accenture, 2008c; Choudary; 2007; Kaplan; 2009; Mahowald, 2009; Pring et al., 2007; Sääksjärvi et al., 2005).
- *More predictable cash flows (in-out cash).* In most cases, the costs of using applications, which is paid by customers, are relatively the same in every period. The

Table 4. The SaaS benefits matrix.

| Benefits | Accenture (2008c) | Choudhary (2007) | Dym (2009) | Geisman (2008) | Hoogvliet (2008) | Kaplan (2009) | Lassila (2006) | Mahowald (2009) | Merchant and Geisman (2006) | Pring et al. (2007) | Rowell (2009) | Sääksjärvi et al. (2005) | Occurrence |
|---|---|---|---|---|---|---|---|---|---|---|---|---|---|
| SaaS enables rapid deployment cycles (development, integration, implementation, and upgrades) | x | - | x | x | x | x | - | x | x | x | x | x | 10 |
| SaaS allows to focus on more strategic projects (core competency) | - | - | x | - | x | x | x | - | - | - | - | x | 5 |
| SaaS provides access with the most recent version of the application | x | x | - | x | x | x | x | x | x | - | x | x | 10 |
| SaaS application is accessible from anytime & anyplace via any devices with internet connection | x | - | x | - | x | - | x | - | - | - | x | x | 6 |
| SaaS requires lower up investment for infrastructure (hardware) & operational management (maintenance staff) | x | x | - | - | - | x | - | x | - | x | - | x | 6 |
| SaaS offers less risky investment | x | - | - | x | x | x | - | - | - | - | - | - | 4 |
| When it is succeeded in integrating into SaaS offering, the bargaining power/competitive differentiation of companies will be increased | - | x | x | - | - | - | x | - | - | - | - | x | 4 |
| More predictable cash flows (in-out cash) | - | x | x | - | x | - | x | - | - | - | x | x | 6 |
| SaaS offers payment flexibility | - | - | - | x | - | x | - | x | x | x | x | x | 7 |
| SaaS users and/or companies can achieve greater profits | x | x | - | - | - | - | - | - | - | - | - | - | 2 |
| Lower initial TCO | - | - | x | - | x | - | - | x | - | x | - | x | 5 |
| Lower production & distribution cost | - | - | x | - | - | - | x | - | - | - | - | x | 3 |
| Lower upgrade & maintenance cost | - | - | - | - | - | - | x | - | x | x | - | x | 4 |
| Lower switching cost | - | - | - | - | - | - | - | - | x | x | - | - | 2 |
| Effective low cost marketing | - | - | x | - | - | - | - | - | - | - | - | - | 1 |
| Increased total available market | - | - | x | - | - | - | - | - | x | - | x | - | 3 |
| Quicker & easier to market | - | - | x | - | x | - | - | - | - | - | - | - | 2 |
| Expands the potential customer base | - | - | - | - | - | - | x | - | x | - | - | - | 2 |
| Improved customer relationship | - | - | x | - | x | - | - | - | x | - | x | - | 4 |
| Faster time to value | - | - | - | x | x | - | - | - | x | - | - | - | 3 |

*continued on following page*

## Software as a Service and the Pricing Strategy for Vendors

*Table 4. Continued*

| | | | | | | | | | | |
|---|---|---|---|---|---|---|---|---|---|---|
| Shortened sales-cycle | - | - | - | - | - | - | x | - | x | 2 |
| Feasible to develop new functionalities/applications for supplementary use | - | x | - | - | - | x | - | - | - | 2 |
| SaaS makes it easier and/or less costly to get the required technical expertise | - | - | - | - | - | - | x | - | x | 2 |
| Ability to switch across providers | x | - | - | - | - | - | - | - | - | 1 |
| Freedom to choose (or better software) | - | x | - | - | - | - | - | - | - | 1 |
| Shared responsibility for support infrastructure | - | - | x | - | - | - | - | - | - | 1 |
| Encourage more standard IT | - | - | - | - | x | - | - | - | - | 1 |
| Simplify sharing system/information | - | - | - | - | x | - | - | - | - | 1 |
| More stable security | - | - | x | - | - | - | - | - | - | 1 |
| Number of benefits issued in each article | 6 | 13 | 5 | 12 | 7 | 10 | 8 | 6 | 7 | 14 |

*- = not applied  x = applied*

same situation applies to SaaS companies (vendors and/or providers), which receive relatively the same revenues. This fact conveys a better overview and more predictable cash flow both for customers and SaaS companies (Choudary, 2007; Dym, 2009; Hoogvliet, 2008; Lassila, 2006; Rowell, 2009; Sääksjärvi et al., 2005).

- *SaaS users and vendors can achieve greater profits.* Customers who make use of SaaS software on their business process can achieve a Return on Investment (ROI) faster since SaaS is easier and quicker to implement than on-premise. Next to that, by having the lower cost of investment, their ROI will automatically be greater as well *(note: for the same net profits compared to on-premise)*. The same situation applies to the SaaS vendors who expend lower costs for the deployment process of their software, which includes the development, the integration, the implementation, and the upgrade processes (Accenture, 2008c; Choudary, 2007).

- *SaaS expands the potential customer base.* Because SaaS applications have been designed as Web-based applications, it makes it easier for the vendors to track and analyze how customers use their applications. For example, by understanding the features that are used more or less by their customers. Such information becomes a vital asset for the vendors to improve their software value in the near future (Lassila, 2006; Sääksjärvi et al., 2005).

The following benefits might be less important for the vendors, providers, or customers. Moreover, they have not been mentioned in any of the literature above. However, we assume that they need to get more attention as well.

- *SaaS helps green environment.* Compared to on-premise software, the multi-tenancy model of SaaS has significantly reduced the number of physical infrastructures needed to support the software, both in data centers and on the customer site. The less infrasctructure is used, the less polution that is created. Besides that, all the contracts, factuurs, and other legal agreements, which were previously printed on paper, are now accessible online and can be more easily analysed. These facts has shown us how SaaS can support the green environment to reduce the effects of global warming. So far, only one book by Schulz (2009) has been found that addresses this issue.

- *SaaS minimizes software piracy.* The transformation from analog to digital in the early 1980s in line with the proliferation of computers and the Internet have made it much easier to make and distribute digital copies. For on-premise software, customers receive the installation file, perhaps on CDs or any other format before they can make use of the software. Since they have access to the software file, it allows 'smart' people to track the code to find ways to illegally duplicate it without paying anything, or even produce and resell their own version of the software. This software piracy has created a huge loss in the software industry. During the five year period (1994 – 1998), the loss has already approached $60 billion (Moores & Dhillon, 2000). Since SaaS applications are installed in data centers where customers do not have direct access to the source code for the software, it can certainly minimize the current piracy activities in the market. So far, only studies like Moore and Mahmoud (2009) discuss this particular benefit.

Table 5 describes the benefits received by vendors, providers, and end users.

## Risks

In reality, the implementation of SaaS has also created some risks or challenges for vendors, providers, and also end users. Therefore, by using a similar approach as above, we have listed several risks or challenges in adopting SaaS in Table 6.

Up until now, there is less literature that discusses the risks of adopting the SaaS model compared with its benefits. Hence, we have included eight additional studies to be further analyzed as listed in Table 6. From the table, and input from several of our respondents, the following risks are commonly discussed:

- *Security and privacy.* Many customers, especially the enterprises, are still vigilant to put their data off premise, which means that they will lose direct control of their data. They are still questioning the privacy and the procedures provided by SaaS vendors to cope with security vulnerabilities such as hardware failure, power failure, or disasters like fire and flood (Caldwell & Eid, 2007; Hayes, 2008; Lin et al., 2009; Merchant & Geisman, 2006; Pring et al., 2007; Sääksjärvi et al., 2005; Tarzey et al., 2007; Wang, 2009).
- *Reliability and performance.* Depending on the technical solution used by the vendor or the provider to deliver their SaaS solutions, the reliability and performance issues often become challenges. For example, the compatibility on different operating systems and Web browsers, and the latency resulted from the hardware used in the infrastructure (Hayes, 2008; Lassila, 2006; Lin et al., 2009; Sääksjärvi et al., 2005).
- *Integration among PaaS platforms and other SaaS applications (interoperability).* Since the SaaS model architecture is a complex network, and users do not have access to the software code, it makes the

*Table 5. The benefits of SaaS for vendors, providers, and end Users*

| Benefits | SaaS vendors | SaaS providers | End Users |
|---|---|---|---|
| SaaS enables rapid deployment cycles (development, integration, implementation, and upgrades) | x | x | x |
| SaaS allows to focus on more strategic projects (core competency) | x | 0 | x |
| SaaS provides access with the most recent version of the application | 0 | 0 | x |
| SaaS application is accessible from anytime & any-place via any devices with internet connection | 0 | 0 | x |
| SaaS requires lower up investment for infrastructure (hardware) & operational management (maintenance staff) | x | x | x |
| SaaS offers less risky investment / failed of deployment | 0 | 0 | x |
| When it is succeeded in integrating into SaaS offering, the bargaining power/competitive differentiation of companies will be increased | x | x | x |
| More predictable cash flows (in-out cash) | x | x | x |
| SaaS offers payment flexibility, which is preferred by the customers | 0 | 0 | x |
| SaaS users and/or companies can achieve greater ROI (because of lower investments) | x | x | x |
| Lower initial Total Cost of Ownership (TCO) | x | x | x |
| Lower production & distribution cost | x | x | 0 |
| Lower upgrade & maintenance cost | x | x | x (free in most cases) |
| Lower switching cost | x | 0 | x |
| Effective low cost marketing | x | 0 | 0 |
| Increased total available market | x | 0 | 0 |
| Quicker & easier to market | x | 0 | 0 |
| Expand the potential customer base | x | x | 0 |
| Improved customer relationship | x | 0 | 0 |
| Faster time to value | 0 | 0 | x |
| Shortened sales-cycle | x | 0 | 0 |
| Feasible to develop new functionalities/ applications for supplementary use | x | 0 | 0 |
| SaaS makes it easier and/or less costly to get the required technical expertise | 0 | 0 | x |
| Ability to switch across providers | x | 0 | 0 |
| Freedom to choose (or better software) | 0 | 0 | x |
| Encourage more standard IT | x | x | 0 |
| Simplify sharing system/ information | x | 0 | x |

*continued on following page*

*Table 5. Continued*

| Benefits | SaaS vendors | SaaS providers | End Users |
|---|---|---|---|
| More stable security | x | x | x |
| SaaS helps green environment | x | x | x |
| SaaS minimizes software piracy | x | 0 | 0 |
| **Number of benefit issues for different roles** | **23** | **12** | **19** |

*0 = not applied    x = applied*

*Note:* When a SaaS vendor uses its own infrastructure to deliver its solutions (do not outsource to third party), then the listed benefits for SaaS providers on the table above will also apply to this vendor. SaaS vendors can also make use of all the listed benefits to help them identify the value propositions for their solutions.

integration between the platforms and other SaaS software much slower and more difficult (Caldwell & Eid, 2007; Hayes, 2008; Lin et al., 2009; Pring et al., 2007; Wang, 2009).

- *Less tailoring options available for customers.* In a one-to-many SaaS model, what users see is what they get. Even though there are still SaaS vendors that allow some customization of data fields and reports, the software code is not accessible to end users (Caldwell & Eid, 2007; Hoch et al., 2001; Merchant & Geisman, 2006; Pring et al., 2007; Sääksjärvi et al., 2005).
- *Required special technical skills for implementing and integrating SaaS applications (difficult to manage complex network).* A SaaS model with the multi-tenancy as its key is a complex architecture. Although there is only one single code base for sharing the resources across multi users, SaaS vendors must put a tremendous amount of effort into guaranteeing that the users' data is differentiable and will not mixed up with other user's data. And for users who demand additional integration for particular SaaS software with their legacy system or with some other SaaS software, they will certainly need the help from special skilled experts, which also means extra costs for these users (Caldwell & Eid, 2007; Lassila, 2006; Sääksjärvi et al., 2005). In the world of SaaS, these skilled experts or companies are more known as *'integrator'* or *'enabler'*. More information about these particular roles can be found in the SaaS Ecosystem section.
- *High initial investment for SaaS vendors or providers.* In order to deliver SaaS software, SaaS companies must have the supported infrastructure ready. Therefore, for SaaS providers and the vendors that make use of their own infrastructure, high initial investment in this infrastructure is inevitable (Lassila, 2006; Sääksjärvi et al., 2005).
- *The break-even point will be reached longer (after few months- years) compared to on-premise model with its up-front paid perpetual license.* Customers do not buy SaaS software. Instead, they rent it. The fee for renting SaaS software is much less compared to the initial payment the customers must pay for on-premise software. Hence, for SaaS vendors, they have less revenue in the first few periods until they have reached the break-even point (BEP). When we look at this from the customer's perspective, we see a different situation. Even though there is a higher up-front cost, it seems after a certain period of time that on-premise software might be cheaper than the SaaS subscription model. But, this

Table 6. The SaaS risks matrix

| Risks / challenges | Hayes (2008) | Lassila (2006) | Lin et al. (2009) | Merchant and Geisman (2006) | Pring et al. (2007) | Sääksjärvi et al. (2005) | Tarzey et al. (2007) | Wang (2009) | Occurrence |
|---|---|---|---|---|---|---|---|---|---|
| Security & privacy | x | - | x | x | x | x | - | x | 6 |
| Reliability & performance | x | x | x | - | - | x | - | - | 4 |
| Data migration | - | - | x | - | - | - | - | - | 1 |
| Service Level Agreements (SLA) | - | - | - | - | - | - | x | x | 2 |
| Integration among PaaS platforms and other SaaS applications (interoperability) | x | - | x | - | x | - | - | x | 4 |
| For software vendor, the financial burden of hardware & software investment, that previously have been the burden of their customers (on-premise) | - | - | - | - | - | - | x | - | 1 |
| Managing the expectation of sales staff & resellers | - | - | - | - | - | - | x | - | 1 |
| For vendors who also offering on-premise, it is important to have the alignment of their on-premise & SaaS pricing | - | - | - | - | - | - | x | x | 2 |
| The break-even point will be reached longer (after few months- years) compared to on-premise with its front-up paid perpetual license | - | - | - | - | - | - | x | - | 1 |
| SaaS companies should offer multiple pricing options to suit different clients | - | - | - | - | - | - | x | - | 1 |
| Billing not as 'utility-based' as customers think. In fact, customers still have to fully pay for some services that they have used in less volume or even not at all. | - | - | - | - | x | - | - | - | 1 |
| Less tailoring available for the customers | - | - | - | x | x | x | - | - | 3 |
| Lack of competitive differentiation among users of the same SaaS application | - | - | - | - | x | - | - | - | 1 |
| For many enterprise requirement, SaaS functionalities will be too simplistic & crude | - | - | - | - | x | - | - | - | 1 |
| The customization of SaaS applications typically incurs extra costs | - | x | - | - | - | x | - | - | 2 |
| A number of 'hidden & unexpected' costs emerging | - | - | - | - | x | - | - | x | 2 |
| Weakened IT management control | - | - | - | - | x | - | - | - | 1 |
| High initial investment for SaaS vendors and/or providers | - | x | - | - | - | x | - | - | 2 |
| The customer is typically bound with a long-term contract (switching costs) in exchange for the lower price. | - | - | - | x | - | - | - | - | 1 |
| Longer-term TCO uncertainties | - | - | - | - | x | - | - | - | 1 |
| Reduce the application turnover when moving to SaaS model (service fees instead of license & consultation fees) | - | x | - | - | - | x | - | - | 2 |
| Requires commitment to a more frequent release/upgrade cycle | - | x | - | - | - | x | - | - | 2 |
| Required special technical skills for implementing & integrating the SaaS applications / difficult to manage complex network | - | x | - | - | x | x | - | - | 3 |
| Number of risks issued in each article | 3 | 6 | 5 | 3 | 10 | 8 | 6 | 6 | |

- = not applied  x = applied

is not going to happen. By referring to the research of Mahowald (2003), Sysmans (2006) from SIIA clearly states that *"when people resources and cost of upgrades are correctly taken in consideration, this break-even point may never be realized."*

In addition, we have also distinguished the risks listed in Table 6 into different groups as follows.

Again, when a SaaS vendor makes use of their own infrastructure to deliver its software, then the listed risks for SaaS providers in Table 7 will also apply for this specific vendor. According to Lin et al. (2009), the risks that currently need significant improvement for SaaS vendors and providers are security, performance, and interoperability issues since they do not currently perform well in these areas.

## SaaS Ecosystem

A software ecosystem has been increasingly used in the last decade. It was first described by Moore (1993) to express the interdependence of the players in business networks. Kittlaus and Clough (2009) also defined software ecosystem in their book as an informal network of (legally independent) units that have a positive influence on the economic success of a software product and benefit from it. A software ecosystem not only consists of the players, but also of more informal networks that emerge around important themes of the software industry such as leading products, system platforms or new technology trends (Economides & Katsamakas, 2005; Kittlaus & Clough, 2009). Jansen et al. (2009) provides the most comprehensive definition of software ecosystem (SECO), which is *"a set of actors functioning as a unit and interacting with a shared market for software and services, together with the relationships among them"*. The scope of a software ecosystem can be quite broad, depending on how we look at it.

In order to get a better overview of how SaaS software is being delivered and the monetary cycle that goes with it, this research has addressed the scope of a software ecosystem for five different key players. These players are the big players that are mostly involved in any situations for delivering SaaS software to the customers. The five players are listed below.

1. *SaaS Vendor.* This is the company who develops and builds the SaaS software. To deliver their software they could use their own infrastructure independently or host it for a selected partner.
2. *SaaS Provider.* This is the external company that provides the vendor with the infrastructure to run and deliver the SaaS software to the customers (Carraro & Chong, 2006). SaaS provider is also mostly resposible for all of the maintenance activities.
3. *End users.* These are the customers who rent and make use of the SaaS software. They could be personal users or enterprises.
4. *Reseller.* Like on-premise software, for better market penetration, SaaS vendors also sell their software using an indirect model via extended sales channels better known as 'resellers' (Murfin, 2005; Schultze, 2007). Having a partnership with a reseller can help the vendor to lower their sales cost, minimize the infrastructure investment, and lower domain expertise requirements. The disadvantages for the vendor would be less profit since they must share it with the reseller and less control over their customer experience (Murfin, 2005). However, according to Chapman (2008), 61% of SaaS vendors in the current market use direct sales with their customers and only 11% make use of a reseller. Depending on the agreement between the reseller and SaaS vendor, a reseller might also sell the software under their own label (better known as *'white labeling'*). So,

*Table 7. The risks of SaaS for vendors, providers, and end users*

| Risks / Challenges | SaaS vendors | SaaS providers | End Users |
|---|---|---|---|
| Security and privacy | x | x | x |
| Reliability and performance | x | x | x |
| Data migration | x | 0 | x |
| Service Level Agreements (SLA) | x | x | x |
| Integration among PaaS platforms and other SaaS applications (interoperability) | x | x | x |
| For SaaS companies, the financial burden of hardware & software investment, that previously have been the burden of their customers (on-premise) | x | x | 0 |
| Managing the expectation of sales staff & resellers | x | x | 0 |
| For vendors who also offering on-premise, it is important to have the alignment of their on-premise & SaaS pricing | x | 0 | 0 |
| The break-even point will be reached longer (after few months- years) compared to on-premise with its front-up paid perpetual license | x | x | 0 |
| SaaS companies should offer multiple pricing options to suit different clients | x | 0 | 0 |
| Billing not as 'utility-based' as customer thinks. In fact, customers still have to fully pay for some services that they have used in less volume or even not at all. | 0 | 0 | x |
| Less tailoring & integration options available for the customers | 0 | 0 | x |
| Lack of competitive differentiation among users of the same SaaS application | 0 | 0 | x |
| For many enterprise requirements, SaaS functionalities will be too simplistic & crude | 0 | 0 | x |
| The customization of SaaS applications typically incurs extra costs | x | x | x |
| A number of 'hidden & unexpected' costs emerging | 0 | 0 | x |
| Weakened IT management control | 0 | 0 | x |
| High initial investment for SaaS vendors and/or providers | x | x | 0 |
| The customer is typically bound with a long-term contract (switching costs) in exchange for the lower price. | 0 | 0 | x |
| Longer-term TCO uncertainties | 0 | 0 | x |
| Reduces the application turnover when moving to SaaS model (service fees instead of license & consultation fees) | x | x | 0 |
| Requires commitment to a more frequent release/upgrade cycle | x | x | 0 |
| Required special technical skills for implementing & integrating the SaaS applications / difficult to manage complex network | x | x | x |
| **Number of risks issued for different roles** | **15** | **12** | **15** |

*0 = not applied   x = applied*

from the customer's point of view, it seems that the reseller is the actual vendor.

5. *Integrator.* This is for customers, especially the enterprises that mostly need additional efforts like customizing the features of SaaS software to meet their business requirements, or integrating it with other software (either on-premise or SaaS) that have been installed on their system. Since this company enables SaaS software to be used by the customers as they want to, this company is also known as 'enabler'. The tasks mentioned are not easy and require special skill to perform it (Lassila, 2006; Pring et al., 2007; Sääksjärvi et al., 2005). The integrator is mostly a third company who has specialized in this area or a SaaS provider that has skill and experience to do so. Some of them even provide consulting services to their customers. On the contrary, in some cases, several IT consulting companies even become the integrator itself.

From this study and the information collected from our respondents we have identified six different scenarios for a SaaS ecosystem. These scenarios depict the current situation in the market for how SaaS software is being delivered from the vendor until it becomes accessible to its customers and also the monetary cycle that goes with it. The first two scenarios (A and B) are the simplest scenarios for SaaS without involving any other parties. These two scenarios are also aligned with the base definition as described earlier. Regarding sales and marketing purposes, it is possible that a vendor applies multiple scenarios.

## Scenario A

The vendor develops, hosts, and operates SaaS software independently by making use of their own infrastructure (data center) and IT staff. The software is being directly delivered and sold to the customer. An example of this scenario would be Bluedog Ltd., which offers its workbench SaaS under the support of their own data center. This workbench is a bundled of SaaS applications, which is comprised of Content Management System, Customer Relationship Management, Project and Activity Management, Workflow and Document Management System, Learning Management System, and a few others (Bluedog, 2009).

## Scenario B

In this scenario, the vendor develops its software and hosts it for the selected partner (provider). The provider supports them with the infrastructure to enable delivering the software as service to customers of the vendor. In most cases, all the maintenance activities such as backup and performance issues become the responsibility of this provider. As an exchange, the vendor pays them for those services. Two examples of this scenario are: 1) *Exact Online* from Exact Software B.V., which is a Dutch SaaS accounting software that is hosted by KPN CyberCenters (KPN, 2009a).

*Figure 3. Scenario A for SaaS ecosystem*

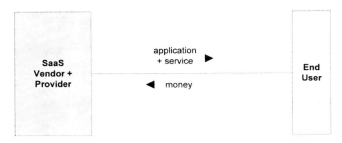

*Figure 4. Scenario B for SaaS ecosystem*

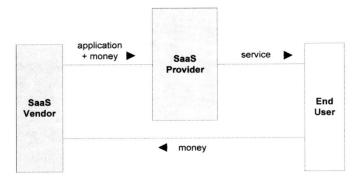

Currently this software is only used for the Belgium-Netherland-Luxemburg (Benelux) market. 2) Intacct, as one of the SaaS leaders for financial management and accounting applications, also cooperates with IBM data center to operate and deliver their software (Intacct, 2009a).

## Scenario C

In this scenario, besides performing its main tasks which are supplying and maintaining the infrastructure to ensure that software from the vendor runs properly, the SaaS provider also acts as a 'reseller'. So, the SaaS provider sells the vendor's software to their own customers, either white labeling or not. In this situation, the provider and vendor share the percentage of the revenue received from customers. The amount of this percentage and the cost burden for the vendors can be different, depending on the agreement between both parties. Here, KPN CyberCenters as the partner of Exact Software, is also being a reseller of *Exact Online*. From KPN (2009b), it is clearly shown that KPN does not white label the Exact Online.

In some cases, the provider could also act as 'integrator' by incorporating SaaS software from different vendors that are hosted on the infrastructure and then sell them to their own customers, either separately or in packages.

## Scenario D

This scenario illustrates the existence of a completely separate reseller to be a partner of a SaaS vendor. By referring to scenario A and B, we distinguish this scenario into two categories, D1 and D2. Scenario D1 means that the vendor who

*Figure 5. Scenario C for SaaS ecosystem*

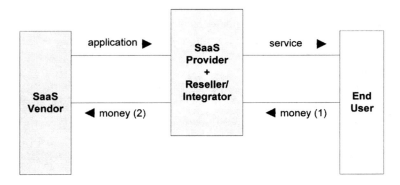

owns and controls their infrastructure also has a partnership with 'reseller' to sell their software (in this situation, the reseller sells under their own label). And for scenario D2 where the vendor has already selected a host partner (SaaS provider), it also partners with third party resellers for their sales and marketing purpose. Of course, it is important to have a clear agreement between the vendor and resellers. A bad agreement may result in loss for both companies and end their relationship in the future. A good example of this situation is the partnership between NetSuite, one of the biggest global SaaS vendors in the accounting field, who had an issue with one of its resellers named Skyytek in the beginning of 2009 (Kanaracus, 2009).

## Scenario E

This scenario exemplifies the existence of an 'enabler', or also known as 'integrator', who is hired by end users. The end users are typically the enterprises who want to employ particular SaaS software and integrate and/or modify it to meet their business requirement (Kittlaus & Clough, 2009). Here, integrating means incorporating the employed SaaS software with the legacy system of the enterprise like existing on-premise and other SaaS software. Modifying means customizing the functionalities of that software with the requisite functionalities from the enterprise. In order to enable this integration and customization process, the integrator needs direct information from both vendor and provider. The required information is depicted as dotted lines in Figure 7. Due to the confidentiality of intellectual properties, additional agreements among the vendor, the provider, and the integrator must be applied as well. Referring to both scenario A and B, we also distinguish this scenario into two categories, E1 and E2. An example of this integrator is a company called 'e-zest'. More information on what they offer to their clients can be found on E-zest (2009).

The integrator's revenue is typically a one-time fee given by the end users as an exchange for the

*Figure 6. Scenario D for SaaS ecosystem*

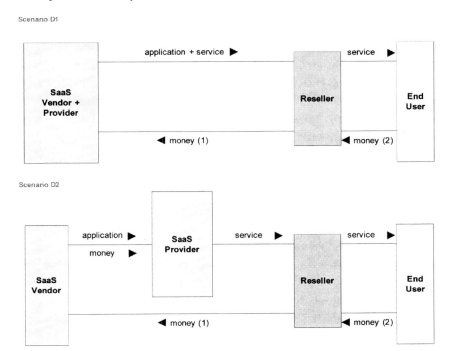

*Figure 7. Scenario E for SaaS ecosystem*

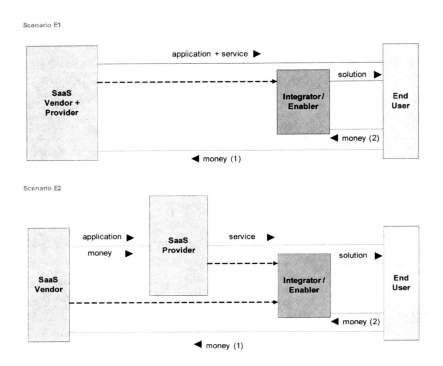

service for modifying the SaaS software to meet the end user's desires.

## Scenario F

Similar to scenario E, this scenario focuses more on the existence of an integrator to help the end users implement SaaS software that meets their business requirements. The only difference is that the integrator already has partnerships with the vendors. Therefore, in order to solve the end users' problem, the integrator will embrace the best suited SaaS solution from his business partners. In most cases, the integrator also gives consulting service to their end users. For example, Accenture works together with Salesforce.com in order to implement CRM solutions for its clients (Salesforce, 2009a; Stiffler & Bois, 2008). In this scenario, the integrator does not only receive the payment from the end users, but also a commission fee from the vendors. Regarding the location of infrastructure used to deliver the SaaS software, we have also divided this scenario into two categories, F1 and F2.

The aforementioned scenarios (A, B, C, D1 and D2, E1 and E2, F1 and F2) only illustrate the relationships among certain players: SaaS vendor, SaaS provider, reseller, integrator, and end users in the current SaaS market. However, it is possible that more and more parties are involved in the near future.

## The Pricing Strategy Guideline Framework for SaaS Vendors

The number of SaaS (Software as a Service) solutions in the market has been rapidly growing in the last few years. There have been several successful pioneer SaaS vendors in many different sectors such as *Salesforce* in the CRM sector, *NetSuite* in the Finance sector, and *Workday* in the HRM sector. These big vendors have good experience

*Figure 8. Scenario F for SaaS ecosystem*

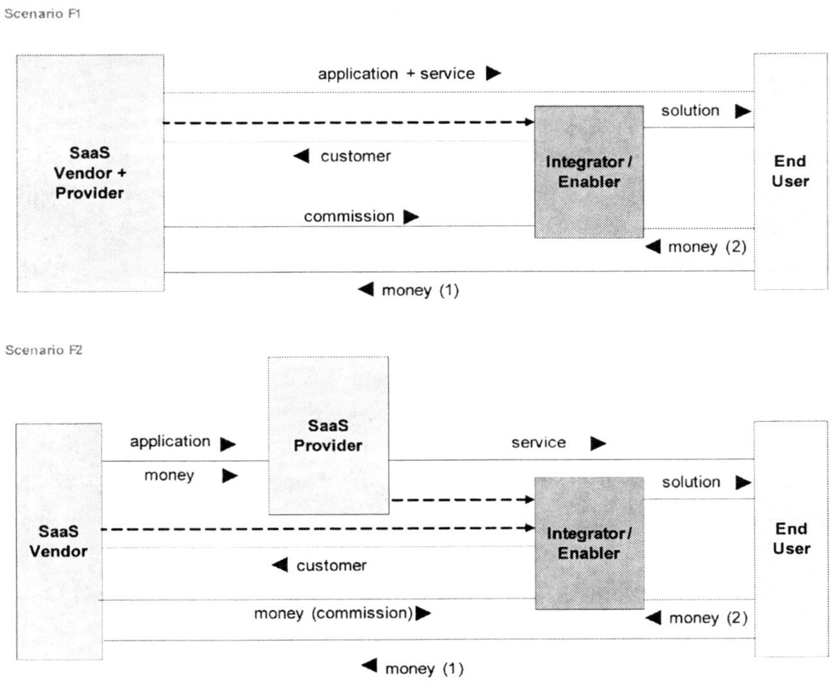

specializing in the kind of software they want to offer, to which markets, and how they are going to sell them. Unfortunately, this is not the case for the small and medium vendors, particularly the startup SaaS vendors. Many of them have emerged and then disappeared again in the market. This is often due to a lack of maturity in their pricing strategy. Of course, there is no single pricing strategy that fits best for all vendors because there are always other factors that also need to be taken into account.

One of the most important benefits of adopting SaaS software is the availability of *having more flexible payment methods* (Geisman, 2008; Kaplan, 2009; Mahowald, 2009; Merchant & Geisman, 2006; Pring et al., 2007; Rowell, 2009; Sääksjärvi et al., 2005). Customers can be offered the finest payment possibilities while still maximizing the vendors' profits, which is not a trivial task to accomplish. There have been various pricing issues faced by SaaS vendors in the market like low sales cycles, low win rates, chaotic pricing, and difficulty to enter new market segment (Jones, 2008; Geisman & Nelson, 2008). Therefore, in order to minimize these issues, the availability and integration of a good pricing strategy as part of the overall corporate strategy for SaaS vendors becomes essential.

SaaS vendors must establish their pricing strategy before conveying their software solution to the market. They must be able to understand all of the potential revenue sources, deployment and distribution costs of their solution, its ability to sell them at prices which will allow for maximum profit, and ensure that the strategy is sustainable over time so that the vendor can continuously achieve its objectives (Kittlaus & Cough, 2009). Hogan and Nagle (2005) also stated that *"a comprehensive pricing strategy is comprised of multiple layers creating a foundation for price setting that minimizes erosion and maximizes profits over time"*. These multiple layers become the core of the *Pricing Strategy Guideline Framework* (PSGF). Until now, there have not been any frameworks in the SaaS research area that provide the vendors with such a pricing guideline. Hence, we have constructed this framework as a solution that *provides SaaS vendors with a guideline to ensure that all the fundamental pricing elements are included in their pricing strategy*. It is intended for small-to-medium SaaS vendors, particularly the startup vendors that tend to have less experience in pricing their SaaS solutions.

The framework has made use of the 'Strategic Pricing Pyramid' by Hogan and Nagle (2005; 2006), 'Five Forces Model' by Porter (1985; 2001), 'Value creation in E-Business' by Amit and Zott (2001), several essential SaaS elements from different literature, and some additional input from our interview respondents. The PSGF framework is shown in Figure 9 and consists of five layers: *Value Creation, Price Structure, Price and Value Communication, Price Policy,* and *Price Level*.

## Value Creation

To run the business successfully, every company – including SaaS vendors, in all different sectors must have a solid corporate strategy. As one of the business units, the Sales and Marketing unit is the company's key source for obtaining more customers who are willing to buy their products or services. In other words, these are the customers that would bring them revenues. For SaaS vendors, the revenue is generated by charging the customers for using their solutions. Therefore, it is important that Sales, Marketing, and Pricing are collaborated closely.

*Figure 9. The pricing strategy guideline framework*

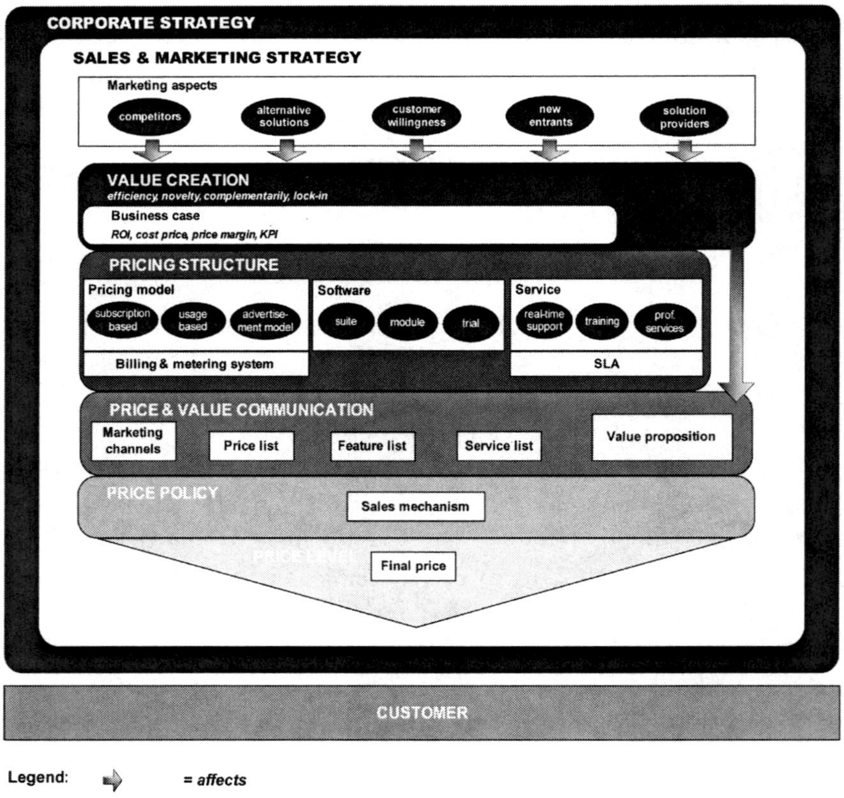

Software vendors always expect to set prices that can capture the value of their solution and that can also maximize their profits (Hogan & Nagle, 2005). Therefore, it is crucial for the vendors to have a good understanding of how much value their SaaS solution can generate for their customers. And this is indeed not a simple task. The vendor must be aware of different aspects that influence the value creation itself, both from internal and/or external aspects. The former aspect can be identified from the corporate strategy of the vendors. And for the latter aspect, the vendors must not neglect the willingness of their customers, the existence of alternative solutions, the situation of their competitors, the new entrants and also the power of the providers (Porter, 1985; 2001). We also see that the corporate strategy is interconnected with the five forces in Porter's model. Within the SaaS market, we also notice that *the bargaining power of the provider* becomes less relevant when the vendors operate their software on their own infrastructure or when the providers themselves do not offer additional value compared to their competitors. For example, a provider who also acts as a reseller and integrator has more value compared to a provider that only provides and maintains the infrastructure. In practice, SaaS vendors hardly take into account all five competitive sources; instead they are more likely to focus on several or even only on one source.

The value delivered from a SaaS solution can also be identified from the four key drivers of e-business and the linkages among them: *efficiency, complementarities, lock-in, and novelty* (Amit & Zott, 2001). In the context of SaaS, efficiency may include lowering the transaction costs, streamlining the inventory management, and simplifying and speeding up the transaction processes. Complementarities are presented when SaaS vendors offer a bundle of solutions together. The value of lock-in can be achieved because the vendors have more control over their customers. Customers have no direct access to the physical data and are dependent on the format and the offered data export features. Finally, SaaS can also innovate the ways in which vendors and customers do their business, called novelty. We notice that the *efficiency* and *lock-in* are the most valued drivers applied to SaaS.

**Business Case**

For SaaS, a business case could be defined as a financial estimation that compares the associated costs of deploying a SaaS solution to the quantified economic benefits or value to be derived from it within a certain period of time (Kittlaus & Clough, 2009). It might include the cost price, price margin, and Return on Investment (ROI) and some other financial measurements (KPIs) such as the number of new customers that sign up every month. We believe that having a concrete business case becomes essential for SaaS vendors before pricing their solution. However, it is true that the business case does not guarantee that the vendor will achieve exactly the same numbers as what has been estimated, but it will certainly give a better financial control for them.

**Price Structure**

The next step after defining the value creation and the business case, is determining the pricing structure for the SaaS solution. It covers how the vendors are going to charge their customers; what kind of metrics they must use – especially when they adopt the usage-based model - and how the metrics and billing processes are going to be measured and tracked; how the software is distributed to the customers; and finally what kinds of services that they want to provide their customers that also is included in the Service Level Agreement (SLA).

Pricing Model

Since SaaS is a new hype in the market, many people are still confused about pricing and licensing for SaaS. Many consider these two terms to be basically the same. In the case of SaaS, the customers do not buy a license but 'rent' the software. Hence, only the term pricing is applied. The pricing models that are adopted by SaaS vendors to charge their customers are mentioned similarly in different literature (Kaplan, 2009; Kittlaus & Clough, 2009; Sessions, 2006, Tarzey et al., 2007). We see that the pricing for SaaS is basically shaped from one or a combination of the following three charging alternatives: *(1) subscription-based* customers are charged the same fixed-price for every month, quarter, or year for an independent usage (Postmus et al., 2008). *(2) usage-based* where customers are charged based on their usage volume from several measurable metrics. This is the reason why SaaS is also typically paid for on a "pay-per-use" basis. 'Price metrics' are variables that drive different prices for a single SaaS solution (Kittlaus & Clough, 2009). The used metrics might diverge for different solutions, e.g. amount of data transferred, the time spent by customers in using the software, number of registered users accessing the software, number of completed transactions while using the solution (Ferrante, 2006; Kittlaus & Clough, 2009). The last two are the most common metrics used in the SaaS world. Nevertheless, there are still many other specific metrics that could also be useful for SaaS vendors, which can be found in Dunham (2009). *(3) advertisement model* where customers pay no cost for using the solution while the vendors earn the revenue from advertisements of third parties that are embedded on their Web pages. This model has rarely been applied by SaaS vendors. The best example is the Google or Yahoo search engine and Google Apps which includes Gmail as well (Kittlaus & Clough, 2009; Sessions, 2006).

The most favorable charging alternative in the current SaaS market is the subscription combined with the usage-based (number of users) model (Herbert et al, 2007). Our interviews show that almost 70% of the respondents have applied a subscription-based model with the combination of number of users as the chosen metric. The remaining respondents are still charging their customers using only a fixed fee subscription model. However, it is important to note that it is possible that the SaaS vendors will come up with new charging alternatives in the near future.

## Billing and Metering System

When formulating the pricing strategy, SaaS vendors also need to keep in mind the billing and metering mechanism for the usage of their solutions. This is a vital element to be able to run their business properly. They must ensure that they have accurately charged the single account of the customers based on the usage amount of the solution, either for current, previous or future transactions. A poor billing and metering system may cause the vendors to lose their customers, as the customers might feel cheated by the vendors. This billing and metering system becomes important for the vendors to track their customers' usage and might also be useful for them to expand the potential of their customer base, which is useful input for their future marketing strategy. Hence, having an accurate metering and billing system that can handle the mechanism and report overall usage is crucial. This system must be integrated with the vendor's solutions to allow the unification of all financial data and system operations (Blokdijk, 2008; Green et al., 2004). The billing and metering system can be built in-house or outsourced to third parties. The in-house systems tend to provide a better alignment of customer resource usage to billing because of their ability to connect to the SaaS platform at more points and collect more information, but it is a lengthy development project in most cases. Outsourcing allows resources to be redirected to the core platform and provides a quicker time to the market, but reaches fewer points than an in-house system (Chaudhuri, 2008; Progress Software, 2008).

## Software

The choice of how a SaaS solution is being distributed to customers is also an important element to be considered by the vendors. Generally, in the case of SaaS, the solution set can be delivered separately or in combination of three types: *(1) suite* or bundled software of several solutions provides more value than separated solutions, which makes it more preferable for the customers (Bakos & Brynjolfsson, 1999; Dewan & Freimer, 2003) and generates increased sale-revenue for the vendors. Nowadays, bundling seems to be the most popular method for the vendors to sell their software. *(2) module*, which is a term used when the vendors sell their solutions separately. In most cases, vendors will only sell them in modules when they are assured that their specific solution has a high-level value and strong competitive advantage. They believe, even though the price of their SaaS solution is relatively more expensive, customers still will be willing to use the software because they cannot find any better alternatives. *(3) trial*. SaaS vendors also offer a no-cost fee for using their solutions. These solutions are usually limited by their features and/or limited by a certain period of time *(trial version)* such as 14 days or 30 days. The goal of offering this free solution is to encourage them to use the the paid solution.

In reality, SaaS vendors can employ the combination between 'bundled' and 'module' options. For example, vendor 'XYZ' sells their SaaS solution in three different packages (A, B, and C) and in total, the solution has ten different modules (1, 2, 3…10). Package A contains module 1, 2, 3, 4 while Package B contains module 2, 3, 5, 6, and 7; and Package C contains module 1, 2, 3, …,7 and 9. Of course, the more value delivered by the modules contained in a package, the more expensive the package would be. Figure 10 provides a better view of this package separation. Vendor 'XYZ' offers freedom for their customers to choose which packages or modules are best suited with their business needs. So, when customer only needs module 2 and 5 to support their business, they can decide to buy the two modules separately or choose package B only. In most cases, the accumulated cost of the choices becomes the main consideration of the customer's decision. Please note that it is possible that some modules are not included in any packages. For example, in Figure 10, when a customer needs package A with the additional of module 8 or 10 since these two modules are not included in any packages. The information about the chosen approach to distribute the software must be clearly determined before it can be communicated to the marketing department.

## Services

By adopting the SaaS model, the way the SaaS vendors provide their services to customers can also be improved (Hoogvliet, 2008). These services are typically included in the fee paid by the users. The purpose of providing these services is to help the users with any problems that may occur when they use SaaS software. From our study and additional input from the respondents, we also classify these SaaS services into three types: *(1) real-time support*. Besides only providing the

*Figure 10. Software packaging example*

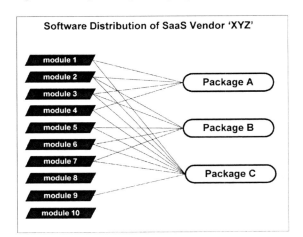

users with a FAQ, the common contact details of the vendors such as phone and fax number, email address, or with a form that needs to be filled out when they want to ask about their problems, SaaS vendors can also provide real-time support such as 'live chat'. This live chat can be built on the vendor's Web site or by taking advantage of the APIs of online VoIP communication tools like Skype. Intacct, one of the leading SaaS vendors in the accounting field, presents a good example by providing different possibilities to their customers to contact its support desk (Intacct, 2009). *(2) training.* In most cases, depending on the features of the SaaS solution, a SaaS vendor might also offer different kinds of training for end users or administrators. The purpose of having this training is to inform the users about the different features of the software and how they can benefit from these features. The training for the administrator users usually covers more aspects and needs more time to complete when compared to the training for the end users. Training can also be performed by providing online video. *(3) professional services.* For customers who want to get help from the vendors such as consultancy service to optimally use the solution to meet their business requirements or technical assistance to extend the functionalities from the vendors, some SaaS vendors also offer professional services. In most cases, these services will convey additional costs to those customers or they are already included in the software package already. A good example of this service is the Customer Success Manager (CSM) provided by Salesforce (Salesforce.com, 2009b), which is responsible for analyzing the users' behavior and provide added value to their users by helping them to be more effective and efficient in using the software according to their business needs. (Salesforce.com, 2009). CSM people visit and have meetings with their customers on a regular basis.

## Service Level Agreements (SLA)

One of the key motivators for moving to SaaS is the reduction of business risk (Tarzey et al., 2007). This is identified by the service level agreement (SLA), which is a contract between a vendor and its users which specifies the level of service that is expected during its period. They can specify the response times for routine and ad hoc queries and response time for problem resolution such as network down and machine failure (Hoch et al., 2001). The aforementioned types of services are mostly covered within the SLA as well. SLAs can be very general or extremely detailed, including the steps to take in the event of a failure. For example, if the problem persists after 30 minutes, a supervisor will be notified; and when the problems stay after one hour, the account representative will be contacted.

## Price and Value Communication

When the pricing structure including all its elements has been determined, it becomes the responsibility of the marketing people of the SaaS vendors to ensure that their prices will be acceptable to the customers whilst maximizing the profit gained. In order to make the price acceptable, it is important that the vendors are able to clearly inform their customers what the value propositions of their solution are. A 'value proposition' is a business statement that summarizes why customers should buy the vendor's SaaS solution. In addition, the vendors may also utilize the matrices of SaaS benefits - risks (Table 4 and 6) to identify the value proposition of their solutions.

When the vendors cannot communicate this value clearly, or when the customers do not completely understand the delivered values of the solution, then customers tend to think that the price offered by the vendors is too high (Hogan & Nagle, 2005). Hence, price and value have a strong relationship. This is also depicted in our framework by the arrow that points from the 'value creation' level to the 'price & value communication' level.

From the marketing research, the marketing people must already have an estimation of *how much the customers are willing to pay for such a solution*. This is a simple question, but a difficult and complex one to answer. Furthermore, they also need excellent approaches to convince their customers to buy their solution. The information related to the features of the solution, the supported services, and information regarding the vendor's background are necessary (Chau, 1995). They need to explicitly list all the included features and supported services in the different prices of their SaaS solution. This information is interrelated with some elements of the pricing structure layer as well. For example, the 'pricing model' is directly related to the listed different prices whereas the 'software' element is linked to the features list and the 'services' element is associated to the services list of this third layer. Examples of the software features, both from the technical and/or non-technical issues are: the software functionalities, the availability of being integrated with existing legacy software, the compatibility and performance issues in different Web browsers, the user friendliness, and the popularity of the software itself. Furthermore, the services like the availability of technical support (24/7 or only during working hours), and end user training (direct or via online demos) should be clearly mentioned as well. The different listed prices, features, and services could also be part of the value proposition of the vendors.

In the case of SaaS, choosing the right marketing channels becomes another important element to consider. Of course, the old marketing channels such as brochures, advertisements on conferences or in magazines, are still used by several vendors. But, we do not see them to be the best solution anymore. New marketing channels like online advertisements via Google Ads and new real-time promotion are becoming more effective approaches to market the vendors' solution. The real-time promotion provides up-to-date information to the users such as the new prices, discounts, or features of SaaS software online. Salesforce.com is a good example who offers this marketing channel for their customers.

For the vendors, it is always important to be honest with their customers about the limitations of their SaaS solution and also for other services that are excluded from the listed prices because this will most definitely affect their reputation (Hoxmeier, 2000). Other additional sources that are necessary for their customers are the references and the past experiences of using that software from other customers. This information is commonly found in a case study provided by SaaS vendors.

## Price Policy

Up until this level, the value propositions, the charging mechanism including the chosen metrics and billing system, and the alignment between prices and value have been defined. It is at the next level when SaaS vendors need to sell and to close the best deal with their customers. In fact, even though the SaaS vendors have published their price list on their website, the prices can still be altered - especially for customers with a large number of users (for example, 500 users or more). These customers may expect and demand additional benefits from a vendor such as free set-up costs and purchasing discount (cheaper price). In order to close the sales and get a signed agreement appropriately including those additional benefits, the expertise of sales people of SaaS vendors is required. These are the people who have good knowledge and understanding of what they are selling, to whom they are selling to, and how they will sell it (Aston, 2009). To accomplish their tasks, they require information from the marketing department as well. This is the reason why a solid relationship between sales and marketing people in an organization is vital.

This relationship can also be seen in the framework from the position where the pricing and value communication level is above and directly connected with the pricing policy.

To sell the SaaS solution, the sales people must use their best approach or *sales mechanism* to reach the customers. There could be different approaches for different customers. However, there must be pricing regulation, also known as *pricing policy* from the vendors to control what may and may not be done by their sales people – e.g. a policy of never giving discounts bigger than 25%. This is important since the sales people always have the tendency to give more and more discounts to customers in order to boost their sales numbers (Hogan & Nagle, 2005). Pricing policy is also needed when the vendors have to deal with pricing in order to sell their solutions. In reality, the vendors may also create price discrimination for different types of customers, depending on the level of their usage (Postmus et al., 2008) and also duration of the contract.

## Price Level

When sales people have reached an agreement with a customer, then the price margin of the SaaS solution of that specific customer can be set and followed by calculating its final unit price. A unit price is basically the sales price of a SaaS solution for a specific customer. For SaaS vendors who do not give additional benefits at all to their customers, because it is part of their pricing policy, for example, the listed prices on their websites are the final unit price and apply to all customers.

Before granting the customer access to their SaaS solution, it is significant for the vendors to clearly outline all information on the contract regarding the prices and any possible additional benefits, the minimum period of the contract and the Service Level Agreements.

## CONCLUSION

Software as a Service (SaaS), one of the hot topics in the software industry today, has been increasingly adopted by many vendors and customers. This chapter provides an in depth overview of the state-of-the-art of SaaS, which covers several aspects of SaaS such as the history, the definition, the key characteristics, the differences between SaaS and on-premise software, its relationship with other technologies, the benefits and risks of SaaS, and also the different scenarios in the SaaS ecosystem. In reality, there is no single pricing strategy that fits all different organizations. From the literature review and several pricing related theories, and also the input and feedback from the interview respondents and experts, we have successfully collected different fundamental pricing elements of SaaS. These elements are the value drivers (efficiency, complementarities, novelty, lock-in) of the SaaS solution, which are affected by the market aspects (external factor) and the corporate strategy (internal factor); a business case containing some relevant KPIs like the number of new customers that sign up every month and ROI; the different SaaS pricing models that include the chosen metrics and billing-metering system to support it; the distribution of the SaaS solution: as a suite and/or in a separate module, also include the option to offer trial version; what kinds of services that are provided to the users on different prices; the creation of SLA; the chosen marketing channels to reach the users; and finally, the sales mechanism that is directly influenced by the applied pricing policies. The aforementioned factors are the elements of the *Pricing Strategy Guideline Framework*, which has been formulated to help SaaS vendors to advance their pricing strategy.

## FUTURE RESEARCH DIRECTIONS

Since the PSGF framework contains several fundamental pricing elements, it opens the door for other studies of each of those elements. For example, research on how to choose the best billing and metering system, research about the methodology of selecting the most profitable metrics of SaaS or the methodology of performing a good SaaS business case, and research about the available marketing channels from different SaaS vendors including the new innovative *Customer Success Manager* from Salesforce.com and many others.

Because the PSGF framework elements are primary, we believe that our framework might apply not only for SaaS, but also for other Cloud-based services such as PaaS and IaaS. Consequently, the incentive to apply the usage-based model for the PaaS or IaaS vendors becomes bigger, which of course implies more complex metrics and the required billing-metering system to support it.

The six scenarios of SaaS have been designed by taking into account only the five big players, which are: vendors, providers, resellers, integrators, and end users. In reality, within a SaaS ecosystem, smaller parties are involved as well. As an example, there are organizations with excellent experience in laws and other legal issues, who specialize in constructing the SLA for different vendors or providers. Therefore, it is definitely possible to extend these six scenarios.

We found that there are only a few studies that focus on the benefits of adopting SaaS such as 'helping the green environment' and 'reducing the software piracy'. Hence, we think that it could be an interesting topic for other researchers to delve deeply into these topics.

The presence of Software + Services (S+S), which allows the users to use the Web-based software even when they have no Internet connection, has arrived about one year ago (Microsoft, 2009). We feel that it is a new interesting topic, which also needs further research. Some aspects to be considered are the security concerns, the synchronization process, and the pricing-licensing issue compared to SaaS.

## REFERENCES

Accenture (2008a). *Software as a Service (SaaS) Practice*. Retrieved February 26, 2009, from *Accenture Collaboration* database.

Accenture (2008b). Traditional licensing vs. SaaS (Software as a Service) vs. ASP (Application Service Provider). Retrieved January 28, 2009, from *Accenture Collaboration* database.

Accenture (2008c). SaaS Overview. Retrieved November 26, 2008, from *Accenture Collaboration* database.

Amit, R., & Zott, C. (2001). Value creation in e-business. *Strategic Management Journal, 22*(6-7), 493–520. doi:10.1002/smj.187

Armbrust, M., Fox, A., Griffith, R., Joseph, A. D., Katz, R. H., Konwinski, A., et al. (2009). *Above the Clouds: A Berkeley View of Cloud Computing*. Retrieved February 19, 2009, from http://www.eecs.berkeley.edu/Pubs/TechRpts/2009/EECS-2009-28.pdf

Aston, G. (2009). *What Is A Sales Strategy?* Retrieved June 28, 2009, from http://www.freshbusinessthinking.com/business_advice.php?AID=2668&Title=What+Is+A+Sales+Strategy

Bakos, Y., & Brynjolfsson, E. (1999). Bundling information goods: Pricing, profits, and efficiency. *Management Science, 45*(12), 1613–1630. doi:10.1287/mnsc.45.12.1613

Barnatt, C. (2008). Explaining Web 2.0. *ExplainingComputers.com*. Retrieved October 9, 2008, from http://www.youtube.com/watch?v=7BAXvFdMBWw

Blokdijk, G. (2008). *SaaS 100 Success Secrets - How Companies Successfully Buy, Manage, Host and Deliver Software as a Service (SaaS)*. Brisbane, Australia: Emereo Pty Ltd.

Bluedog (2009). *Workbench "Always on the Job!"*. Retrieved March 17, 2009, from http://www.bluedog.net/wb/93/wo/90kVL48YV9Ww5MvGIx7Naw/0.5

Buyya, R., Yeo, C. S., Venugopal, S., Broberg, J., & Brandic, I. (2009, June). Cloud computing and emerging IT platforms: Vision, hype, and reality for delivering computing as the 5th utility. *Future Generation Computer Systems, 25*(6), 599–616. doi:10.1016/j.future.2008.12.001

Caldwell, F., & Eid, T. (2007). Is SaaS safe for financial governance, risk, and compliance solutions. *Gartner*, Article G00150913. Retrieved November 13, 2008, from Gartner database.

Carr, N. (2006). *Here comes HaaS*. Retrieved October 18, 2008, from http://www.roughtype.com/archives/2006/03/here_comes_haas.php

Carraro, G., & Chong, F. (2006). *Architecture strategies for catching the long tail*. Microsoft. Retrieved October 4, 2008, from http://msdn.microsoft.com/en-us/library/aa479069.aspx

Chapman, M. R. (2008). *The 2008 Softletter SaaS Report*. Retrieved March 25, 2009, from *Softletter.com* database.

Chau, P. Y. K. (1995). Factors used in the selection of packaged software in small businesses: Views of owners and managers. *Information & Management, 29*(2), 71–78. doi:10.1016/0378-7206(95)00016-P

Chaudhuri, S. (2008). SaaS pricing and metering. Retrieved May 20, 2009, from http://sumanchaudhuri.wordpress.com/2008/02/28/saas-pricing-and-metering/

Chou, D. C., & Chou, A. Y. (2008). Software as a service (SaaS) as an outsourcing model: An economic analysis. *Proceeding of the 39th Southwest Decision Science Institute Conference,* Houston, Texas. Retrieved October 3, 2008, from http://www.swdsi.org/swdsi08/paper/SWDSI%20Proceedings%20Paper%20S469.pdf

Choudary, V. (2007). Software as a service: Implications for investment in software development. *Proceedings of the 40th Hawaii International Conference on System Sciences,* Waikoloa, Hawaii.

Clayton, S. (2009). *Google talks software plus services*. Retrieved August 16, 2009, from http://blogs.msdn.com/stevecla01/archive/2009/01/28/google-talks-software-plus-services.aspx

Creeger, M. (2009). Cloud computing: An overview. *ACM Queue; Tomorrow's Computing Today, 7*(5), 1–5.

Danielson, K. (2007). *Confusing SaaS with Web 2.0*. Retrieved August 28, 2009, from http://www.ebizq.net/blogs/saasweek/2008/05/confusing_saas_with_web_20/

Dewan, R. M., & Freimer, M. L. (2003). Consumers prefer bundled add-ins. *Journal of Management Information Systems, 20*(2), 99–111.

Dunham, M. (2009). SaaS metrics – Saasonomics-101. Retrieved April 1, 2009, from http://blog.sciodev.com/2009/02/10/saas-metrics-saasonomics-101/

Dym, R. (2009). Why software as a service? Helping our customers reduce costs and increase revenue. *OpSource - The Business of Web Operations* database. Retrieved March 22, 2009, from http://www.opsource.net/saas/wp_why_saas.pdf

E-zest. (2009). E-zest: Company Overview. Retrieved August 1, 2009, from http://www.e-zest.net/technical_consulting.html

Economides, N., & Katsamakas, E. (2005). Linux vs. Windows: a comparison of application and platform innovation incentives for open source and proprietary software platforms. New York University, Law and Economics Research Paper No. 05-21. Retrieved August, 12, 2009, from http://papers.ssrn.com/sol3/papers.cfm?abstract_id=822894

Ferrante, D. (2006). Software licensing models: What's out there? *IT Professional, 8*(6), 24–29. doi:10.1109/MITP.2006.147

Geisman, J. (2008). *SaaS pricing for prosperity* [Webinar]. *MarketShare, Inc.* Retrieved April 16, 2009, from http://www.softwarepricing.com/readingroom/Content/OpSource%20Pricing%20for%20Prosperity%20Webinar.pdf

Geisman, J., & Nelson, B. (2008). *Pricing a SaaS product – what's the big deal?* [Webinar]. *PragmaticMarketing.com*. Retrieved June 01, 2009, from http://www.pragmaticmarketing.com/resources/archived-webinars/pricing-a-saas-product-2013-what2019s-the-big-deal

Giurata, P. (2008). SaaS, *PaaS, cloud computing, on-demand - what do they all mean?* Retrieved October 15, 2008, from http://www.catalystresources.com/saas-blog/saas_paas_cloud_computing_on_demand_what_do_they_all_mean/

Google. (2009). Run your Web apps on Google's infrastructure. Retrieved December 5, 2008, from http://code.google.com/appengine/

Green, K., Klemenhagen, B., & Hoch, F. (2004). Software as a service: Changing the paradigm in the software industry. *Software & Information Industry Association (SIIA) and TripleTree*. Retrieved January 23, 2009, from http://www.pangeafoundation.org/pdf/saas_aug04.pdf

Greschler, D., & Mangan, T. (2002). Networking lessons in delivering 'software as a service' – Part II. *International Journal of Network Management*, *12*(5), 317–321. doi:10.1002/nem.446

Hayes, B. (2008). Cloud computing. *Communications of the ACM*, *51*(7), 9–11. doi:10.1145/1364782.1364786

Herbert, L., Ross, C.F., Thresher, A., & Bartolomey, F. (2007). The components of SaaS pricing and negotiations. *Forrester Research*, Document # 43581. Retrieved February 20, 2009, from Forrester database.

Hoch, F., Kerr, M., & Griffith, A. (2001). *Software as a service: strategic backgrounder*. Retrieved October 3, 2008, from http://www.siia.net/estore/ssb-01.pdf

Hogan, J., & Nagle, T. (2005). What is strategic pricing? *Strategic Pricing Group Insights*. Retrieved March 15, 2009, from http://www.monitor.com/Portals/0/MonitorContent/documents/Monitor_What_Is_Strategic_Pricing.pdf

Hogan, J., & Nagle, T. (2006). *The Strategy and Tactics of Pricing: A Guide to Growing More Profitably* (4th ed.). Upper Saddle River, NJ: Prentice Hall.

Hoogvliet, M. T. (2008). SaaS interface design. Retrieved April 28, 2009, from http://one3rd.nl/whitepaper_maartenhoogvliet_saasinterfacedesign.pdf

Hoxmeier, J. A. (2000). Software preannouncements and their impact on customers' perceptions and vendor reputation. *Journal of Management Information Systems*, *17*(1), 115–139.

IBM (2008). Making sense of SOA and today's IT innovations. *IBM Smart SOA solutions*. Retrieved November 20, 2008, from IBM database.

Intacct (2009). Intacct Support. Retrieved on August 10, 2009, from http://us.intacct.com/services/support.php

Jansen, S., Brinkkemper, S., & Finkelstein, A. (2009). Business network management as a survival strategy: A tale of two software ecosystem. *Proceedings the International Workshop on Software Ecosystem (IWESECO 2009)*, Virginia, USA.

Jones, D. (2008). The five qualities of good software pricing. *Forrester Research*, Document # 46218. Retrieved February 20, 2009, from Forrester database.

Kanaracus, C. (2009). NetSuite, ex-reseller locked in ugly legal battle. *Techworld.* Retrieved August 1, 2009, from http://www.techworld.com.au/article/304143/netsuite_ex-reseller_locked_ugly_legal_battle

Kaplan, J. M. (2009). SaaS movement accelerating. *Business Technology Trends & Impacts Advisory Service Executive Update, 8*(22).

Kittlaus, H., & Clough, P. N. (2009). *Software Product Management and Pricing: Key Success Factors for Software Organizations.* Berlin: Springer.

Knol, P., Spruit, M., & Scheper, W. (2008). Web 2.0 revealed - business model innovation through social computing. *Proceedings of the Seventh AIS SIGeBIZ Workshop on e-business (WeB 2008),* Paris, France.

KPN. (2009a). *Exact Online.* Retrieved on February 26, 2009, from http://zakelijk.kpn.com/web/file?uuid=d386cdd0-b010-4bd3-b66d-2e183510674a&owner=3c096f1f-64ae-471a-a72b-161373c8b70b

KPN. (2009b). *KPN Online Boekhouden.* Retrieved on February 26, 2009, from http://zakelijk.kpn.com/business/meer-diensten/softwareonline/alle-software-online/online-boekhouden.htm

Langholz, G., Kandel, A., & Mott, J. L. (1998). *Foundations of digital logic design.* Singapore: Word Scientific Publishing Co. Pte. Ltd.

Lassila, A. (2006). Taking a service-oriented perspective on software business: How to move from product business to online service business. *IADIS International Journal on WWW/Internet, 4*(1), 70-82.

Lenk, A., Klems, M., Nimis, J., Tai, S., & Sandholm, T. What's inside the cloud? An architectural map of the cloud landscape. *ICSE Workshop on Software Engineering Challenges of Cloud Computing 2009,* Vancouver, Canada.

Lin, G., Fu, D., Zhu, J., & Dasmalchi, G. (2009). Cloud computing: IT as a service. *IT Professional, 11*(2), 10–13. doi:10.1109/MITP.2009.22

Luit Infotech. (2008). *Difference between the ASP model and the SaaS model.* Retrieved November 12, 2008, from http://www.luitinfotech.com/downloads/saas-asp-difference.pdf

Mahowald, R. P. (2003). Do service providers deliver value and reduce enterprise costs? *IDC.* Retrieved November 5, 2008, from IDC database.

Mahowald, R. P. (2009). SaaS, PaaS, and cloud: choices for success. *Proceedings of the IDC SaaS Summit Spring 2009,* Document # 217935. Retrieved April 5, 2009, from IDC database.

Maximilien, E. M., Ranabahu, A., & Gomadam, K. (2008). An online platform for Web APIs and service mashups. *IEEE Internet Computing, 12*(5), 32–43. doi:10.1109/MIC.2008.92

Merchant, N., & Geisman, J. (2006). Solving the puzzle: pricing, licensing and business models. *Rubicon Consulting, Inc & Market Share, Inc.* Retrieved March 5, 2009, from http://rubiconconsulting.com/downloads/whitepapers/Rubicon_Solving-Puzzle.pdf

Microsoft. (2009a). *Cloud computing infrastructure.* Retrieved August 12, 2009, from http://www.microsoft.com/virtualization/en/us/cloud-computing.aspx

Microsoft. (2009b). Software + services. *The Architecture Journal, 13.* Retrieved August 15, 2009, from *MSDN Architecture Center* database via http://msdn.microsoft.com/en-us/architecture/bb906058.aspx

Moore, B., & Mahmoud, Q. H. (2009). A service broker and business model for SaaS applications. *Proceedings of the IEEE/ACS International Conference on Computer Systems and Application,* Rabat, Morocco.

Moore, J. F. (1993). Predators and prey: A new ecology of competition. *Harvard Business Review*, *71*(3), 75–86.

Moores, T., & Dhillon, G. (2000). Software piracy: A view from Hong Kong. *Communications of the ACM, 43*(12), 88–93. doi:10.1145/355112.355129

Murfin, J. (2005). Business case for entering SaaS. Retrieved August 3, 2009, from *Microsoft Solution for Hosted Messaging and Collaboration* database.

Obannon, I. (2009). What Web 2.0, SaaS and cloud computing mean for tax & accounting professionals. Retrieved August 28, 2009, from http://getanewbrowser.com/2006/05/web-20-soa-saas/

Papazoglou, M. (2003). Service-oriented computing: Concepts, characteristics and directions. *Proceedings of the Fourth International Conference on Web Information System Engineering (WISE '03)*, Rome, Italy.

Porter, M. E. (1985). *Competitive Advantage: Creating and Sustaining Superior Performance*. New York: Free Press.

Porter, M. E. (2001). Strategy and the Internet. *Harvard Business Review*. Retrieved March 26, 2009, from http://www.soum.com.br/sonaomuda/ecommerce/Arquivos/Artigos/Strategy_and_the_internet.pdf

Postmus, D., Wijngaard, J., & Wortmann, H. (2009). An economic model to compare the profitability of pay-per-use and fixed-fee licensing. *Information and Software Technology, 51*(3), 581–588. doi:10.1016/j.infsof.2008.08.004

Pring, B., Desisto, R.P., & Bona, A. (2007). The cost and benefits of SaaS vs. on-premise deployment. *Gartner*, Article G00151171. Retrieved November 13, 2008, from Gartner database.

Pring, B., & Stahlman, M. (2009). Sizing the cloud: The world's first comprehensive cloud computing services forecast [Webinar]. *Gartner*. Retrieved March 19, 2009, from Gartner database.

ProgressSoftware. (2008). SaaS billing & metering. *Progress Software Corporation*. Retrieved May 20, 2009, from http://communities.progress.com/pcom/servlet/JiveServlet/download/12057-2-11235/SaaS_BillingMetrics.pdf

Rayner, N. (2008). The impact of SOA and SaaS on financial systems. *Gartner*, Article G00157191. Retrieved November 13, 2008, from Gartner database.

Rowell, J. (2009). A step-by-step guide to starting up SaaS operations. *OpSource – The SaaS Delivery Experts* database. Retrieved March 22, 2009, from http://www.opsource.net/saas/starting_up_saas_operations.pdf

Rust, R. T., & Kannan, P. K. (2003). E-service: A new paradigm for business in the electronic environment. *Communications of the ACM, 46*(6), 36–42. doi:10.1145/777313.777336

Sääksjärvi, M., Lassila, A., & Nordstrom, H. (2005). Evaluating the software as a service business model: From CPU time-sharing to online innovation sharing. *Proceedings of the IADIS International Conference e-Society 2005*, Qawra, Malta.

Salesforce.com. (2008). Developer Force: Apex Code. Retrieved December 3, 2008, from http://wiki.developerforce.com/index.php/Apex

Salesforce.com. (2009a). Contact Partner. Retrieved on August 5, 2009, from http://www.salesforce.com/partners/opportunities/consulting-partners/profiles/a0x30000000CfKu.jsp

Salesforce.com. (2009b). Job detail of Customer Success Manager. Retrieved on August 20, 2009, from http://www.salesforce.com/company/careers/locations/a0800000000Ab42AAC/a017000000AIvVJ.jsp

Schultze, A. (2007). *Channel Excellence*. USA: Axel Schultze.

Schulz, G. (2009). *The Green and Virtual Data Center*. Boca Raton, FL: Auerbach Publications. doi:10.1201/9781420086676

Sessions, R. (2006). Software as a service: Another perspective. *The ObjectWatch Newsletter, 52*. Retrieved October 3, 2008, from http://www.objectwatch.com/newsletters/ObjectWatchNewsletter052.pdf

Stiffler, D., & Bois, R. (2008). Consulting in the cloud: The emerging SaaS consulting, product development, and outsourcing ecosystem (Excerpt). *AMR Research, Inc*. Retrieved December 2, 2008, from http://www.accenture.com/NR/rdonlyres/D2BC2C93-38BB-41AB-9F3E-48B26C91D26C/0/ConsultingintheCloudEmerginSaaSConsulting.pdf

Sysmans, J. (2006). Software as a service: A comprehensive look at the total cost of ownership of software applications. *Software & Information Industry Association (SIIA)* database. Retrieved April 3, 2009, from via http://www.siia.net/estore/ssb-01.pdf

Takeda, H., Veerkamp, P. J., Tomiyama, T., & Yoshikawa, H. (1990). Modeling design processes. *AI Magazine, 11*(4), 37–48.

Tarzey, B., Longbottom, C., & Stimson, T. (2007). On-premise to on-demand: The software as a service opportunity for independent software vendors. *Quocirca Insight Report*. Retrieved March 22, 2009, from Quaocirca database.

Tukey, J. W. (1958). The teaching of concrete mathematics. *The American Mathematical Monthly, 65*(1), 1–9. doi:10.2307/2310294

Vaquero, L. M., Rodero-Merino, L., Caceres, J., & Linder, M. (2009). A break in the clouds: Towards a cloud definition. *ACM SIGCOMM Computer Communication Review, 39*(1), 50–55. doi:10.1145/1496091.1496100

Vouk, M. A. (2008). Cloud computing – issues, research and implementations. *Proceedings of the 30th International Conference on Information Technology Interfaces*, Dubrovnik, Croatia.

Walsh, K. R. (2003). Analyzing the application ASP concept: Technologies, economies, and strategies. *Communications of the ACM, 46*(8), 103–107. doi:10.1145/859670.859677

Wang, R. (2009). Shape your apps strategy to reflect new SaaS licensing and pricing trends. *Forrester Research*, Document #46602. Retrieved February 28, 2009, from Forrester database.

Wilson, D., & Basiliere, P. (2008). The flavors of e-procurement extend beyond software as a service and on-premise. *Gartner*, Article G00146768. Retrieved November 13, 2008, from Gartner database.

Wordreference.com. (2009). *WordNet 2.0*. Princeton, NJ: Princeton University. Retrieved July 5, 2009, from http://www.wordreference.com/definition/software

York, J. (2008). Contrasting software-as-a-service and enterprise software business models. Retrieved January 27, 2009, from http://chaotic-flow.com/2008/09/02/contrasting-software-as-a-service-and-enterprise-software-business-models-2/

Youseff, L., Butrico, M., & Da Silva, D. (2008). Toward a unified ontology of cloud computing. *Proceedings on Grid Computing Environments Workshop (GCE) 2008*, Austin, Texas.

Zhen, J. (2008). Defining SaaS, PaaS, IaaS, etc. Retrieved November 15, 2008, from http://cloudfeed.net/2008/06/03/defining-saas-paas-iaas-etc/

## ADDITIONAL READING

Biddick, M. (2010, January 18). Time for a SaaS strategy. *InformationWeek, 1254*, 27–32.

Braude, E. (2008). Software-as-a-service and offshoring. *International Journal of Business Insights & Transformation, 2*(1), 93–95.

Campbell-Kelly, M. (2009). Historical reflections: The rise, fall, and resurrection of software as a service. *Communications of the ACM, 52*(5), 28–30. doi:10.1145/1506409.1506419

Choudhary, V. (2007). Comparison of software quality under perpetual licensing and software as a service. *Journal of Management Information Systems, 24*(2), 141–165. doi:10.2753/MIS0742-1222240206

Concha, D., Espadas, J., Romero, D., & Molina, A. (2010). The e-HUB evolution: From a custom software architecture to a software-as-a-service implementation. *Computers in Industry, 61*(2), 145–151. doi:10.1016/j.compind.2009.10.010

Cusumano, M. A. (2007). The changing labyrinth of software pricing. *Communications of the ACM, 50*(7), 19–22. doi:10.1145/1272516.1272531

Cusumano, M. A. (2008). The changing software business: Moving from products to services. *Computer, 41*(1), 20–27. doi:10.1109/MC.2008.29

Elfatatry, A. (2007). Dealing with change: Components versus services. *Communications of the ACM, 50*(8), 35–39. doi:10.1145/1278201.1278203

Fan, M., Kumar, S., & Whinston, A. B. (2009). Short-term and long-term competition between providers of shrink-wrap software and software as a service. *European Journal of Operational Research, 196*(2), 661–671. doi:10.1016/j.ejor.2008.04.023

Fox, L. (2009). Integrating SaaS and legacy apps: 5 steps for success. *NetworkWorld Asia, 5*(1), 30.

Greer, M. B. (2009). *Software as a Service Inflection Point: Using Cloud Computing to Achieve Business Agility*. Bloomington, IN: iUniverse.

Hatch, R. (2008). *SaaS Architecture, Adoption and Monetization of SaaS Projects using Best Practice Service Strategy, Service Design, Service Transition, Service Operation and Continual Service Improvement Processes*. Brisbane, Australia: Emereo Pty Ltd.

Hill, S. Jr. (2008). SaaS seems to favor users more than vendors. *Manufacturing Business Technology, 26*(1), 48.

Jaisingh, J., See-To, E. W. K., & Tam, K. Y. (2008). The impact of open source software on the strategic choices of firms developing proprietary software. *Journal of Management Information Systems, 25*(3), 241–275. doi:10.2753/MIS0742-1222250307

Keller, E. (2007). How software application pricing models are likely to change. *Manufacturing Business Technology, 25*(1), 42–43.

Lamont, J. (2010). SaaS: Integration in the cloud. *KM World, 19*(1), 12–22.

Murphy, S., & Samir, W. (2009). 'In the cloud' IT creates new opportunities for network service providers. *Journal of Telecommunications Management, 2*(2), 107–120.

Orr, B. (2007). Microsoft begins its radical shift to software as a service. *ABA Banking Journal, 99*(12), 46–47.

Postmus, D., Wijngaard, J., & Wortmann, H. (2009). An economic model to compare the profitability of pay-per-use and fixed-fee licensing. *Information and Software Technology, 51*(3), 581–588. doi:10.1016/j.infsof.2008.08.004

Poston, R. S., Kettinger, W. J., & Simon, J. C. (2009). Managing the vendor set: Achieving best pricing and quality service in IT outsourcing. *MIS Quarterly Executive, 8*(2), 45–58.

Raghu, T. S., Sinha, R., Vinze, A., & Burton, O. (2009). Willingness to pay in an open source software environment. *Information Systems Research*, *20*(2), 218–236. doi:10.1287/isre.1080.0176

Sainio, L.-M., & Marjakoski, E. (2009). The logic of revenue logic? Strategic and operational levels of pricing in the context of software business. *Technovation*, *29*(5), 368–378. doi:10.1016/j.technovation.2008.10.009

Susarla, A., Barua, A., & Whinston, A. B. (2009). A transaction cost perspective of the 'software-as-a-service' business model. *Journal of Management Information Systems*, *26*(2), 205–240. doi:10.2753/MIS0742-1222260209

Thompson, J. K. (2009). Business intelligence in a SaaS environment. *Business Intelligence Journal*, *14*(4), 50–55.

Turner, M., Budgen, D., & Brereton, P. (2003). Turning software into a service. *Computer*, *36*(10), 38–44. doi:10.1109/MC.2003.1236470

Vorisek, J., & Feuerlicht, G. (2004). Is it the right time for the enterprise to adopt software-as-a-service model? *Information & Management*, *17*(3-4), 18–21.

Waters, B. (2005). Software as a service: A look at the customer benefits. *Journal of Digital Asset Management*, *1*(1), 32–39. doi:10.1057/palgrave.dam.3640007

## KEY TERMS AND DEFINITIONS

**Apex:** It is a strongly-typed programming language that executes on the platform used for Salesforce.com.

**Enabler (or Integrator):** A role of IT specialist in the implementation of integrated solutions and effective support of retail systems. It is also known as Integrator.

**Key Performance Indicator (KPI):** KPI is a performance measurement that is commonly used by an organization to define and to evaluate the success of their product or service in the market.

**Platform as a Service (PaaS):** PaaS is the delivery of a computing platform and solution stack as a service.

**Pricing Strategy Guideline Framework (PSGF):** PSGF is the tentative framework resulted from the conducted research on this report.

**Provider:** In SaaS, the provider refers to the organization which supplies the vendor with the infrastructure to operate the vendor's software. Usually the provider is responsible for the maintenance and support.

**Reseller:** Reseller is an organization that purchases goods or services with the intention of reselling rather than consuming or using them. They are an intermediary between software producers (vendor) and end users.

**SaaS Ecosystem:** A set of actors functioning as a unit and interacting with a shared market for software and services, together with the relationships among them.

# Chapter 11
# The Private Copy Issue:
## Piracy, Copyright and Consumers' Rights

**Pedro Pina**
*Polytechnic Institute of Coimbra, Portugal*

## ABSTRACT

*Digital copyrights involve a combination of technology and law that seek to provide full control of the work by the rightholder. Managing rights over digital copyrighted contents through the use of consumers' technological protection measures may however jeopardize some freedoms that copyright law has traditionally recognized, such as the private copy. In the present chapter, the author describes the conflict between the exclusive right to the exploitation of the work and the private copy issue; how modern copyright obstructs private copying and recent proposals regarding the conciliation between rightholders' and consumers' interests.*

## DIGITAL COPYRIGHT MANAGEMENT AND CONSUMERS CONCERNS

It is today unquestionable that, in a digital environment, technology and law are complementary realities. Nevertheless, sometimes, technology allows uses that may override individuals' rights and protected interests. The relationship between digital copyright and consumers' rights is precisely one of the fields where both the expressed complementarity and conflict can be clearly perceived.

In fact, the emergence of digital technology, principally the Internet, has created the possibility of a free and global flow of informational contents that was reflected in the metaphor of the information highway. Soon, the economic facet of this inter-relational digital structure was revealed and online markets and e-commerce were developed. However, informational cybermarkets present some problems concerning the immaterial and intellectual nature of its products (for example, music, movies, digital books or software).

Immaterial goods are public, non-rivaled and non-excludable goods. In fact, the consumption of an informational good by one person doesn't

DOI: 10.4018/978-1-61692-877-3.ch011

exclude the possibility of consumption by others and, without regulation no one can be excluded from using the good. These characteristics are emphasized in the digital world as the positive externalities created by the free flow of copyrighted content information which may result in a disadvantage for content creators and rightholders. That is the general justification for public legal regulation of the intellectual creations' market where intellectual property law is presented as an instrument that can be used to create scarcity, since it gives the holders the economic exclusive rights of the works' exploitation, excluding others from using it without proper authorization.

Nevertheless, considering the global geographical distribution of Internet users, the differences between intellectual property national legislations and the merely reactive judicial enforcement mechanisms, holders felt that law, by itself, was no longer sufficient to ensure their rights and to protect their economically legitimate interests.

For that reason, the answer to the problems left to rightholders by technology was sought in technology itself: as Clark (1996) stated, the answer to the machine seemed to be in the machine (p. 139). In fact, one of the main characteristics of modern copyright law is the interplay between legal and consumers technological protection measures (TPM), such as steganography, encryption or electronic agents. TPM may not be mere defensive electronic fences. Indeed, digital rights management (DRM) systems based on the combination of different TPM allow rightholders to potentially control all the utilizations of digital copyrighted works, including, *inter alia*, access to contents or even to some personal data of the users.

The idea of a technological perfect control (Lessig, 2006, p. 183) of digital contents made some rightholders think that they could dismiss or to underrate copyright law's protection. Combining TPM and contracts, they unilaterally set the conditions for end users to access protected contents.

As Guibault and Helberger (2005) state,

*"DRM systems [...] create an environment in which various types of use, including copying, are only practically possible in compliance with the terms set by the right holders. Therefore, they usually do not deny access but rather manage access to content by combining technical measures with a payment mechanism. DRM-based business models ensure that consumers pay for actual use of content, and that the content is protected and cannot be accessed by unauthorized users" (p. 9).*

Moreover, normally, consumers of digital works don't have the means or the power to bargain; they can only accept or decline the terms of the end user license agreement that is presented to them on a "take-it-or-leave-it" basis. Code and contract seemed to be a perfect combination to ensure holders' interests and to enable the management of protected works in even stronger terms than copyright law did. In fact, if copyright law traditionally represented a compromise between holders' and collective interests as a means to promote cultural, artistic and scientific creation, this privatization of copyright (McManis, 1999) could make holders escape from free usages that reflected public interests, like the ones recognized in the cases of fair use, in the USA, or the copyright exemptions recognized by European legislations. Nevertheless, reality showed that the use of TPM and DRM did not stop copyright infringements and illegal file-sharing, especially through peer-to-peer (P2P) networks, as circumvention mechanisms were developed.

Because of that, copyright law was recovered and adapted. To fully protect holders' self help and private enforcement systems, copyright law was completed with provisions prohibiting the use and the creation of circumvention mechanisms.

The Digital Millennium Copyright Act, in the USA, and the Directive 2001/29/EC of the European Parliament and of the Council of May

22, 2001 on the harmonization of certain aspects of copyright and related rights in the information society (InfoSoc Directive), in the European Union, were the main legal instruments that punished circumvention acts and the creation or the commercialization of circumvention devices. Both pieces of legislation were enacted to implement the World Intellectual Property Organization (WIPO) Copyright Treaty and the WIPO Performances and Phonograms Treaty. According to Article 11 of the former treaty,

*"Contracting Parties shall provide adequate legal protection and effective legal remedies against the circumvention of effective technological measures that are used by authors in connection with the exercise of their rights under this Treaty or the Berne Convention and that restrict acts, in respect of their works, which are not authorized by the authors concerned or permitted by law".*

As the mentioned provisions came into force,

*"copyright owners have three levels of cumulative protection: the first is the legal protection by copyright. The second level is the technical protection of works through measures protection techniques. The third level is the new legal protection against circumvention of consumer's technological protection measures introduced by the WIPO Treaties"* (Werra, 2001, p. 77)

The use of TPM and DRM is, therefore, legally recognized and inserted inside copyright law, though they regard more the enforcement question than the substance of copyright itself. To distinguish these technological enforcement provisions from substantive copyright, they are often called *paracopyright* (Jaszi, 1988) or *übercopyright* (Helberger & Hugenholtz, 2007). The use of TPM and DRM is largely criticized as the fields of public domain and free uses (that traditionally were considered by copyright law) have diminished. Copyright has expanded and strengthened, but its internal limitations remained the same, as if digitization didn't exist. As a consequence, balance between holders' and users' rights inside copyright law has vanished, which made users seek protection outside the boundaries of copyright law. One of the branches of law that started changing copyright law was consumers' law. In fact, DRM systems raise some concerns on consumers related to their privacy and anonymity, to interoperability, to fairness of contracts, or to the free usages that traditional copyright recognized and that are inhibited by DRM. Amongst the latter concerns, stands the private copy issue.

## PROBLEMS CONCERNING THE DIGITAL PRIVATE COPY

Stéphane P., a French consumer, bought a DVD of David Lynch's movie *Mulholland Drive*. When he wanted to make a copy of the movie in VHS format to watch at his parents' house, he realized that embedded TPM prevented him from doing the referred format-shifting.

Backed by a French consumer organization, Stéphane P. sued the rightholder and argued that the technical device was unlawful as it didn't allow him to exercise his right of private copying. The *Court de Cassation de Paris* declared that the private copy does not constitute a user's right, but a mere exemption to the traditional principle that prohibits the reproduction, in full or in part, of a protected work without the consent of the rightholders. As a consequence, a consumer that has legally acquired a copyrighted work, doesn't have the right to circumvent a TPM to make a private copy and, therefore, he/her has no legal basis for a claim when such devices are embedded in the purchased product. Nevertheless, as it constitutes a statutory exemption, one who makes a private copy cannot be held liable for copyright infringement.

This seminal decision, though it seems to be completely disappointing for consumers interests, alerts rightholders that they cannot just simply ignore the private copy exception and use TPM arbitrarily.

Actually, the complete blocking of any possibilities of making private copies was an impermissible behavior under French copyright law and is not at the disposal of private parties (Helberger, 2005). Indeed, it's up to legislators to establish the limits of the private copy exception. As Helberger (2005) states,

*"the decision of the Court of Appeals confirms, however, once more that the rules on the legal protection of technological measures in copyright law are still in many respects flawed and incomplete. It concludes that it is task of the legislator to bring more light in the complicated relationship between private copying and the usage of technological measures".*

In fact, although current legislations around the world foresee a private copy exception or a fair use defense, they don't clarify sufficiently the meaning or the extent of the exemption.

At the international level, reproducing a copyrighted work without the holders' authorization is considered lawful if it has a non-commercial or non-profitable purpose. Nevertheless, that conclusion can only be achieved if the three-step test rules that were introduced by article 9.2. of the Berne Convention for the Protection of Literary and Artistic Works are respected. According to the identified provision, a private reproduction is only free in certain special cases that don't conflict with a normal exploitation of the work and don't unreasonably prejudice the legitimate interests of the author. This solution is also predicted in article 10 of the WIPO Copyright Treaty, in article 13 of the Agreement on trade-related aspects of intellectual property rights (TRIPS).

Senftleben (2004) points out that, in its origins, the three-step test formula reflected a compromise between the formal and harmonized recognition of the holders' reproduction right and the preservation of existing limitations in different national legislations. The option that was then taken consisted not in enumerating exhaustively a list of existing free uses, but in the formulation of a general clause and abstract criteria that, "due to its openness, […] gains the capacity to encompass a wide range of exceptions and forms a proper basis for the reconciliation of contrary opinions" (p. 51).

In this conception, the three-step test presents some similarities with the general clause of fair use that is provided by the § 107 of the USA Copyright Act, like its flexibility. Under this provision, in determining whether the use made of a work in any particular case is a fair use, the following four factors must be considered:

*"(1) the purpose and character of the use, including whether such use is of a commercial nature or is for nonprofit educational purposes; (2) the nature of the copyrighted work; (3) the amount and substantiality of the portion used in relation to the copyrighted work as a whole; and (4) the effect of the use upon the potential market for or value of the copyrighted work".*

However, with the massive increase in copyright infringement in the digital world, the flexible interpretation of the three-step test tended to be substituted and replaced by a rigid and narrow one.

In fact, in the European Union, the three-step test was introduced by article 5.5. of the InfoSoc Directive, but in a rather curious manner: the test was presented as a restriction to the exhaustive list of limitations of the exclusive rights over the work granted to the holders.

Limitations on exclusive reproduction rights are foreseen in article 5.2. of the Directive in the following cases:

*"(a) in respect of reproductions on paper or any similar medium, effected by the use of any kind of photographic technique or by some other process having similar effects, with the exception of sheet*

*music, provided that the rightholders receive fair compensation; (b) in respect of reproductions on any medium made by a natural person for private use and for ends that are neither directly nor indirectly commercial, on condition that the rightholders receive fair compensation which takes account of the application or non-application of technological measures referred to in Article 6 to the work or subject-matter concerned; (c) in respect of specific acts of reproduction made by publicly accessible libraries, educational establishments or museums, or by archives, which are not for direct or indirect economic or commercial advantage; (d) in respect of ephemeral recordings of works made by broadcasting organisations by means of their own facilities and for their own broadcasts; the preservation of these recordings in official archives may, on the grounds of their exceptional documentary character, be permitted; (e) in respect of reproductions of broadcasts made by social institutions pursuing non-commercial purposes, such as hospitals or prisons, on condition that the rightholders receive fair compensation".*

Additionally, the European legislature provides that Member States may foresee the following limitations on the same reproduction right and also to right of communication to the public of works and right of making available to the public other subject-matter:

*"(a) use for the sole purpose of illustration for teaching or scientific research, as long as the source, including the author's name, is indicated, unless this turns out to be impossible and to the extent justified by the non-commercial purpose to be achieved; (b) uses, for the benefit of people with a disability, which are directly related to the disability and of a non-commercial nature, to the extent required by the specific disability; (c) reproduction by the press, communication to the public or making available of published articles on current economic, political or religious topics or of broadcast works or other subject-matter of the same character, in cases where such use is not expressly reserved, and as long as the source, including the author's name, is indicated, or use of works or other subject-matter in connection with the reporting of current events, to the extent justified by the informatory purpose and as long as the source, including the author's name, is indicated, unless this turns out to be impossible; (d) quotations for purposes such as criticism or review, provided that they relate to a work or other subject-matter which has already been lawfully made available to the public, that, unless this turns out to be impossible, the source, including the author's name, is indicated, and that their use is in accordance with fair practice, and to the extent required by the specific purpose; (e) use for the purposes of public security or to ensure the proper performance or reporting of administrative, parliamentary or judicial proceedings; (f) use of political speeches as well as extracts of public lectures or similar works or subject-matter to the extent justified by the informatory purpose and provided that the source, including the author's name, is indicated, except where this turns out to be impossible; (g) use during religious celebrations or official celebrations organised by a public authority; (h) use of works, such as works of architecture or sculpture, made to be located permanently in public places; (i) incidental inclusion of a work or other subject-matter in other material; (j) use for the purpose of advertising the public exhibition or sale of artistic works, to the extent necessary to promote the event, excluding any other commercial use; (k) use for the purpose of caricature, parody or pastiche; (l) use in connection with the demonstration or repair of equipment; (m) use of an artistic work in the form of a building or a drawing or plan of a building for the purposes of reconstructing the building; (n) use by communication or making available, for the purpose of research or private study, to individual members of the public by dedicated terminals on the premises of establishments referred to in paragraph 2(c) of works and other subject-matter not*

*subject to purchase or licensing terms which are contained in their collections; (o) use in certain other cases of minor importance where exceptions or limitations already exist under national law, provided that they only concern analogue uses and do not affect the free circulation of goods and services within the Community, without prejudice to the other exceptions and limitations contained in this Article".*

The transcript limitations apply as a whole to the analogical world. Nevertheless, in a digital environment, where TPM may be used, the list of limitations is shorter in practice. Although the Directive predicts that it must be ensured that

*"rightholders make available to the beneficiary of an exception or limitation [...] the means of benefiting from that exception or limitation, to the extent necessary to benefit from that exception or limitation and where that beneficiary has legal access to the protected work or subject-matter concerned",*

the limitations admitted for the referred purposes are significantly less than those foreseen for the analogical world, since the European legislator has only previewed the possibility of circumvention with the rightholders' collaboration for the hypotheses of Article 5(2)(a), (2)(b), (2)(c), (2)(d), (2)(e), (3)(a), (3)(b) or (3)(e). And, as Ascensão (2008, p. 63) emphasizes, "amongst the sacrificed limitations is the quotation right! That's to say that the most important limitation, the one imposed by the necessity of cultural and social dialogue, ceases to exist online".

Moreover, article 6.4. of the InfoSoc Directive expressly admits the possibility of escaping from copyright law to contract law, since Member States only have to ensure that rightholders take appropriate measures to make available to beneficiaries the means of benefiting from those exceptions or limitations in the absence of voluntary measures, including agreements between rightholders and other parties concerned at the limitations.

This clause permits the private copying exception to co-exist with the anti-circumvention protection of technical measures, normally through the exemption of a limited number of reproductions that is found compatible with the notion of strictly personal use (Mazziotti, 2008, p. 207).

Thus, European Union legislation has narrowed the three-step test's range of application, especially in the digital world. Nevertheless, the Directive leaves space for national legislators to develop the solution. The problem is that there is no harmonized solution over the private copy issue in national legislations, except for the fact that the beneficiary of the limitation must be a natural person.

For instance, determining the exact meaning of each of the three steps has not yet found a harmonized solution. The task presents relevant difficulties considering how the social and economic environment has changed decades after the test was introduced by the Berne Convention. In fact, as Vicente (2004) recognizes,

*"technological developments, together with the impossibility of an effective control of the reproduction for private use by rightholders and collecting societies, have put into crisis one of the of premises that traditionally formed the basis for legal private copy: its small relevance to the commercial exploitation of protected works" (pp. 5-6).*

Currently, with the emergence of P2P file-sharing networks and the technical possibility of making perfect copies, that premise must be considered completely obsolete. Consequently, relevant questions remain controversial: can a digital file containing copyrighted content that was downloaded from a P2P network for personal use be considered a private copy? Does it comply with the "no conflict with normal exploitation" step?

As there is no authoritative interpretation, some courts have decided in favor of Internet users when they prove the non-commercial purpose of downloading. In France, the Tribunal de Grande Instance de Paris declared that Anthony G., a user of Kazaa P2P network who kept 1,875 MP3 and DIVX files containing copyrighted content stored in his computer's hard drive, was acting within the limits of the private copy exception as he had not downloaded and uploaded files for profit. Although the decision was reversed by a higher court, it echoed throughout P2P communities. In Spain, Judge Aldecoa ruled P2P music download legal as in those networks there is no talk of money or any other compensation beyond the sharing of material available among various users. These decisions represent, however, a minority current, as, generally, courts tend to rule that making copyrighted works available for the public without the holders' authorization is an infringement and affects the normal exploitation of the works. Thus, downloading digital copyrighted contents from P2P networks is, generally, not considered a private copy or a fair use of the works. That doesn't mean that consumers' rights must be kept out of the equation in the P2P issue, as privacy concerns may arise when rightholders use intrusive mechanisms to identify users' eventual online infringements (Cohen, 1996; Katyal, 2004a; 2004b).

Nevertheless, eventually, considering the current legal solutions, downloading protected content from a P2P network may be lawful in some specific cases. Imagine the case of an individual that has purchased a CD from his favorite band and wants to make a copy of it to listen in his car. As the original CD had copy-protection technology, it can be considered legal to download it from a P2P network, since that action is consistent with the three-step test and the fair use factors.

That leads us to the problem of private copies' enforcement and its non-harmonized solutions. Regarding the possibility of safeguarding the limitation when TPM are being used, in Germany, according to § 95b of the *Urheberrechtsgesetz*, only analogical private copies are enforceable. In other words, according to this solution, when TPM preventing copies are enabled, digital private copies are simply not admitted, even when the copy serves a legitimate end, like parody, teaching or familiar use. In other countries, like Portugal or France, the opposite solution is in place, as the beneficiary who wants to make a digital copy may ask an administrative regulatory authority to decide *ad hoc* if he/her may have access to the necessary means to copy the work.

From the summarily described circumstances, we may ask: if among scholars, judges or legislators there is no consensus over what is lawful or not, how can a non-lawyer determine his/her behaviors regarding the private copy issue? The line between private copy and piracy is getting thinner and uncertain and that is harmful and insecure for all the stakeholders' interests.

In the current inconclusive copyright paradigm, as Koelman (2006) remarks,

*"the three-step test has evolved from an intentionally vague criterion that would never really need interpretation, to a test that actually has to be applied by the judiciary. However, the test is not fit to deal with the issue of which usage should exclusively be controlled by the right-holder and which should not. This is because, even though it was clear from the beginning that exemptions are needed in order to serve the public interest, the three-step test does not give judges sufficient latitude for considering other interests than the right-holders" (p. 2).*

In the absence of legislators' intervention, a balanced interpretation of the three-step test provisions has become a crucial goal. For that reason, a group of some of the most important copyright scholars in the EU from the Max Planck Institute for Intellectual Property, Competition and Tax Law outlined some guiding principles to take into account when interpreting and applying the Three-Step Test.

According to their advised opinion, not disregarding that

*"account should be taken of the interests of original rightholders, as well as of those of subsequent rightholders [...], the Three-Step Test should be interpreted in a manner that respects the legitimate interests of third parties, including interests deriving from human rights and fundamental freedoms; interests in competition, notably on secondary markets; and other public interests, notably in scientific progress and cultural, social, or economic development" (Geiger et al, 2008, p. 5).*

For that purpose,

*"the Three-Step Test requires a comprehensive overall assessment, rather than the step-by-step application that its usual, but misleading, description implies. No single step is to be prioritized. As a result, the Test does not undermine the necessary balancing of interests between different classes of rightholders or between rightholders and the larger general public. Any contradictory results arising from the application of the individual steps of the test in a particular case must be accommodated within this comprehensive, overall assessment" (p. 2).*

Considering that the three-step test, as defined by the Berne Convention, must be interpreted by national legislators, when foreseeing limitations on exclusive rights, and by the judiciary, interpretation under these principles prevents reductionist results that are achieved when one focuses restrictively on one of the steps, especially on the second. In fact, reducing copyright only to its economic reflex implies forgetting its social and collective function and its public interest based justification, converting it into a body of law that merely seeks to protect investments.

If, on the one hand, a broad interpretation of what is a private non-commercial use may also lead to the unfair economic loss of rightholders; on the other hand, as, potentially, each private copy diminishes the possibilities of the economic exploitation of a copyrighted work, a narrow interpretation of the meaning of the second step clearly creates a disadvantaged position for general public interest. As Mazziotti (2008) notes,

*"the requirement protecting the normal exploitation of a copyrighted work does not intend to reserve all of the possible market segments to the copyright owner. Instead, the purpose of this requirement is to ensure that free uses of the work do not enter into economic competition with forms of exploitation of the work that have, or are likely to acquire, considerable economic importance" (p. 209).*

Considering the *Mulholland Drive* case, particularly the fact that Stéphane P. had purchased an original DVD and intended to make a copy for personal and non-commercial uses, under the Max Planck Institute's proposal, it becomes evident that the copy "would not have entered into competition with the exploitation of the DVD film" (Mazziotti, 2008, p 209).

## NEW PERSPECTIVES OVER THE PRIVATE COPY ISSUE

The private copy issue is a good example of how the consumer has been placed in a secondary status in the copyright paradigm.

Copyright is meant to serve the public interest by promoting creativity and innovation through a reward mechanism, *i.e.*, the recognition of the holders' exclusive economic right to exploit the works. As it was thought that what was good for creators was good for society, copyright law was centered in the author's figure. Only the usages made in the private and familiar sphere of the user and that had no relevant economic meaning were out of rightholders' control. Thus, the user was left a mere passive and consumptive role and he/she didn't need to have positive rights against the authors.

While works were available only in analog formats, and private copies had less quality than the original support, rightholders didn't have relevant economic losses from private copying. But, with digitization, the scenario has dramatically changed and, today, usages made in the private sphere, such as P2P file-sharing, may conflict with normal exploitation of the works. Copyright law has strengthened leaving even less space for the user and for private copying. If, in the past, copyright law held internally the balance between divergent rights and interests, the current legal and digital management systems "may override copyright's escape valves – the idea-expression dichotomy, fair use, statutory exemptions – which are as much a part of copyright as are the exclusive right's themselves" (Goldstein, 2003, p. 170).

Nonetheless, "it is an historical constant that when internal limitations are missing, external limitations emerge" (Ascensão, 2008, p. 55). That idea reflects the growing recognition of external limitations on copyright as the expansion of this branch of law has put it on a collision course with other fundamental rights of similar or greater importance. If users can't find protection inside the boundaries of copyright law, they will seek it outside, like in consumers' law.

Nevertheless, as Helberger & Hugenholtz (2007) pointed out, if consumer law provides for a number of legal instruments that may legitimize the interests of consumers in making private copies, such as the rules of conformity of products with reasonable consumer expectations, the test of fairness of contractual terms, and rules on consumer information, it "also suffers from a lack of legal certainty and other deficiencies" (p. 1063). Indeed,

*"although consumer law norms generally apply to the supply of goods and services, they have not been designed with consumers of digital content in mind. Even if consumers may successfully claim a right to make private copies under a variety of consumer protection doctrines, the lack of legal certainty that the existing copyright framework [regarding what must be considered a reasonable consumer expectation] has to offer is mirrored in the framework of consumer law" (p. 1097).*

Though bridges can be crossed between copyright and consumers' law, the private copy issue remains fundamentally a copyright problem and internal balance must be found within the boundaries of copyright law.

Some proposals have been made to integrate consumers' interests in copyright law and to restore or to reinvent the lost balance.

Lessig (2001), Netanel (2003) and Fisher (2004) propose an alternative compensation system based in levies for non commercial use. According to this system, a levy on digital media and online services is foreseen to compensate rightholders from the loss of the right to control non-commercial uses of their works. This scheme replaces a proprietary system by a remuneration rights system and does not impede P2P file sharing.

In France, a mandatory collective management system with a global license for P2P file sharing was discussed but rejected by a commission of the High council of Literary and Artistic Property. Respecting acts of downloading, this proposal was similar to the non-commercial use levy on digital media, but regarding downloading plus uploading acts, the proposal foresaw the remuneration of rightholders through a collective management collecting society. Those who only wanted to download files would be safeguarded by the levy; those who wanted to truly share would need the license from the collecting society on behalf of the rightholders.

Geiger (2008b) proposed the recognition of a true right for private copying for creative purposes, or in other words, for "uses that serve the creation of future works" (p. 123). That right should be enforceable against rightholders using anti-copy TPM.

Mazziotti (2008) proposes the creation of a true users' rights infrastructure as copyright law should be revised and extended to include consumer protection issues, recognizing the existence of subjective rights of the users and not mere exemptions.

Other approaches are possible: the solution may not be found exclusively in the law, but also or mainly in the market, as new business and content distribution models based on the P2P networks can be created. Future years will bring clarity to discussion. What is certain, for now, is that a repressive legal solution that only protects rightholders' economic interests may create insecurity and lack of confidence of online consumers; and that an open solution based on a broad interpretation of private use may seriously affect the normal exploitation of an intellectual digital work, to the disadvantage of creators and of society. Balance in copyright law means that reductionist perspectives must be abandoned: copyright law can't be neither a mere authors' law nor a mere users' law, and must be directed to benefit society as a whole.

For now, until new balanced legal provisions are foreseen, the Max Planks Institute's interpretation proposal for the three-step tests seems to be a good start to find reasonable solutions regarding the private copy issue. Furthermore, co-regulation mechanisms combining legitimate TPM and the recognition and the enforcement of the private copy must be developed, using as a model the French and the Portuguese legal solution of an administrative authority provided with the legal and technical means to access the work when rightholders impede the consumer from making legitimate copies.

## REFERENCES

Ascensão, J. (2008). Sociedade da informação e liberdade de expressão. In *Direito da Sociedade da Informação* (pp. 51–73). VII.

Clark, C. (1996). The answer to the machine is in the machine. In Hugenholtz, P. B. (Ed.), *The Future of Copyright in a Digital Environment*. The Hague: Kluwer Law International.

Cohen, J. E. (1996). A right to read anonymously: A closer look at 'copyright management' in cyberspace. *28 Conn. L. Rev 981*. Retrieved October 15, 2009, from http://ssrn.com/abstract=17990

Fisher, W. W. (2004). *Promises to Keep. Technology, Law, and the Future of Entertainment*. Palo Alto, CA: Stanford University Press.

Geiger, C. (2008b). The answer to the machine should not be the machine: Safeguarding the private copy exception in the digital environment. *European Intellectual Property Review*, *30*(4), 121–129.

Geiger, C., Hilty, R. M., Griffiths, J., & Suthersanen, U. (2008). A balanced interpretation of the "three-step test" in copyright law. *Max Planck Institute for Intellectual Property, Competition and Tax Law*. Retrieved October 15, 2009, from http://www.ip.mpg.de/shared/data/pdf/declaration_three_step_test_final_english.pdf

Goldstein, P. (2003). *Copyright's Highway: From Gutenberg to the Celestial Jukebox*. Palo Alto, CA: Stanford University Press.

Guibault, L., & Helberger, N. (2005). Copyright Law and Consumer Protection. *European Consumer Law Group*. Retrieved October 15, 2009, from http://www.ivir.nl/publications/other/copyrightlawconsumerprotection.pdf

Helberger, N. (2005, August 26). Not so silly after all – new hope for private copying. *INDICARE Monitor*. Retrieved October 15, 2009, from http://www.indicare.org/tiki-read_article.php?articleId=132

Helberger, N., & Hugenholtz, P. B. (2007). No place like home for making a copy: Private copying in European copyright law and consumer law. *Berkeley Technology Law Journal*, *22*, 1061–1098.

Jaszi, P. (1998). Intellectual property legislative update: Copyright, paracopyright, and pseudo-copyright. Paper presented at the Association of Research Libraries conference: The Future Network: Transforming Learning and Scholarship. Eugene, Oregon, May 13-15, Retrieved from http://www.arl.org/resources/pubs/mmproceedings/132mmjaszi~print.shtml

Katyal, S. (2004a). The new surveillance. *Case Western Law Review, 54*(297). Retrieved September 25, 2009, from http://ssrn.com/abstract=527003

Katyal, S. (2004b). Privacy vs. piracy. *Yale Journal of Law & Technology, 7*. Retrieved September 25, 2009, from http://ssrn.com/abstract=722441

Koelman, K. J. (2006). Fixing the three-step test. *European Intellectual Property Review, 8*, 407–412.

Lessig, L. (2001). *The Future of Ideas: The Fate of the Commons in a Connected World*. New York: Random House.

Lessig, L. (2006). *Code 2.0*. New York: Basic Books.

Mazziotti, G. (2008). *EU Digital Copyright Law and the End-User*. New York: Springer.

McManis, C. R. (1999). The privatization (or shrink-wrapping) of American copyright law. *California Law Review, 87*, 173–190. doi:10.2307/3481006

Netanel, N. W. (2003). Impose a Noncommercial Use Levy to Allow Free Peer-to-Peer File Sharing. Harvard Journal of Law & Technology, *17*. Retrieved October 15, 2009, from http://ssrn.com/abstract=468180

Senftleben, M. (2004). *Copyright, Limitations and the Three-Step Test. An Analysis of the Three-Step Test in International and EC Copyright Law*. The Hague: Kluwer Law International.

Vicente, D. M. (2004). Cópia privada e sociedade da informação. Retrieved October 15, 2009, from http://www.apdi.pt/APDI/DOUTRINA/c%C3%B3pia%20privada%20e%20sociedade%20da%20informa%C3%A7%C3%A3o.pdf

Werra, J. (2001). Le régime juridique des mesures techniques de protection des oeuvres selon les Traités de l'OMPI, le Digital Millennium Copyright Act, les Directives Européennes et d'autres legislations (Japon, Australie). *Revue Internationale du Droit d'Auteur, 189*, 66–213.

## ADDITIONAL READING

Abbott, R. (2009). The reality of modern file sharing. *Journal of Internet Law, 13*(5), 3–8.

Aplin, T. (2005). *Copyright Law in the Digital Society: The Challenges of Multimedia*. Oxford, UK: Hart Publishing.

Beckerman, R. (2009). Content holders vs. the Web: 2008 US copyright law victories point to robust Internet. *Journal of Internet Law, 12*(7), 16–21.

Berkeley Technology Law Journal. (2009)... *Annual Review, 24*(1), 363–404.

Billington, H. B. (2006). The podcasting explosion: US and international law implications. *Intellectual Property & Technology Law Journal, 18*(11), 1–5.

Bond, C., Paramaguru, A., & Greenleaf, G. (2007). Advance Australia fair? The copyright reform process. *The Journal of World Intellectual Property, 10*(3/4), 284–313. doi:10.1111/j.1747-1796.2007.00324.x

Brown, B. (2008). Fortifying the safe harbors: Reevaluating the DMCA in a Web 2.0 world. *Berkeley Technology Law Journal*, 2008 Annual Review, *23*(1), 437-467.

Brown, E., & Beckham, B. (2008). Internet law in the courts. *Journal of Internet Law, 12*(5), 17–20.

Calandrillo, S. P. (2008). Dangers of the Digital Millennium Copyright Act: much ado about nothing? *American Law & Economics Association Papers, 41*, 1–63.

Charn, W. W. (2008). The reform of copyright protection in the networked environment: A Hong Kong perspective. *The Journal of World Intellectual Property, 11*(5/6), 498–526.

Chen, A. C. (2007). Copy locally, share globally: A survey of P2P litigation around the world and the effect on the technology behind unauthorized file sharing. *Intellectual Property & Technology Law Journal, 19*(9), 1–4.

Cloak, T. (2007). The digital Titanic: The sinking of YouTube.com in the DMCA's safe harbor. *Vanderbilt Law Review, 60*(5), 1559–1597.

Coats, W. S., Lerner, J. L., & Krause, E. (2010). Preventing illegal sharing of music online: The DMCA, litigation, and a new graduated approach. *Journal of Internet Law, 13*(7), 3–7.

Crum, J. (2008). The day the (digital) music died: Bridgeport, sampling infringement, and a proposed middle ground. *Brigham Young University Law Review*, (3): 943–969.

Dannenberg, R., & Gerk, D. R. (2009). DMCA copyright protections: Uniquely American or common & uniform abroad? *Intellectual Property & Technology Law Journal, 21*(5), 1–7.

Geach, N. (2009). The future of copyright in the age of convergence: Is a new approach needed for the new media world? *International Review of Law Computers & Technology, 23*(1/2), 131–142. doi:10.1080/13600860902742588

Ginsburg, J. C. (2007). The pros and cons of strengthening intellectual property protection: Technological protection measures and Section 1201 of the United States Copyright Act. *Information & Communications Technology Law, 16*(3), 191–216. doi:10.1080/13600830701680453

Herman, B. D. (2008). Breaking and entering my own computer: The contest of copyright metaphors. *Communication Law and Policy, 13*(2), 231–274. doi:10.1080/10811680801941276

Holsapple, C. W., Iyengar, D., Jin, H., & Rao, S. (2008). Parameters for software piracy research. *The Information Society, 24*(4), 199–218. doi:10.1080/01972240802189468

Kleve, P., & De Mulder, R. (2007). Anomalies in Internet law. *International Review of Law Computers & Technology, 21*(3), 305–313. doi:10.1080/13600860701701587

Lysonski, S., & Durvasula, S. (2008). Digital piracy of MP3s: Consumer and ethical predispositions. *Journal of Consumer Marketing, 25*(3), 167–178. doi:10.1108/07363760810870662

Mak, C. (2008). Fundamental rights and the European regulation of iConsumer contracts. *Journal of Consumer Policy, 31*(4), 425–439. doi:10.1007/s10603-008-9084-3

Masango, C. A. (2009). Understanding copyright in support of scholarship: Some possible challenges to scholars and academic librarians in the digital environment? *International Journal of Information Management, 29*(3), 232–236. doi:10.1016/j.ijinfomgt.2008.10.003

Myers, K. The RIAA, the DMCA, and the forgotten few webcasters: A call for change in digital copyright royalties. *Federal Communications Law Journal, 61*(2), 431–456.

Prasad, A., & Agarwala, A. (2008). Armageddon on the digital superhighway: Will Google's e-library project weather the storm? *Computer Law & Security Report, 24*(3), 253–260. doi:10.1016/j.clsr.2008.01.002

Rott, P. (2008). Download of copyright-protected Internet content and the role of (consumer) contract law. *Journal of Consumer Policy, 31*(4), 441–457. doi:10.1007/s10603-008-9081-6

Sawyer, M. S. (2009). Filters, fair use, & feedback: User-generated content principles and the DMCA.

Sinha, R. K., & Mandel, N. (2008). Preventing digital music piracy: The carrot or the stick? *Journal of Marketing, 72*(1), 1–15. doi:10.1509/jmkg.72.1.1

Stokes, S. (2009). *Digital Copyright: Law and Practice*. Oxford, UK: Hart Publishing.

Tschmuck, P. (2009). Copyright, contracts and music production. *Information Communication and Society, 12*(2), 251–266. doi:10.1080/13691180802459971

Waterman, D., Sung, W. J., & Rochet, L. R. (2007). Enforcement and control of piracy, copying, and sharing in the movie industry. *Review of Industrial Organization, 30*(4), 255–289. doi:10.1007/s11151-007-9136-x

Wu, S., & Chen, P. (2008). Versioning and piracy control for digital information goods. *Operations Research, 56*(1), 157–172. doi:10.1287/opre.1070.0414

## KEY TERMS AND DEFINITIONS

**Copyright:** The set of exclusive moral and economic rights granted to the author or creator of an original intellectual work, including the right to copy, distribute and adapt the work.

**Digital Rights Management (DRM):** A copyrighted works' management system based on digital technology that, amongst other powers, allows copyright holders to control access to works or to prevent unauthorized copies.

**Private Copy:** A copy of a copyrighted work that is made for personal and non-commercial use.

**Technological Protection Measures (TPM):** Digital technology-based tools idealized to control third parties access to works or subsequent unauthorized uses.

# Chapter 12
# Service Systems as Digital Products

**Hsin-Lu Chang**
*National Chengchi University, Taiwan*

**Michael J. Shaw**
*University of Illinois at Urbana-Champaign, USA*

**Feipei Lai**
*National Taiwan University, Taiwan*

## ABSTRACT

*In this research, the authors study service systems and assess one emerging service innovation model: services as digital products. The focused area of application is to make the remote healthcare platform developed at National Taiwan University Hospital (the U-Health Service System) a comprehensive and effective "Service System as a Digital product"—made possible by new technology but in need of service designs and innovations. In addition to studying service delivery and innovation for the U-Health Service System, our research aims to stimulate attention toward the promising research opportunities for information systems scholars in several domains: (1) the development of service systems as digital products, (2) the development of service value models based on service processes, service delivery, service metrics and service outcomes, (3) the development of service metrics and the fit between service models, perceived value, and service metrics, and (4) the management of service systems and how to make them sustainable operationally and economically.*

## INTRODUCTION

Service systems have become an important source of new digital products. Take Kindle, the e-book device developed and sold by the online retailer Amazon, as an example. The "product" being sold is not just the Kindle device itself. Amazon is aiming to market the whole *system* and *services* that deliver books in digital formats—In other words, Amazon is selling a "service system" for consumers to purchase, download, and read books digitally. The same service system can be used for selling other products as well, such as newspapers. In that sense it will not be surprising if Amazon's Kindle in the future competes with Apple's iPod in selling digital music services because both are

DOI: 10.4018/978-1-61692-877-3.ch012

service systems selling digital products. There are many other examples of service systems as digital products. One particular example we will focus on in this chapter is a service system for healthcare delivery, also based on a mobile device that can be used in the comfort of homes.

In this chapter we address the following primary research issues concerning digital products:

1. The development of service systems as digital products.
2. The service value models based on service processes, service delivery, service metrics and service outcomes.
3. The service metrics and the fit between service models, perceived value, and service metrics.
4. The management of service systems and how to make them sustainable operationally and economically.

## BACKGROUND

### The Concept of Service Systems

The concept of service systems have been defined from different perspectives in recent years. We can categorize them into the following four viewpoints:

- **Front-back view:** A service system includes visible front stage and invisible back stage. Front stage is where "final assembly" of service elements takes place and service is delivered to customers. Back stage is the technical core where inputs are processed and service elements created (Teboul, 2006; Lovelock et al., 2008).
- **Provider-client view:** A service system comprises service providers and service clients working together to co-produce value (Spohrer et al., 2007). While front-back view emphasizes the interactions with human customers, this view suggests value co-production by providers and clients as the essence of service.
- **Work system view:** A service system is a work system in which human participants or machines perform work using information, technology, and other resources to produce products and services for internal and external customers (Alter, 2008). In contrast to emphasize customer or producer concerns, this view analyzes the system from the business perspective and focuses on how to produce products and services for customers.
- **Value network view:** A service system is a value network where the value is driven and determined by the end consumer and delivered through a complex web of direct and indirect relationship between network actors (Basole & Rouse, 2008). This view is based on the premise that a service system is not merely operated in a dyadic provider-client relationship, but is deeply embedded in complex economic systems consisting of numerous inter-organizational relationships.

Based on our analysis of how other researchers have modeled service systems, we derive an integrative model of service systems that comprise elements in four levels: (1) front and back stage; (2) provider and client; (3) work system; and (4) value network (Figure 1).

The integrative model of service systems in Figure 1 focuses on customer or business concerns without describing how the system evolves when information technology (IT) is introduced. However, IT is a very important part in today's service systems. The past literature has recognized that IT plays a central role in reducing complexity for consumers by providing greater levels of value network integration, information visibility, and

*Figure 1. The integrative model of service systems*

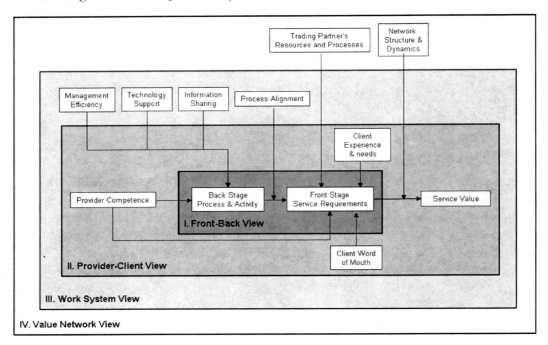

means to manage and anticipate change. Without IT, service systems would find coordination difficult, leading to missed opportunities for innovation or efficiency gains. Besides, IT may reform the service system by replacing human providers or human-interacted work systems to provide variety of self-services to customers.

## Service as Digital Products

IT also provides significant opportunities for digitization of services in response to contemporary customer requirements (Rai & Sambamurthy, 2006). There are several reasons for developing digitized services. First, the globalization and rapid product and services innovations require digitized services for process integration and business intelligence. Second, many firms rely on service innovations for growth. Service digitization offers them a great chance to provide a niche service or a total solution for logically related activities.

Service can be digitized in different contexts. For example, the "Wine Snob" application provides a digitally facilitated service encounter that enriches the interactions between sommeliers and customers in a restaurant (WineSnob, 2009). Amazon's "Kindle" represents a digitized service platform, which allows users to download digital versions of books, newspapers, and magazines (Amazon Kindle, 2008). Salesforce.com's Web-based software provides a full customer relationship management (CRM) service offering online. Customers start to consume the service as soon as they sign on to its website (Davila, 2003). General Motors' (GM) in-vehicle service system "OnStar" creates a digital connection between subscribing drivers and GM caring advisors, providing real-time information and assistance on the road.

Emphasizing the service as a digital product makes it easier to design and understand digitally enabled service systems. In the following sections, we will use a case study for the design of a "u-health service" to illustrate this concept.

## THE U-HEALTH SERVICE SYSTEM

### Problems in the Current Service Systems

The current remote healthcare service systems can be roughly categorized by three types: life caring, medical caring, and long-term caring. All of them adopt the passive, request-and-respond model (Chen et al., 2001). That is, patients or their families have to specify their service requirements by themselves. Due to the inadequate integration of healthcare information systems, however, most patients do not have sufficient information to decide which services suit them the best. In addition, resources are concentrated on the urban hospital sectors, the healthcare manpower in the rural areas is significantly insufficient, and thus the care provided tends to be episodic and fragmented. Moreover, the healthcare systems of different medical institutions can not communicate with each other, and thus the patient care records are not shared between institutions and most are paper-based and not well organized (Lin et al., 2001). In meeting these challenges, National Taiwan University Hospital (NTUH) is undergoing health service reforms, combining home healthcare service network and sensors network to provide a continuous healthcare service for remote patients.

### Development of U-Health Service in NTUH

The U-Health service system is comprised of four services which will be rolled out in an integrated manner (please refer to Figure 2). Through the system, the patients upload their daily biometric information to the database. If they have any question, they can call their care managers at call centers to receive care and guidance. Care managers call out the patients through video conferencing or phone interview twice or thrice a week (5 to 10 minutes for each time). In each conference or interview, care managers review the biometric records and provide advice and counseling. If care managers find any symptoms that require immediate attention, they will provide immediate guidance and alert the patient's physician. Patients' biometric information can be assessed by both care managers and intensive care unit (ICU). In any emergency case that the patients can not find their care managers, they can directly contact ICU for advice.

Patients receive care in different settings based on their diseases and symptoms. For each setting, there are six digitized services to be chosen:

1. **Tele-education:** This service item provides current health information and knowledge that will allow for the empowerment of individuals. This service will enhance the preventive approach and motivate individuals to take the responsibility to lead a healthy life.
2. **Tele-diagnosis:** The service item provides patients immediate treatment on the basis of transmitted tele-monitoring data or tele-consultation.
3. **Tele-monitoring:** The care managers at call centers proactively contact patients to monitor their progress and ensure they are following the plan of care set by their physician.
4. **Tele-consultation:** This service item provides healthcare consultations from a distance. Patients can receive consultation through videoconferencing which incorporates digitization and transmission of patients' relevant medical records including text, diagnostic images and medical charts.
5. **Tele-therapy:** This service item provides psychological counseling and rehabilitation treatments through high-quality live videos.
6. **Lifetime health:** This service item provides personalized, proactive and prospective lifetime health plans to achieve a continuum

*Figure 2. The service components of u-health service system*

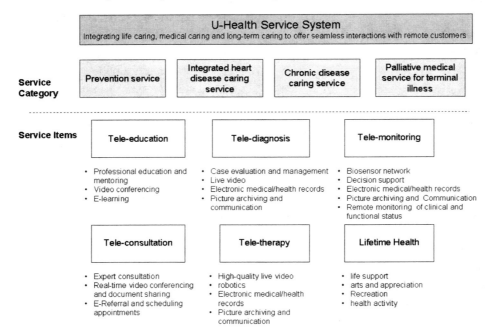

of care to keep the individual in the highest possible state of health.

## Finding the Fit for the U-Health Service System

According to Figure 1, the targeted customer segments of U-Health Service are (1) elderly, (2) patients with heart diseases, (3) patients with chronic diseases, and (4) patients with terminal illness. To ensure the service propositions designed in U-Health service system can offer the proper value to each individual customer segment, the hospital (NTUH), and partners, we propose a research framework to evaluate the fitness between service propositions and stakeholders (customers, NUTH, and partners). The service propositions are examined across six P's (product, price, place, promotion, process, and people). The framework is shown in the Figure 3.

## Service Expectation

First, we need to consider what customers value the U-Health service. The value can be seen as the differences between perceived benefits and perceived costs. One obvious cost is the willingness to pay the price (Teboul, 2006). Derived from SERVQUAL (Jiang, 2002; Keettinger et al., 2005), eight dimensions of customer value propositions for U-Health service are proposed below:

- *Tangibles*: physical facilities, equipment, and personnel are visually appealing
- *Reliability*: the information and knowledge provided by the service is dependable (always available when needed) and accurate
- *Responsiveness*: the service is provided promptly (i.e. timely information and knowledge can be provided)
- *Assurance*: the service can inspire trust and confidence of customers to manage their health and to be more involved in their own care.

*Figure 3. Research framework*

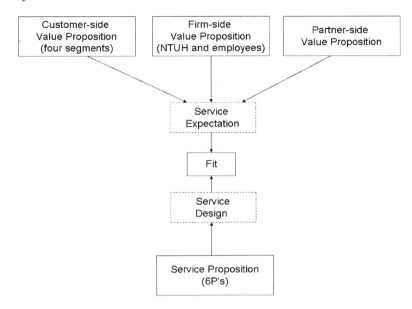

- *Empathy*: the service can provide caring and individualized attention to customers
- *Convenience*: the service can fulfill customer's demands for convenience (e.g., shorter waiting time for treatment, convenient scheduling for treatment, and reduced paperwork)
- *Control*: the service can enhance customers' perception of taking control of their own health
- *Choice*: the service can provide a variety of choices and information for healthcare

Second, we consider the value to NTUH and its employees. For NTUH, there are three dimensions: operational improvement, patient-physician relationship improvement, and revenue growth. The key is to help NTUH build a strong and sustainable competitive position.

- Operational Improvement
  - Productivity improvement:
    - Minimize labor intensity
    - Improve clinical decision-making
    - Reduce the incidence of medical error
    - Reduce the cost of treatment
    - Save time and paper through automated process
    - Efficient encounter
    - Reduce traffic in the waiting room
    - Reduce clinic visits
- Patient Relationship Improvement
  - Improve patent ability to access health care
    - Reduce patient travel time and cost to access healthcare service
  - Improve patient-physician communication
  - Improve wellness index
- Revenue Growth
  - Increase patient acceptance
  - Increase Physician participation

Since employees, including physicians and care managers, are involved in producing, delivering, and marketing the service, their participation determines the success of the system. Therefore, besides studying the value to NTUH, it is also important to evaluate the impact of the service proposition on employees' willingness to participate. We adopt the technology acceptance model (Davis, 1989; Venkatesh & Davis, 2000) to study their expectations to the services. Two value propositions are proposed below:

- Usefulness
    - Improve my performance in my job
    - Increase my productivity
    - Enhances my effectiveness in my job
- Ease of Use
    - The interface is clear and understandable
    - The operation is easy for me

Finally, we study the values to the partnering firms. There are potentially four types of partnering firms in this U-Health service system: community service (such as insurance, restaurants, travel agencies), caring service (such as daycare center), healthcare service (such as clinics and pharmacy), and IT service providers (such as IT platform providers and communication service providers). According to Merchant and Schendel (2000), three types of advantages from partnering are summarized:

- Task-related advantage
    - Increase scale and scope economics
    - Generate superior competitive insights
        - Ability to better anticipate, comprehend, and adapt to emerging threats and opportunities
    - Raise entry barriers
- Partner-related advantage
    - Improve communication between partners
    - Reduce transaction costs/risks of partnering
    - Improve productivity of partnering
    - Reduce managerial costs of partnering
- Institution-related advantage
    - Increase incentives to participate
    - Improve coordination and control

## Service Design

The six elements of the service propositions should be designed to maximize benefits and minimize the costs. The most significant decisions in the U-Health service system are summarized as follows:

- **Product:** There are six service items integrated to offer life caring, medical caring, and long-term caring.
- **Price:** Device rental fees can be paid monthly ($600), quarterly, and annually. There are additional charge for extra tele-consulting (e.g., more than twice a week) and weekly health report.
- **Place:** The service is provided at patients' homes or nursing houses.
- **Promotion:** Physicians play the major role in marketing the service. They identify the specific needs of patients and then suggest suitable services to meet those needs. Care managers and case managers become part-time marketers. They are responsible for the contact with patients.
- **People:** 45 to 100 case managers and caring managers are expected to recruit in charge of the whole experience.
- **Process:** Highly-trained case managers and caring managers proactively contact participants to monitor their progress and ensure they are following the plan of care set by their physician.

## Service Fitness

A winning service proposition has to achieve a strong fit and a good balance of value perceived by the three stakeholders: NTUH, patients, and partners. The fitness is analyzed with the following matrixes, depicted in the Figure 4. These four simple matrixes show how a firm should manage the service as a digital product. It also highlights the fact that a breakthrough digitized service has to achieve a strong fit and a good balance of value perceived by its stakeholders.

## FUTURE RESEARCH DIRECTIONS

In this research we study and assess service systems as digital products. The focused area of application is to make the remote healthcare platform developed at NTUH (the U-Health Service System) a comprehensive and effective "Service System"—made possible by new technology but in need of service designs and innovations. There are several suggestions for the future work:

1. The development of service systems as digital products. Just as the way Amazon developed the Kindle© system as a digital platform for delivering digital books as a service to book readers, developing the service as a digital product provides a rich context for IS research to address SSME issues and models that go beyond abstract concepts.

2. The development of service value models based on service processes, service delivery, service metrics and service outcomes. To make the service system viable economically, it is important to focus on value propositions from stakeholders' perspectives. Such a 'going back to the basics' approach can keep enterprises considering the initial business objectives in mind while designing their service system, therefore ensuring the service proposition is well aligned with the business strategies.

3. The development of service metrics and the fit between service models, perceived value, and service metrics. In this research we take the approach that the operational strategy of a service system should consist

*Figure 4. Finding the fit for u-health service system*

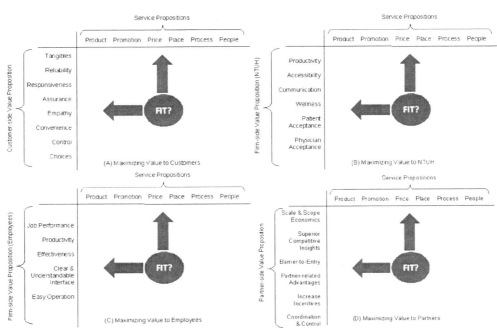

of defining the key elements of the service mix to maximize value to customer, value to employee, value to firm, and value to partners.

4. The management of service systems and how to make them sustainable operationally and economically. In making service systems work, it is critical for enterprises to encourage the continuous improvement of service practices. Value and service propositions may change with different industrial and product focus. Enterprises need to monitor and adjust the service proposition regularly to maintain the sustainable profits and growth.

## CONCLUSION

Healthcare service is being transformed by advances in information technology (IT) and by the empowered, computer-literate public. The healthcare service of the future can be considered as a model of "service systems as digital products," which is patient-centered and more integrated and networked to allow for greater sharing of information and resources between healthcare providers and clients. The digitization also makes the service available around home and community, accessible for remote patients, and customizable in delivery. While these innovations have great potential in the future, this chapter addresses the issues concerning with service propositions, fitness, as well as cost and benefit of the service systems. We hope that this chapter stimulates further thinking in how service systems can be examined as digital products and ultimately advances the field of service science.

## REFERENCES

Alter, S. (2008). Service system fundamentals: Work system, value chain, and life cycle. *IBM Systems Journal*, *47*(1), 71–85. doi:10.1147/sj.471.0071

Basole, R. C., & Rouse, W. B. (2008). Complexity of service value networks: Conceptualization and empirical investigation. *IBM Systems Journal*, *47*(1), 53–70. doi:10.1147/sj.471.0053

Chen, H., Guo, F., Chen, C., Chen, J., & Kuo, T. (2001). Review of telemedicine projects in Taiwan. *International Journal of Medical Informatics*, *61*(2-3), 117–129. doi:10.1016/S1386-5056(01)00134-4

Davila, T. (2003). Salesforce.com: The evolution of marketing systems. *Harvard Case #E-145*, March 18. Wine Snob. Retrieved November 28, 2009, from http://www.iwinesnob.com/

Davis, F. D. (1989). Perceived usefulness, perceived ease of use, and user acceptance of information technology. *Management Information Systems Quarterly*, *13*(3), 319–339. doi:10.2307/249008

Jiang, J. J., Klein, G., & Carr, C. L. (2002). Measuring information system service quality: SERVQUAL from the other side. *Management Information Systems Quarterly*, *26*(2), 145–166. doi:10.2307/4132324

Kettinger, W. J., & Lee, C. C. (2005). Zones of tolerance: Alternative scales from measuring information systems service quality. *Management Information Systems Quarterly*, *29*(4), 607–623.

Kindle, A. (2008)... *Technology Review*, *111*(2), 94–95.

Lin, C., & Chen, H. (2001). Chen, C., & Hou, S. Implementation and evaluation of a multifunctional telemedicine system in NTUH. *International Journal of Medical Informatics*, *61*(2-3), 175–187. doi:10.1016/S1386-5056(01)00140-X

Lovelock, C., Wirtz, J., & Chew, P. (2008). *Essentials of Service Marketing*. Pearson Edition.

Merchant, H., & Schendell, D. (2000). How do international joint ventures create shareholder value? *Strategic Management Journal*, *21*(7), 723–737. doi:10.1002/1097-0266(200007)21:7<723::AID-SMJ114>3.0.CO;2-H

Spohrer, J., Maglio, P. P. Bailey, J., & Gruhl, D. (2007). Steps toward a science of service systems. *IEEE Computer Society*, January, 71-77.

Teboul, J. (2006). *Service is Front Stage: Positioning Services for Value Advantage*. Insead Business Press.

Venkatesh, V., & Davis, F. D. (2000). A theoretical extension of the technology acceptance model: Four longitudinal field studies. *Management Science*, *46*(2), 186–204. doi:10.1287/mnsc.46.2.186.11926

## ADDITIONAL READING

Bath, P. A. (2008). Health informatics: Current issues and challenges. *Journal of Information Science*, *34*(4), 501–518. doi:10.1177/0165551508092267

Chae, Y. M., Lee, J. H., Ho, S. H., Kim, H. J., Jun, K. H., & Won, J. U. (2001). Patient satisfaction with telemedicine in home health services for the elderly. *International Journal of Medical Informatics*, *61*(2-3), 167–173. doi:10.1016/S1386-5056(01)00139-3

Chang, H. (2010). A roadmap to adopting emerging technology in e-business – An empirical study. *Information Systems and e-Business Management*, *8*(2), 103-130.

Chang, H., & Lue, C. (2009). An exploratory study of risk factors for implementing service-oriented IS projects. *Lecture Notes in Business Information Processing*, *22*, 83–95. doi:10.1007/978-3-642-01256-3_8

Chen, H. (2008). New trends in telehealth. *NTU Medical Informatics*. Retrieved December 20, 2009, from http://www.netown.net.tw/Download

Chia, E. (2008). Digitizing healthcare. *Enterprise Innovation*, *4*(4), 26–27.

Costa-Font, J., Mossialos, E., & Rudisill, C. (2009). When is the Internet a valued communications device for health information in Europe? *Economics of Innovation and New Technology*, *18*(5/6), 429–445. doi:10.1080/10438590802547159

CyberDialogue. (2000). Online health information seekers growing twice as fast as online population. Retrieved December 20, 2009, from http://www.cyberdialogue.com

Goldschmidt, P. G. (2005). HIT and MIS: Implications of health information technology and medical information systems. *Communications of the ACM*, *48*(10), 68–74. doi:10.1145/1089107.1089141

Goldzweig, C. L., Towfigh, A., Maglione, M., & Shekelle, P. G. (2009). Costs and benefits of health information technology: New trends from the literature. *Health Affairs*, *28*(2), 282–293. doi:10.1377/hlthaff.28.2.w282

Hensing, J., Dahlen, D., Warden, M., Van Norman, J., Wilson, B. C., & Kisiel, S. (2008). Measuring the benefits of IT-enabled care transformation. *Healthcare Financial Management*, *62*(2), 74–80.

Kudyba, S. P. (2010). *Healthcare Informatics: Improving Efficiency and Productivity*. Boca Raton, FL: CRC Press. doi:10.1201/9781439809792

Li, J., & Shaw, M. J. (2004). Protection of health information in data mining. *International Journal of Healthcare Technology and Management*, *6*(2), 210–222. doi:10.1504/IJHTM.2004.004977

Lin, F., Shaw, M., & Chuang, M. Y. (2005). A unified framework for managing Web-based services. *Information Systems and e-Business Management*, *3*(3), 299-322.

Rada, R. (2008). *Information Systems and Healthcare Enterprises*. Hershey, PA: IGI Publishing.

Raghupathi, W., & Tan, J. (2008). Information systems and healthcare XXX: Charting a strategic path for health information technology. *Communications of the AIS*, *2008*(23), 501-522.

Spetz, J., & Keane, D. (2009). Information technology implementation in a rural hospital: A cautionary tale. *Journal of Healthcare Management*, *54*(5), 337–348.

Suleiman, A. B. (2000). The untapped potential of telehealth. *International Journal of Medical Informatics*, *61*(2-3), 103–112. doi:10.1016/S1386-5056(01)00132-0

Tan, J. (2008). *Healthcare Information Systems and Informatics: Research and Practices*. Hershey, PA: IGI Publishing.

Themistocleous, M., Mantzana, V., & Morabito, V. (2009). Achieving knowledge management integration through EAI: A case study from healthcare sector. *International Journal of Technology Management*, *47*(1-3), 114–126. doi:10.1504/IJTM.2009.024117

Topacan, U., Basoglu, A. N., & Daim, T. U. (2010). Exploring the adoption of technology driven services in the healthcare industry. *International Journal of Information Systems in the Service Sector*, *2*(1), 71–93.

## KEY TERMS AND DEFINITIONS

**Service Design:** Service propositions designed across six dimensions: product, place, price, promotions, process, and people.

**Service Expectation:** The value customers expect from consuming the service.

**Service Fitness:** A service proposition achieves a good balance of value perceived by its stakeholders.

**Service Propositions:** Service offerings designed to maximize benefits and minimize the costs of clients.

**Service Systems:** Comprising service elements in four levels: (1) front and back stage; (2) provider and client; (3) work system; and (4) value network.

**Service Systems as Digital Products:** Digitally-enabled service systems.

**U-Health Service System:** Combining home healthcare service network and sensors network to provide a continuous healthcare service for remote patients.

# Section 4
# Visions for the Future

# Chapter 13
# Transitioning to Software as a Service:
## A Case Study

**Dave Sly**
*Proplanner.com, USA*

## ABSTRACT

*Managing a digital product company in 2010, and beyond, is a very unique and challenging experience when compared to the traditional product development and deployment models of the late 20th century taught in many MBA schools today. This chapter presents the benefits, challenges, and lessons learned of a CEO who is transforming a traditional "Software as a Product" engineering software firm that he founded in the mid 1980's to a Web-based, "Software as a Service" firm that is positioned for 2010 and beyond.*

## A CASE STUDY IN MAKING THE TRANSITION FROM SOFTWARE PRODUCT SALES TO SOFTWARE AS A SERVICE

### Background

Proplanner.COM started out as a company called Cimtechnologies Corporation, which was a university spinoff of graduate student research from the Industrial Engineering department at Iowa State University in 1987. The company sold PC-based manufacturing planning software applications tailored to the design of factories and their supporting materials handling systems. The software applications were priced from around $1,000 per license up to $15,000 per license and included a number of modules.

Products were sold in the US from the combination of a direct, and indirect, sales force. The indirect sales force comprised of independent dealers of software from two different vendors (Autodesk and Production Modeling Corporation). These dealers were geographically located throughout the world. Software presentations were all conducted face-to-face, typically at the customer's factory site. It would often take three or four presentations over the course of two to

DOI: 10.4018/978-1-61692-877-3.ch013

three months in order to gain approval for the purchase order process, and that process could take another one to two months.

The software applications performed tasks to diagram and calculate the flow of materials in the factory, as well as, create 2D and 3D layout drawings within the AutoCAD graphics application. As such, the features and benefits of the software were fairly easy to communicate, and were largely well understood by the prospect. In addition, customers were able to quickly learn and use the applications with a minimum of training and support.

Since about 2003, the products expanded their focus to include tasks such as Assembly Line Balancing, Shop Floor Work Instructions Authoring/Publishing, Worktask Time Estimation (Observed and Predetermined), FMEA/Control Plan Quality Management, Engineering Change Management and finally, the integrated workflows and data flows between the associated modules.

The current distribution model was providing diminishing results in terms of both units sold and sales revenue. In addition, direct sales for products in this price range were very expensive, and it took too long to find, train and support new territory representatives. Since around 1998, the company had been expanding in its use of the Web to support customers. By the end of 2001, it was proposed to begin offering software sales and product downloads via the Web.

In 2002 the Proplanner.COM company was formed and within two years the company was offering products for sale, download, and subscribed use via the Web. The new company was founded in order to provide financial resources and technical focus on the Web-based model. The original Cimtechnologies Corporation firm had been acquired by a very large IT firm, and this firm wanted a clean break between the two technologies and sales models. In part, this was due to a lot of internal resistance by the IT firm's sales staff, to the prospect of converting to a direct Web-based model for sales and support.

## Desired Benefits

One of the critical benefits that the company was seeking involved a consistent revenue base afforded by subscription sales. The traditional software model experienced significant revenue swings of four to six times on a monthly basis. This variability created challenges for budgeting of staff and managing cash flow.

It was also hoped that the sales effort could be reduced by a commensurate reduction in the significant up-front cost for the software purchase. Basically, the higher the up-front cost, without proven benefits of the software, then the more effort that needs to be expended in sales to convince the client that their decision is wise, and the more decision makers are likely to need to be involved. Instead, if the client can pay-as-they-go, then the initial successes drive increased usage (more licenses) within the company, and this increases the monthly revenue, without a commensurate increase in sales activity. In many ways, software as a service (SaaS) provides an opportunity for a traditional model "pilot project" to transition smoothly into a full deployment sale, without introducing a significant front-loaded one-time cost which will require multi-level management approvals and perhaps even future budgeting cycles.

Finally, the steady revenue and rapid low-cost sales growth of the SaaS model was heavily desired by Venture Capital firms and this created much greater corporate valuation multipliers (three to six times revenue) versus that of the traditional sales model which was experiencing reduced multipliers down to one to two times revenue by 2002.

## Challenges Experienced

The company has experienced many challenges in making the transition to the SaaS model, and still today in 2010, the company only receives about 30% of its revenue via subscription sales, and none of its customers have chosen to have the applications hosted externally.

One of the principle issues that the company experienced from the beginning was related to data security. A majority of clients were very concerned with their critical production information being managed and stored at an external Internet accessible site. This issue was mainly raised by the IT staff, and it was clear that some of the justification was related to the IT department's associated control of their data. Interestingly, the manufacturing engineering departments saw the SaaS model as a benefit over the traditional model, as it was common for them to prefer product and data center support from an external firm dedicated to the application, rather than their in-house general services provider which would need to learn and support a non-native system.

As a result of this concern, Proplanner introduced the ability for the software to be installed within the client's data center and intranet, while still allowing them to pay for the software on a subscription basis, rather than via an upfront purchase. In this situation, customers would essentially pay an annual software fee equivalent to approximately 40% of what competitors were charging for their upfront software purchases. It is important to note that most companies in the manufacturing software field were charging a 20% annual fee for product updates and phone support, therefore this subscription fee was established to cover the software costs over a three to five year period (not including time value of money).

This SaaS pricing approach is the current model being offered by Proplanner, and generates approximately 30% of the annual revenues, although that percentage has been increasing over the past two years. Proplanner has contracted with an external data center which specializes in very secure hosting of data and applications. Proplanner offers this external hosting service for a nominal fee (approximately $200/month) which represents less than 5% of the monthly product subscription fee. Currently, no firm has yet subscribed to this service (aside from pilot projects of the software), due to the aforementioned "control" issues which continue to be raised by the IT organizations within these large manufacturing firms.

Clearly the sales models between face-to-face selling of software products, or services, and Web-based "viral marketing" of downloadable products and services are significantly different. A major challenge encountered by Proplanner was that of trying to provide enough information on the website to a prospective client, that they could get their pre-sale questions answered, download the software and implement it - all without speaking to a sales person. In addition, it was likely that the client was in a face-to-face selling position with a competitor who was providing a traditional software product sale. This sales situation quickly proved impossible, and the company was forced to return to a face-to-face, albeit Web-enabled, sales model. Currently, the company performs client prequalification and intermediate Q&A via Web conferencing, and yet still relies on one to three face-to-face meetings prior to securing a sale.

A major contributing factor to the slow Web-based approach has been the complexity associated with the data and workflow between the various software modules. While the Time Estimation, Line Balancing, Flow Analysis, and to some extent, Work Instruction modules are fairly easy to comprehend, the integration between them, and the associated client-configuration options, have not been. As such, the current sales process involves multiple meetings (face and Web) with IT, Product Engineering, Manufacturing Engineering and even Accounting/Upper Management to educate the client and secure the sale. As such, it became clearly evident that any product, or implementation complexity (even a little) will severely affect the viral (automated Web sale) nature of the sales process.

The company mainly markets its products via the Google search engine, and some targeted trade shows and mailings (mostly email). These marketing techniques are used to drive prospects to the company's website which is rich in downloadable content, as compared to other competing firms in the industry. In particular, Proplanner offers a discussion forum, detailed white papers, and downloadable brochures and software manuals. The Line Balancing, Time Estimation, and Flow Planning modules can also be downloaded, on a trial basis, by simply asking the prospect to provide their contact information, which is subsequently validated by the Proplanner internal sales staff. Prior to providing the free trial downloads, the company was receiving very few client requests for information. These requests increased four times, immediately upon offering the free trial software. Of course, the free trial is only applicable to the stand-alone modules of the application, and not the full data integrated application, due to the complexity associated with installation and configuration.

Proplanner is offering the SaaS model via its direct internal sales staff. This staff receives the vast majority of their compensation via a commission on billed revenues. As a result, the sales staff does make more up-front money if they sell a product, versus if they offer products on a subscription basis. On the other hand, they may risk losing or delaying a sale if the upfront purchase option is simply too expensive for the client to approve. The company is currently implementing a lower commission for software sales versus software subscription, however even with this change the sales staff will receive more up-front payment for a software sale versus a subscription sale.

A final concern raised by other SaaS providers, but not yet experienced by Proplanner, was the lack of customer loyalty and the rapid nature at which firms can switch between SaaS providers when there is not a personal contact and where the transition effort and cost is very low. Once again, since Proplanner has not achieved a true level of impersonal interaction, and there is a lot of configuration required to implement the Proplanner system at a client site, this is not likely to be a foreseeable problem.

## Technology Issues

Proplanner chose to develop in the Microsoft (MS) MS.NET environment as a thick client, client-server application. The primary reason for this choice was that the applications were computationally intensive and required a very intuitive interface that was highly interoperable with MS Excel, and other MS applications. In addition, the company wanted the flexibility to deploy the applications on a traditional network, as well as a fully web-deployed model. This flexibility proved beneficial, as customers have primarily deployed internal to their network and evaluated the software over the Web (thereby using both options).

Addressing problems with firewalls, internal corporate security filters, and incompatibilities between client installed Internet browsers was initially one of the most challenging development issues encountered. Further complicating this situation was the fact that all of the externally hosted deployments were used for pre-purchase pilot projects, and so a slow and complicated start-up process would often hinder the sale and raise unwanted attention from the IT staff, prior to the engineering department making a decision to push for the purchase of the software. Luckily, subsequent versions of MS.NET (mainly 3.0 and above) have greatly reduced these problems.

Other development environments, such as Google and Flash apps for the Web have become very popular since Proplanner launched its development effort on the MS.NET platform. Those environments clearly make for ideal platforms in a 100% hosted "cloud computing" solution, however they still suffer from a relatively weak integration with other MS Office apps, and of course a very difficult "in-house" hosting option. As such, Proplanner still strongly believes that the MS.NET based solution is superior to other solutions available as of 2010.

## Benefits Experienced

So far, the greatest SaaS benefit that the company has experienced is with the SaaS pricing model. With nearly 1/3 of monthly revenues generated from monthly subscriptions, cash flow and staffing decisions are much easier to manage. In addition, the SaaS sales have nearly all been made in significantly less time and effort than their up-front purchase counterparts.

Technically, the greatest benefit of SaaS has been with the ease of the one-click client installation and update process. In addition, the externally hosted (include intranet) capability has been perceived as a very valuable installation option for the client - even though no client has yet chosen that option. Basically, the fact that the products can be hosted externally provides the customer with future hosting flexibility.

Finally, client satisfaction has been greatly enhanced with the use of Web-based collaboration technologies which allow the Proplanner staff to assist their customers with technical support, product updates and even data entry and analysis services. In fact, the company sees a potential future business opportunity with assisting their clients with remote engineering services.

## SOLUTIONS AND RECOMMENDATIONS

The Proplanner case study further validates the premise that online content must involve a mature process and be easily learned and quickly and painlessly implemented. As such, while there are still benefits for more sophisticated applications and implementations to use the SaaS model, software firms should take care to ensure that the investment in SaaS is focused on their most simple of processes and products.

Finally, SaaS does not imply Web-based selling. Companies with sophisticated and technical products will need to continue their direct sales efforts and should be very careful with their use of Web technologies in the early stages of the sales process where it is critical for the sales staff to establish personal relationships and learn about customer pains and problems - firsthand.

## REFERENCES

Aramand, M. (2008). Software products and services are high tech? New product development strategy for software products and services. *Technovation*, *28*(3), 154–160. doi:10.1016/j.technovation.2007.10.004

Baca, F. (2009). Considering HR outsourcing? Consider SaaS. *Financial Executive*, *25*(8), 59–60.

Biddick, M. (2010, January 18). Time for a SaaS strategy. *InformationWeek*, *1254*, 27–32.

Blokdijk, G. (2008). *SaaS 100 Success Secrets - How Companies Successfully Buy, Manage, Host and Deliver Software as a Service (SaaS)*. Brisbane, Australia: Emereo Pty Ltd.

Braude, E. (2008). Software-as-a-service and offshoring. *International Journal of Business Insights & Transformation*, *2*(1), 93–95.

Campbell-Kelly, M. (2009). Historical reflections: The rise, fall, and resurrection of software as a service. *Communications of the ACM*, *52*(5), 28–30. doi:10.1145/1506409.1506419

Chia, E. (2008). Marketing firm adopts software-as-a-service path. *Enterprise Innovation*, *4*(1), 23.

Chong, B. (2008). Stop data leaks through SaaS. *NetworkWorld Asia*, *4*(8), 4.

Choudhary, V. (2007). Comparison of software quality under perpetual licensing and software as a service. *Journal of Management Information Systems*, *24*(2), 141–165. doi:10.2753/MIS0742-1222240206

Concha, D., Espadas, J., Romero, D., & Molina, A. (2010). The e-HUB evolution: From a custom software architecture to a Software-as-a-Service implementation. *Computers in Industry*, *61*(2), 145–151. doi:10.1016/j.compind.2009.10.010

Crosman, P. (2009). No slow down for SaaS. *Wall Street & Technology*, *27*(8), 22.

Cusumano, M. A. (2007). The changing labyrinth of software pricing. *Communications of the ACM*, *50*(7), 19–22. doi:10.1145/1272516.1272531

Cusumano, M. A. (2008). The changing software business: Moving from products to services. *Computer*, *41*(1), 20–27. doi:10.1109/MC.2008.29

Demirkan, H., Kauffman, R. J., Vayghan, J. A., Fill, H.-G., Karagiannis, D., & Maglio, P. P. (2008). Service-oriented technology and management: Perspectives on research and practice for the coming decade. *Electronic Commerce Research and Applications*, *7*(4), 356–376. doi:10.1016/j.elerap.2008.07.002

Elfatatry, A. (2007). Dealing with change: Components versus services. *Communications of the ACM*, *50*(8), 35–39. doi:10.1145/1278201.1278203

Fan, M., Kumar, S., & Whinston, A. B. (2009). Short-term and long-term competition between providers of shrink-wrap software and software as a service. *European Journal of Operational Research*, *196*(2), 661–671. doi:10.1016/j.ejor.2008.04.023

Fox, L. (2009). Integrating SaaS and legacy apps: 5 steps for success. *NetworkWorld Asia*, *5*(1), 30.

Gold, N., Knight, C., Mohan, A., & Munro, M. (2004). Understanding service-oriented software. *IEEE Software*, *21*(2), 71–77. doi:10.1109/MS.2004.1270766

Greer, M. B. (2009). *Software as a Service Inflection Point: Using Cloud Computing to Achieve Business Agility*. Bloomington, IN: iUniverse.

Guptill, B., & McNee, W. S. (2008). SaaS sets the stage for 'cloud computing'. *Financial Executive*, *24*(5), 37–44.

Hatch, R. (2008). *SaaS Architecture, Adoption and Monetization of SaaS Projects using Best Practice Service Strategy, Service Design, Service Transition, Service Operation and Continual Service Improvement Processes*. Brisbane, Australia: Emereo Pty Ltd.

Hill, S. Jr. (2008). SaaS seems to favor users more than vendors. *Manufacturing Business Technology*, *26*(1), 48.

Jaisingh, J., See-To, E. W. K., & Tam, K. Y. (2008). The impact of open source software on the strategic choices of firms developing proprietary software. *Journal of Management Information Systems*, *25*(3), 241–275. doi:10.2753/MIS0742-1222250307

Keller, E. (2007). How software application pricing models are likely to change. *Manufacturing Business Technology*, *25*(1), 42–43.

Kruff, J. (2008). Ocimum benefits from SaaS model. *Enterprise Innovation*, *4*(5), 22–23.

Lamont, J. (2010). SaaS: Integration in the cloud. *KM World*, *19*(1), 12–22.

Menken, I. (2008). *SaaS - The Complete Cornerstone Guide to Software as a Service Best Practices Concepts, Terms, and Techniques for Successfully Planning, Implementing and Managing SaaS Solutions*. Brisbane, Australia: Emereo Pty Ltd.

Murphy, S., & Samir, W. (2009). 'In the cloud' IT creates new opportunities for network service providers. *Journal of Telecommunications Management*, *2*(2), 107–120.

Orr, B. (2006). SaaS just may be the end of software as we know it. *ABA Banking Journal*, *98*(8), 51–52.

Orr, B. (2007). Microsoft begins its radical shift to software as a service. *ABA Banking Journal, 99*(12), 46–47.

Postmus, D., Wijngaard, J., & Wortmann, H. (2009). An economic model to compare the profitability of pay-per-use and fixed-fee licensing. *Information and Software Technology, 51*(3), 581–588. doi:10.1016/j.infsof.2008.08.004

Poston, R. S., Kettinger, W. J., & Simon, J. C. (2009). Managing the vendor set: Achieving best pricing and quality service in IT outsourcing. *MIS Quarterly Executive, 8*(2), 45–58.

Raghu, T. S., Sinha, R., Vinze, A., & Burton, O. (2009). Willingness to pay in an open source software environment. *Information Systems Research, 20*(2), 218–236. doi:10.1287/isre.1080.0176

Sainio, L.-M., & Marjakoski, E. (2009). The logic of revenue logic? Strategic and operational levels of pricing in the context of software business. *Technovation, 29*(5), 368–378. doi:10.1016/j.technovation.2008.10.009

Susarla, A., Barua, A., & Whinston, A. B. (2009). A transaction cost perspective of the 'software-as-a-service' business model. *Journal of Management Information Systems, 26*(2), 205–240. doi:10.2753/MIS0742-1222260209

Thompson, J. K. (2009). Business intelligence in a SaaS environment. *Business Intelligence Journal, 14*(4), 50–55.

Turner, M., Budgen, D., & Brereton, P. (2003). Turning software into a service. *Computer, 36*(10), 38–44. doi:10.1109/MC.2003.1236470

Vorisek, J., & Feuerlicht, G. (2004). Is it the right time for the enterprise to adopt software-as-a-service model? *Information & Management, 17*(3-4), 18–21.

Waters, B. (2005). Software as a service: A look at the customer benefits. *Journal of Digital Asset Management, 1*(1), 32–39. doi:10.1057/palgrave.dam.3640007

Wong, K. (2008). SaaS vendors buying innovation rather than developing it themselves. *NetworkWorld Asia, 4*(8), 35.

# Chapter 14
# Digital Media:
## Future Research Directions

**Anthony Hendrickson**
*Creighton University, USA*

**Trent Wachner**
*Creighton University, USA*

**Brook Mathews**
*Creighton University, USA*

## ABSTRACT

*Few would argue that digital technology has impacted nearly every industry, especially media and related firms. Media's initial reaction to digital technology was reactive in nature: How can we convert traditional processes to fit digital technologies, mostly in the form of distribution (e.g., traditional newspapers making content available online)? In this chapter the authors argue that digital technologies have now permeated virtually every aspect of the value chain and are forcing traditional firms to rethink long held business models. The authors identify five areas of potential inquiry: (1) What is a resource in the digital age and why does this matter? (2) Where does value creation fit in today's horizontal business models? (3) What is the demarcation between consuming content and creating content (e.g., user-generated content)? (4) What mechanisms can be used to assess quality in a world where anyone can publish? (5) What is the role of regulation and changing business models in the world of digital technology? They do not claim to have the answers, but they hope to at least create dialogue that encourages future research.*

## INTRODUCTION

Today's ongoing advancement of technology continues to verify the relevance of Gordon Moore's (1965) comments that led to the now famous Moore's Law. Computer and Electrical Engineering Science have proven amazingly robust in its delivery of astonishing productivity gains for nearly five decades. However, somewhat overlooked aspects of this continued technological eruption are the challenges involved in utilizing some of the technologies and in creating business models that allow us to monetize these burgeoning technologies.

In this chapter, we explore how technology challenges our current business models and processes. From this brief reflection, we pose a number

DOI: 10.4018/978-1-61692-877-3.ch014

of themes that warrant further examination of the implications technological innovations have on the global workplace.

## NEW RESEARCH DIRECTIONS FOR DIGITAL PRODUCT MANAGEMENT

First, it is necessary to consider what we mean by *digital*. Digital data is distinguished from analog data in that the datum is represented in discrete, discontinuous values, rather than the continuous, wavelike values of analog (Tocci, Windmer & Moss, 2006). Thus, the *digitization* of data refers to the conversion of information into binary code, allowing for more efficient transmission and storage of data. A key differentiator of our current age from prior human history is that, as of the last decade, we not only *convert* data to a digital format, but we also *create* data in a digital format. Thus, we now have the digital product, a concept defined by Scupola (2005) as "a product whose complete value chain can be implemented with the use of electronic networks. It can be produced and distributed electronically, and paid for electronically" (p. 2563).

Since its inception, the Internet has continued to change the game for IT (Feller et al., 2008; Ghose, 2009; Moon & Sproull, 2008; Smith & Telang, 2009) and allows business to meet today's "qualitative and quantitative diversification of demand" (Theodorou, 2006, p. 70). Once hidden behind corporate walls and offering immense competitive advantages that could be leveraged into profitability, technology is now openly accessible on the Internet (Majchrzak, 2009; Raghu et al., 2009). Competitors--even small start-ups--can adopt it and level the playing field. As a result, technology itself no longer offers competitive advantage and higher profit. Instead, the way it is applied to or combined with new information or technologies (i.e., the *network effect*) creates advantage and profitability (Liebowitz & Margolis, 1994).

Not surprisingly, the same evolution has occurred with digital products, as they are the result of the application of technology. The Internet now makes the production and distribution of digital products available to a wide audience, whether regulated or not. "We have all become potential publishers" (Guadamuz, 2009, p. 3). The material we publish may be our original work, a copy of someone else's work, or a combination of the two with virtually no formal quality control.

In the following sections, we highlight these themes that lend themselves to further study of digital products:

1. Digital products from a Resource Perspective
2. Value Creation in Production and Distribution
3. Integrating User-generated Content (UGC)
4. Quality Assessment and User Perceptions
5. Regulation and Monetization

### Digital Products from a Resource Perspective

Vargo and Lusch (2004) discussed service-dominant logic to bridge the gap between goods and services. Their main argument was that service, not goods, is the basis of exchange. They defined service as "the application of specialized competences (knowledge and skills), through deeds, processes, and performances for the benefit of another entity or the entity itself" (p.283). This approach represents a significant change in the understanding of value and exchange that dates as far back as the 1800s. Vargo and Lusch assert that value and exchange are about people, not products: goods are simply empty shells until acted upon to produce a service.

Vargo and Lusch (2004) refer to knowledge and skills as operant resources, meaning they are used to operate on another resource to produce an effect. Operant resources are infinite, whereas operand resources are finite and usually take the form of production inputs.

The very act of changing a product into a digital format raises questions about the product from a resource perspective, as it goes from being a finite resource to an infinite resource, suggesting a move from operand to operant. However, the term *operant* also suggests that the resource can be used on another resource to produce an effect. Thus, it is important to address the following questions:

- If digital products are operant resources, how are they being used to act upon other resources?
- Do digital products suggest a need to revisit the definitions of operand and operant resources, or perhaps create a third class of resources?
- Are digital products proof that operant resources can exist outside the human mind and body?

## Value Creation in Production and Distribution

*"Users have become publishers, a fact that blurs the boundaries between consumers and producers" (Guadamuz, 2009, p. 2).*

Before the advent of digital products and the Internet, firms controlled production and distribution. By their physical nature, products were protected from unauthorized duplication and distribution. Digitization freed these products from physical form and made them easy to replicate and distribute, thus posing a serious threat to their ability to create value for the firm.

Another threat to the firm's ability to create value with digital products is self-publishing technology. As discussed above, demand has diversified, and firms are not always able to meet customer expectations. Empowered by technology, customers can now produce their own work or re-mix the copyrighted work of firms. This trend has begun the user-generated content movement that has popularized wikis, blogs, video- and photo-sharing Web sites, and many other social sharing networks.

Lessig (2008) offers a historical perspective on the user-generated content movement in terms of a read-only culture versus a read/write culture. Lessig says that for much of human history, creation was a communal process. However, when western copyright laws were put into place, creation was taken out of the community and given to sole creators. As a result, a read-only culture developed. The proliferation of technology has shifted creation back to the community, despite copyright laws, and allowed people the opportunity to read, write, and respond. Once again, a read/write culture is blossoming (Guadamuz, 2009; Lessig, 2008).

The communal creation and distribution of content raises interesting questions about the value derived from such activities.

- If people and firms alike can create their own content, how is value being created?
- How will performance be differentiated?
- Will content become a commodity, much like technology, where the only way to create value is through the way it is combined and integrated?

According to Guadamuz (2009), "Open source software, Creative Commons, wikis, and the Free Culture movement have arisen as a response to the needs of content creators who are often not motivated by commercial gain" (p. 2). If this is the case, then what exactly is the motivation?

## Integrating User Content

The rise of user-generated content has forced firms to sit up and take notice. Guadamuz (2009) pointed out that "in the digital environment the roles are decreasingly clear, as users are empowered by the technology to place their content online" (p. 2).

Firms are realizing that customers want to be part of the production and distribution process.

Allowing such participation may enrich the body of knowledge already being put forth, thus increasing value to the firm without the additional burden of more employees (Majchrzak, 2009; Moon & Sproull, 2008; Te'eni, 2009). In contrast, when firms do not integrate user content, they miss opportunities to enrich their product offering, often to the detriment of their customers.

While many firms are trying to figure out how to integrate user-generated content, they remain skeptical about whether that content will enhance or erode firm value. The debate over content co-creation is especially heated in the newspaper industry. A study of British newspapers revealed that many are "experiencing problems with incorporating user media into professional journalism structures due to concerns about reputation, trust and legal issues" (Hermida & Thurman, 2007, p. 25). To deal with these issues, newspapers are developing a filtered model where comments are sorted before being posted. The comprehensiveness of such filtered models varies and is often driven by economics. There is a need to develop industry standards and best practices around this trend.

Copyright issues sit squarely in the middle of the debate over how firms will integrate user content. Interestingly, Guadamuz (2009) pointed out:

*There is growing evidence the growing numbers of copyright owners are located in the peer production sector. It would be useful if policy was no longer designed with the idealized struggling creator in mind. More often than not, the creator will be a hobbyist, never expecting a monetary return for her troubles. (p. 17)*

From the user's perspective, Da Lio (2005) pointed out difficulties experienced by wiki contributors. We believe those same difficulties may apply to users who want to contribute content to a firm. User difficulties include frustration over modification of submitted content (e.g., by the firm, such as a newspaper comment moderator), fear of criticism from other users, reluctance to give up ownership and anonymity, fear of destructive input, hesitation to edit the materials of others, and uneasiness with taking responsibility for unfinished work that will be made public (Lessig, 2008). These scenarios may actually deter users from submitting content to firms who are actively soliciting this type of participation. This issue speaks to four challenges faced by firms that are trying to tap into what Moon and Sproull (2008) call the "Internet-based volunteer workforce."

Not only is content being created by users, but whole information systems (ISs) are being created by user-developers who have little IT expertise. This is known as User Systems Development (USD). Anecdotal evidence is showing that USD may be more efficient in some instances than Traditional Systems Development (TSD). Advic (2002) explained that a user-developer has primary business knowledge and limited IS knowledge, whereas the IT specialist has primary IS knowledge and limited business knowledge. The tools to develop systems are becoming user-friendly enough to bridge the IS knowledge gap for user-developers, making it easier for them to develop systems themselves (Avdic, 2002).

Firms, like newspapers, are struggling with how to manage and control USD because they exist outside the normal processes. They must address the following two key questions:

- Is it better to control tightly these activities or set up an infrastructure to train employees on USD?
- How will the role of the IT specialist change, given the USD trend?

Other items for consideration, based on research presented in this section include a) how copyright law will change to accommodate the creator as a hobbyist, b) how such a change will affect firms, and c) how the interests of both may be balanced.

More research is necessary to explore models that will help both firms and newspapers efficiently incorporate user content into the product cycle and business processes, and that will help determine what motivates the volunteer workforce to make content contributions to the firm.

## Quality Assessment and User Perceptions

Organizations face numerous problems as more information is converted and created in digital form. The volume of digital products has become cumbersome and difficult to manage. Gistics, a California-based research firm, estimates that 30% of digital assets within an organization are misplaced, and then reworked or duplicated. The number of digital assets an organization may manage has exploded. Organizations struggle to reduce cycle time, despite the need to maintain brand consistency and coordinate cross-media publishing and one-to-one marketing (Subramanian & Yen, 2002).

All of these problems can lead to poor data quality, which can cost a firm dearly. In a 1998 study, Redman found that poor data could result in a cost increase of 8 to 12% for an organization, and that service organizations could face a 40 to 60% increase in expenses due to poor data quality.

With so much at stake for organizations, more research is necessary to determine how the quality of digital products is assessed. Wang et al. (1996) detailed four areas of data quality: (a) intrinsic data quality (e.g., accuracy, objectivity, believability, and reputation); (b) contextual data quality (e.g., relevance, timeliness, and appropriate length); (c) representational data quality (e.g., conciseness, consistency, ease of interpretation and comprehension); (d) accessibility.

The proliferation of user-generated content and co-created content raises some interesting questions about how quality is assessed and perceived. Borchers (2003) referenced journalism literature, which has produced studies comparing the "believability" of content between print and online media. However, this is not an ideal assessment of quality, and the comparison of print and online media is not comprehensive enough.

More research is necessary to determine how to measure actual data quality. One possibility might be to examine multiple digital product formats based on a framework, such as that presented by Wang et al. (1996).

The concept of perceived data quality also presents such issues as (a) whether or not consumers are verifying data quality, (b) how they are verifying data quality, (c) how data quality standards are enforced in online communities where co-creation is the norm, and (d) whether or not consumers perceive data quality differently, depending on the digital format in which it is presented. Accurate methods to measure perceived quality must also be developed.

## Regulation and Monetization

Old business models no longer produce value, and digital products are part of that phenomenon, as they lead the trend towards consumer demand for free products. This phenomenon is due in part to piracy and in part to the proliferation of compelling user-generated content. Regulation and protectionist tactics have helped firms maintain the value of their digital products. However, firms must walk a fine line to protect this value. A 2009 study by Raghu et al. revealed that consumers tend to lash back at firms that try to limit open source software projects, while they are unfazed by firms' use of piracy controls, despite the availability of free open source alternatives. Firms whose whole value is derived from digital products face serious challenges with profitability in an environment where so much is being given away for free--hence the closing of newspapers across the nation and the declines in profits for record labels, film production companies, and others. The effectiveness of regulatory and protective measures as a means to protect profits is likely to continue to decline, and digital product businesses will have to change their business models to remain competitive.

The digital products industry is experiencing an unfortunate lack of consistency among business models' best practices. The *Wall Street Journal* recently catapulted to the top for online newspapers, though charging for access to much of its content; in contrast, the music industry has seen decreases in revenue using a similar fee-for-access model. While Facebook announced profitability in September 2009, YouTube has yet to make money with a similar ad revenue scheme. Twitter is the most interesting example, with CEO Evan Williams' claim that the company currently focuses on creating value rather than creating revenue. Obviously, firms should not pin their hopes on finding the right answer with these models.

The evolution of IT research reveals many of the same issues facing digital product profitability models. "Studies have struggled to prove the business value or profitability of IT" (Barua, et al., 2000, p. 26). Due to the dynamic nature of business today, technology has shortened product cycles to the point that competitive advantage is difficult to sustain. IT literature has shown that a firm's ability to integrate technology into its value chain sets it apart from its competitors. This idea is supported by the network effects model (Liebowitz & Margolis, 1994), which states that the more activities of the value chain a firm conducts online, the more likely the business value of e-commerce will be optimized. The answer to the profitability question may lie in how a firm integrates resources, as indicated by Vargo and Lusch (2004). Thus, the following questions must be answered with regard to monetization and regulation of digital products:

- What factors make certain digital products more successful than others?
- Can we create a model for profitability with digital products that works in a culture that demands free over paid content?
- What can we learn from the evolution of IT research on profitability that we can apply to the study of digital products?
- What can we learn from marketing research, specifically S-D logic, that we can apply to the study of digital products?

## CONCLUSION

We have attempted to address the implications of technological advancement on human information exchange and business practice. While we are often enamored with the functions that technology provides, we struggle to identify its true usefulness and value. Often this challenge is due to our own limits of cognition, as we tend to define possibilities within the context of things we already know. Thus, we struggle to integrate the potential of new technologies into our knowledge base because we naturally attempt to force them into a predetermined context.

As with any endeavor that seeks to expand our horizons of thought, we are limited by our own creativity as well. In this chapter we have posed questions that are somewhat obvious but are derived from consideration and integration of prior literature. However, a multitude of questions are certainly left unanswered. The beauty of seeking answers to questions is that no end is ever in sight. And at times when we appear to be approaching omniscience with respect to technology, innovations once again present us with a whole new world of challenges and questions. We believe the questions posed here will keep us quite busy in the near future.

## REFERENCES

Avdic, A. (2002). User and developer-user systems development using a spreadsheet program. In Khosrow-Pour, M. (Ed.), *Issues and Trends of Information Technology: Management in Contemporary Organizations* (*Vol. 1*). Hershey, PA: Idea Group Publishing.

Barua, A., Konana, P., Whinston, A., & Yin, F. (2000). Making e-business pay: Eight key drivers for operational success. *IT Professional*, *2*(6), 22–30. doi:10.1109/6294.888013

Borchers, A. (2003). Intrinsic and contextual data quality: The effect of media and personal involvement. In Grant, G. (Ed.), *ERP & Data Warehousing in Organizations: Issues and Challenges*. Hershey, PA: Idea Group Publishing.

Da Lio, E., Fraboni, L., & Leo, T. (2005, October). *TWiki-based facilitation in a newly formed academic community of practice*. Proceedings of the 2005 International Symposium on Wikis. San Diego: ACM.

Feller, J., Finnegan, P., Fitzgerald, B., & Hayes, J. (2008). From peer production to productization: A study of socially enabled business exchanges in open source service networks. *Information Systems Research*, *19*(4), 475–493. doi:10.1287/isre.1080.0207

Ghose, A. (2009). Internet exchanges for used goods: An empirical analysis of trade patterns and adverse selection. *Management Information Systems Quarterly*, *33*, 263–291.

Guadamuz, A. (2009, May). *If you build it, they won't come: Placing user-generated content in context of commercial copyright policy*. Presented at the Mashing-Up Culture: The Rise of User-generated content Workshop, Uppsala University, Sweden.

Hermida, A., & Thurman, N. (2007, March). *Comments please: How the British news media is struggling with user-generated content*. Presented at the 8th International Symposium on Online Journalism, University of Texas, Austin.

Lessig, L. (2008). *Remix: Making art and commerce thrive in the hybrid economy*. New York: Penguin Press.

Liebowitz, S. J., & Margolis, S. E. (1994). Network externality: An uncommon tragedy. *The Journal of Economic Perspectives*, *8*(2), 133–150.

Majchrzak, A. (2009). Comment: Where is the theory in wikis? *Management Information Systems Quarterly*, *33*(1), 18–20.

Moon, J. Y., & Sproull, L. S. (2008). The role of feedback in managing the Internet-based volunteer work force. *Information Systems Research*, *19*, 494–515. doi:10.1287/isre.1080.0208

Moore, G. (1965). Cramming more components onto integrated circuits. *Electronics*, *38*(8), 114–117.

Raghu, T. S., Sinha, R., Vinze, A., & Burton, O. (2009). Willingness to pay in an open source software environment. *Information Systems Research*, *20*, 218–236. doi:10.1287/isre.1080.0176

Redman, T. C. (1998). Impact of poor data quality on the typical enterprise. *Communications of the ACM*, *41*(2), 79–82. doi:10.1145/269012.269025

Scupola, A. (2005). Strategies of e-commerce business value optimization. In Khosrow-Pour, M. (Ed.), *Encyclopedia of Information Science and Technology*. Hershey, PA: Idea Group Reference.

Smith, M. D., & Telang, R. (2009). Competing with free: The impact of movie broadcasts on DVD sales and internet piracy. *Management Information Systems Quarterly*, *33*, 321–338.

Subramanian, R., & Yen, M. Y. (2002). Digital asset management: Concepts and issues. In Gangopadhyay, A. (Ed.), *Managing Business with Electronic Commerce: Issues and Trends*. Hershey, PA: Idea Group.

Te'eni, D. (2009). Comment: The wiki way in a hurry--The icis anecdote. *Management Information Systems Quarterly*, *33*, 20–22.

Theodorou, P. (2006). A DSS model that aligns business strategy and business structure with advanced information technology: A case study. In Khosrow-Pour, M. (Ed.), *Cases on Information Technology and Organizational Politics & Culture*. Hershey, PA: Idea Reference Group.

Tocci, R., Windmer, N., & Moss, G. (2006). *Digital Systems: Principles and Applications* (10th ed.). Upper Saddle River, NJ: Prentice Hall.

Vargo, S. L., & Lusch, R. F. (2004). Evolving to a new dominant logic for marketing. *Journal of Marketing*, *68*, 1–17. doi:10.1509/jmkg.68.1.1.24036

Wang, R. Y., Strong, D. M., & Guarascio, L. (1996). Beyond accuracy: What data quality means to data consumers. *Journal of Management Information Systems*, *12*(4), 5–34.

## ADDITIONAL READING

Anderson, C. (2008, February 25). Free! Why $0.00 is the future of business. *Wired Magazine*. Retrieved from http://www.wired.com/print/techbiz/it/magazine/16-03/ff_free

Bakos, J. Y. (1991). A strategic analysis of electronic marketplaces. *Management Information Systems Quarterly*, *15*(3), 294–310. doi:10.2307/249641

Batini, C., & Scannapieco, M. (2006). *Data Quality: Concepts, Methodologies and Techniques*. Berlin: Springer.

Brousseau, E., & Penard, T. (2007). The economics of digital business models: A framework for analyzing the economics of platforms. *Review of Network Economics*, *6*(2), 81–114. doi:10.2202/1446-9022.1112

Chakravorti, B. (2004). The new rules for bringing innovations to market. *Harvard Business Review*, *82*, 59–67.

Gallaugher, J., & Wang, Y.-M. (2002). Understanding network effects in software markets: Evidence from Web server pricing. *Management Information Systems Quarterly*, *26*, 303–327. doi:10.2307/4132311

Kannan, P. K., Pope, B. K., & Jain, S. (2009). Pricing digital content product lines: A model and application for the National Academies Press. *Marketing Science*, *28*(4), 620–636. doi:10.1287/mksc.1080.0481

Katz, M., & Shapiro, C. (1985). Network externalities, competition, and compatibility. *The American Economic Review*, *75*, 424–440.

Katz, M., & Shapiro, C. (1986). Technology adoption in the presence of network externalities. *The Journal of Political Economy*, *94*, 822–841. doi:10.1086/261409

Kauffman, R. J., McAndrews, J., & Wang, Y.-M. (2000). Opening the black-box of network externalities in network adoption. *Information Systems Research*, *11*(1), 61–82. doi:10.1287/isre.11.1.61.11783

Klein, B. D. (2001). User perceptions of data quality: Internet and traditional text sources. *Journal of Computer Information Systems*, *41*(4), 9–15.

Kraut, R., Patterson, M., Lunmark, V., Kiesler, S., Mukopadyaya, T., & Sherlis, W. (1998). Internet paradox: A social technology that reduces social involvement and psychological well being. *The American Psychologist*, *53*, 1011–1031. doi:10.1037/0003-066X.53.9.1017

Lang, K. R., Shang, R. D., & Vragov, R. (2009). Designing markets for co-production of digital culture goods. *Decision Support Systems*, *48*(1), 33–45. doi:10.1016/j.dss.2009.05.010

Lee, Y. W., Pipino, L. L., Funk, J. D., & Wang, R. Y. (2006). *Journey to Data Quality*. Cambridge, MA: The MIT Press.

Payne, A. F., Storbacka, K., & Frow, P. (2008). Managing the co-creation of value. *Journal of the Academy of Marketing Science, 36*(1), 83–96. doi:10.1007/s11747-007-0070-0

Rayport, J. F., & Sviokla, J. J. (1995). Exploiting the virtual value chain. *Harvard Business Review, 73*(6), 75–85.

Rogers, E. (1995). *Diffusion of Innovations* (4th ed.). New York: The Free Press.

Saloner, G., & Shepard, A. (1993). Adoption of technologies with network effects: An empirical examination of the adoption of automated teller machines. *The Rand Journal of Economics, 26*(3), 479–501. doi:10.2307/2555999

Schilling, M. A. (1998). Technological lockout: An integrative model of the economic and strategic factors driving technology success and failure. *Academy of Management Review, 23*(2), 267–284. doi:10.2307/259374

Schilling, M. A. (2002). Technology success and failure in winner-take-all markets: The impact of learning orientation, timing, and network externalities. *Academy of Management Journal, 45*(2), 387–398. doi:10.2307/3069353

Shapiro, C., & Varian, H. (1998). *Information Rules*. Boston, MA: Harvard Business School Press.

Shaw, M. J., Gardner, D. M., & Thomas, H. (1997). Research opportunities in electronic commerce. *Decision Support Systems, 21*(3), 149–156. doi:10.1016/S0167-9236(97)00025-0

Shy, O. (1996). Technology revolutions in the presence of network externalities. *International Journal of Industrial Organization, 14*, 785–800. doi:10.1016/0167-7187(96)01011-9

Song, J., & Walden, E. (2007). How consumer perceptions of network size and social interaction influence the intention to adopt peer-to-peer technologies. *International Journal of E-Business Research, 3*(4), 49–66.

Srinivasan, R., Lilien, G., & Rangaswamy, A. (2004). First in, first out? The effects of network externalities on pioneer survival. *Journal of Marketing, 68*(1), 41–58. doi:10.1509/jmkg.68.1.41.24026

Strader, T. J., Ramaswami, S. N., & Houle, P. A. (2007). Perceived network externalities and communication technology acceptance. *European Journal of Information Systems, 16*(1), 54–65. doi:10.1057/palgrave.ejis.3000657

Strong, D. M., Lee, Y. W., & Wang, R. Y. (1997). Data quality in context. *Communications of the ACM, 40*(5), 103–110. doi:10.1145/253769.253804

Thurman, N. (2008). Forums for citizen journalists? Adoption of user generated content initiatives by online news media. *New Media & Society, 10*(1), 1–30. doi:10.1177/1461444807085325

Vargo, S. L., & Lusch, R. F. (2006). *The Service-Dominant Logic of Marketing: Dialog, Debate, and Directions*. Armonk, NY: M. E. Sharpe.

Viswanathan, S. (2005). Competing across technology-differntiated channels: The impact of network externalities and switching costs. *Management Science, 51*(3), 483–496. doi:10.1287/mnsc.1040.0338

Windrum, P., & Birchenhall, C. (2005). Structural change in the presence of network externalities: A co-evolutionary model of technological successions. *Journal of Evolutionary Economics, 15*, 123–148. doi:10.1007/s00191-004-0226-8

## KEY TERMS AND DEFINITIONS

**Business Model:** A set of planned activities designed to produce a profit in a marketplace. The primary components include the value proposition, revenue source, intended market, competitive environment, competitive advantage source, strategy, organizational structure, and management personnel.

**Data Quality:** Quality of data goes beyond simple accuracy measures. Data quality can be viewed from four different perspectives including intrinsic, contextual, representational, and accessibility data quality.

**Network Externalities (Network Effect):** Positive network externalities are associated with the phenomenon where the value of a product or service increases as the number of users increases. For example, an online auction site's value would increase as the size of the participant base increases because there would be more potential buyers and sellers with which to potentially trade.

**Operant Resources:** Resources that are used to operate on another resource to produce an effect. Knowledge and skills could be considered to be operant resources. Digital assets may also be described as operant resources.

**User Systems Development (USD):** Development of information systems by user/developers who may have little information technology expertise.

# Chapter 15
# Digital Technology in the 21st Century

Troy J. Strader
*Drake University, USA*

## ABSTRACT

*Have digital technologies reached their full potential? It seems pretty clear that the answer is no. All aspects of digital technology continue to evolve as scientists make discoveries, managers incorporate these discoveries into their products and services, government agencies utilize new technologies to improve service provision and information management, and social scientists investigate the impact these new technologies may have on social interaction, the global economy, and society as a whole. This chapter describes recent developments in a range of digital technology areas including input devices, output methods, storage technology, process technology, and various applications including enhancement of the sensory rich environment enabled by digital technology, deep Web search, online language translation, improved security methods, automated payment systems, and interplanetary Internet. Impacts on broader societal institutions such as healthcare, government services, higher education, political campaigns, cybercrime law enforcement, and life at home are also identified. Digital technology trends and implications for digital product managers are discussed as well as directions for future research.*

## INTRODUCTION

Existing digital technology is capable of storing and processing vast amounts of data as well as enabling user-friendly interaction between machines and people, but there is still a lot of room for improvement. Digital technology is based on electromagnetic storage and processing using billions of switches that can be set to one of two positions – on or off. This basic concept has been around since the development of the first electromagnetic computers prior to World War II and has enabled input, processing, output, storage, and telecommunications, and the development of numerous applications that support a wide range of human activities.

DOI: 10.4018/978-1-61692-877-3.ch015

In many of the previous chapters the focus has been on developments in digital technology, applications, business strategy, and related societal issues that have occurred in the past decade or two. The primary purpose for this chapter is to present and discuss technologies that are currently under development and will most likely be available in the early part of the 21st century. These new and potential technologies will enable new applications and also create opportunities and threats for existing digital product companies. They are also likely to impact our global society because they will enable new forms of computing, applications, sensory-rich machine/user interaction, and electronic communications.

This chapter is organized as follows. First, basic digital technologies available during the past few decades are presented to provide a baseline for identifying areas for improvement across the spectrum of present technological capabilities. In the main section for this chapter several examples of interesting new digital technologies are presented and discussed to identify some of the potential improvements in this area that may be adopted in the 21st century and provide opportunities for new digital products, digital services, and improved information search and communications.

Next, impacts on broader societal institutions such as healthcare, government services, higher education, political campaigns, cybercrime law enforcement, and life at home are identified. The chapter concludes with a discussion of trends and directions for future research associated with new digital technologies and applications.

## BACKGROUND

Most digital technologies fall into one of the following categories: input technology, process technology, output technology, storage technology, telecommunications technology, multimedia protocols, software applications, and Internet/Web protocols and applications. Figure 1 describes the relationship between these digital technologies and lists some common examples available at the end of the previous century.

Each of these examples is a big improvement over the earliest digital technologies available in the 1940s and 1950s (Turban & Volonino, 2010). Early input and storage methods involved punch cards and magnetic tapes. They were slow, took a lot of space, and only supported the simplest forms of sequential file processing. Early proces-

*Figure 1. Common digital technologies in the late 20th century*

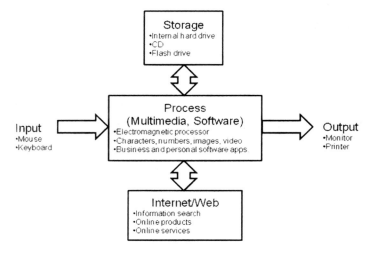

*Digital Technology in the 21st Century*

sors used vacuum tubes that were large, hot, unreliable, and used a lot of electricity. Output may have used a slow teletype-like printout. And in this earliest digital age, there was no Internet so there were no applications such as e-mail or the Web. Networking of individual computers was very rare and involved costly, slow, unreliable connections. Given the explosion of individual computing activities that occurred after the commercialization of the Internet in the early to mid 1990s, most people understand the basic capabilities and limitations of current digital technology as shown in Figure 1. People use digital technologies not only in computers, but also televisions, cameras, phones, music players, and many other devices. What is not as well known is that many exciting new technologies are currently being developed which will impact our lives in the early part of the 21st century.

## Advanced Digital Technologies for the Future

Today, and in the future, individual digital technologies will be developed to better perform various tasks. Input devices convert human activity into a digital representation. Smaller and faster process technology performs basic numerical and logic tasks on input data to produce some form of output. Output technology transforms digital representations into a format that a human can see or hear. Storage technologies place digitized data into denser, more efficiently organized, devices. Telecommunications transmits larger and larger volumes of data more quickly, safely, and with fewer transmission errors. And on top of these hardware devices and protocols a wide range of novel applications can be developed to gather and search vast amounts of information, replicate real world sensory-rich environments, and provide many different services.

An interesting question is why we even need newer digital technologies. One of the reasons is that new technologies can improve all aspects of society including the work environment, economy, governance, home life, and nearly every other human activity. Another reason we need improved digital technologies is to manage the explosion of online activity that is already taking place. Some interesting statistics illustrate just how big the digital explosion has been (Mearian, 2008). By 2011, it is forecast that there will be 1,800 exabytes of electronic data in existence. This is 1.8 million terabytes, or 1.8 billion gigabytes. The compound annual growth rate is almost 60% and most forecasts from the 1990s and 2000s underestimated the growth rate because they did not anticipate the explosion of the user-generated content. Improvements in digital technology are a must because something will be needed to store the data, and even more importantly, to find it and use it. Digitally enabled personal interaction through sites such as Facebook is another area that provides some interesting statistics (Facebook, 2010). In early 2010 there were more than 400 million active Facebook users. Half of them log on in any given day. In a month they upload three billion photos, and more than five billion pieces of content are shared each week. And each of these numbers continues to grow. Obviously, new digital technologies will also need to support growth in these applications through better interfaces and telecommunications.

In this section a number of recent real-world digital technology developments are discussed. The examples are summarized in Figure 2. The examples are not intended to illustrate all possible future digital technologies, but they are sufficient to show that developments are occurring in a wide range of hardware and software areas.

## Brain-Computer Interfaces

Inputting data into a computer typically requires the user to perform some manual task such as typing or moving and clicking a mouse. Bar code readers and point-of-sale systems make data entry easier, but still require hardware devices and some human involvement. Imagine if a person was able to manipulate a computer or machine just by thinking about what they want it to do. Scientists are currently working on several forms of brain-computer interfaces (BCI) that will enable users to interact with machines through thought. Every time we think, move, feel, or remember, signals are generated in our brains through differences in electric potential carried by ions on the membranes of our neurons (Grabianowski, 2009a). A BCI interprets these signals and can send them to a computer just as if the user was typing a command or clicking a mouse. One of the biggest challenges is to create the interface. The least invasive interface uses a set of electrodes attached to the user's scalp, but it may not be able to read all signals and some signals are distorted. To improve the signal, electrodes can be directly implanted into a user's brain itself. The downside is obviously that this method is far more invasive and scar tissue may develop around the implant which reduces its capabilities over time (Grabianowski, 2009a).

A brain-computer interface has a number of potential applications. Users could control games or remote controls through their thoughts. They could also enter data into a computer or select a link when using an Internet browser just by using their thoughts. More importantly, a BCI could allow disabled users to interact with computers and machines in ways that are not available using current digital technologies. BCI technology has enormous potential, but also has some drawbacks. The brain is complex and signals are weak so some brain activity will not be able to be interpreted correctly. And current BCI equipment often involves hardware that is not easily portable or must be wired between the brain interface and computer (Grabianowski, 2009a).

NeuroSky has developed a brain-computer interface peripheral device that works with most current personal computers (NeuroSky, 2009). The MindSet resembles a pair of headphones with an electrode-laden arm that is in contact with a user's forehead allowing interpreted signals to be transmitted to common entertainment, education, home, and work-related software applications. It may be some time before BCI is widely adopted, but if it provides cheap and reliable interaction with computers and machines through thought rather than manual effort on the part of the user, then it may replace many of the input devices the people rely upon today.

*Figure 2. Example new digital technologies for the 21$^{st}$ century*

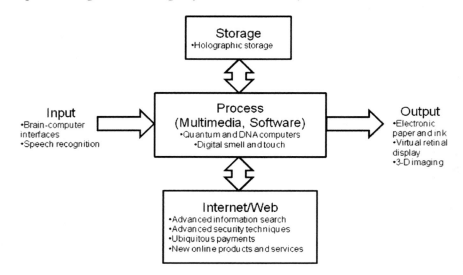

## Speech Recognition

Another technique for efficient data input is speech recognition (Grabianowski, 2009b). Some common uses are in automated telephone customer service systems and inputting data into word processing documents. Telephone systems used to require that a number be pressed on the phone to select an option. In many instances this has now been replaced by speaking a number or word. Personal computers can include input devices that translate speech into text rather than requiring a user to type on the keyboard. Speech, or voice, recognition is available today, but it is not widely used. Most interaction with computers is still through the traditional mouse and keyboard.

The process to translate speech into words on a screen involves several steps (Grabianowski, 2009b).

1. An analog-to-digital converter captures the sound wave and translates it into a digital representation.
2. A system filters the digitized sound to remove noise that will make conversion more difficult in later steps.
3. The digital sound is adjusted to a constant volume level.
4. The signal is divided into discrete segments and the segments are matched to know phonemes. Phonemes are consonant sounds that are unique to the language. English has roughly 40 unique sounds.
5. The phonemes are examined in context and a statistical model is used to identify the most likely word or phrase being spoken.
6. The result is either typed on the screen, or a command may be issued to the computer.

It is apparent that there is a lot of room for error when using a speech recognition system. Even humans misinterpret words and phrases when talking to other people. Systems using artificial intelligence may be able to learn how a particular person speaks which could improve its ability to identify specific words or phrases. There are a number of potential advantages if these systems can be perfected. First, interaction with machines will mimic interaction with other people to bring us closer to a world where machines and people can work together more efficiently. People with disabilities will also benefit from these systems because they will be able to type or issue commands to a machine by only having to speak.

One company that provides speech recognition software for many of the most common platforms is Nuance (http://www.nuance.com/naturallyspeaking/). It is an efficient and useful tool because people can typically speak three times faster than they can type. Many keyboard tasks can be accomplished by translating speech into keystrokes. The program can be used in a range of situations where people do not have easy access to a desktop computer and need to document their activities. Some examples are sales, insurance claims, healthcare, and law enforcement.

## Electronic Paper and Ink

Paper and ink is a simple way to record human thoughts and activities and it has been around for nearly two thousand years. Coupled with printing presses, paper and ink has had a profound impact on people and organizations. Even in the digital world people continue to print out Web pages, receipts, and e-mail messages, but could there be a method that captures the best aspects of paper (portability, relatively low cost, user acceptance) with the best aspects of digital output (less resource use, reusability)? One possibility is electronic paper and electronic ink. Scientists are working on both of these ideas but their use is currently limited.

Electronic ink (e-ink) uses millions of tiny microcapsules that are filled with white and dark chips or balls (Bonsor, 2009). When an electrical charge is applied to the microcapsules, the chips will either rise to the top to display one color or be pulled to the bottom to display the alternative color. Each microcapsule is similar to a pixel on a computer monitor that can each independently display a different color which will result in an overall image that the user can see. Only using positive and negative charges will limit the display to either white or dark unlike the millions of colors available using the red-green-blue (RGB) format that is used in computer monitors. E-ink can be spread over a paper or thin plastic surface and the display can be altered by changing the arrangement of positive and negative charges applied to the surface. With enough microcapsules arranged close together on a page, a person will not be able to tell the difference between traditional printed pages or those displayed using electronic ink.

Electronic ink has several advantages (Bonsor, 2009). It can be used on a wide variety of surfaces and it provides a readable output so users will not have to dramatically alter the ways they read books or newspapers. It also could eliminate the need for many tons of paper. One set of pages in a book could theoretically be used to store any book content and it could easily be changed to another book's content with a wireless download.

An alternative to electronic ink is electronic paper. Color electronic paper (e-paper) looks like regular paper, but it can display more than one million different colors using ambient light (Nusca, 2009). Researchers at Philips have developed this technology that can be used for changing colors on a wall or regulating the amount of sun coming through a window. The technology uses colored charged particles that are controlled by an electrode to generate a spectrum of color combinations similar to the colors available on a television or compute r monitor. The technology is relatively inexpensive and can save time and money when changing the color of an object. Currently its use is very limited, but it has the potential to be used commercially in the future.

## Virtual Retinal Display

Some of the downsides for current computer display technology are that they can take a lot of space and they are not very portable. Future output technology could overcome these problems by projecting images directly into the eye. The idea for virtual retinal displays (VRD) has been around for two decades and despite technology breakthroughs it will be some time before it is available commercially (Wolk, 2009). A VRD uses ultra low power lasers to paint an image onto the human retina using a combination of red, green, and blue light. These displays would not block normal vision, but would overlay images in a person's line of sight without requiring them to move their head. The benefits are that VRD is takes far less space than a typical computer monitor and uses far less energy. Early applications were for the military, but there are numerous commercial and home applications that will be developed in the next decades. Some roadblocks have slowed development of VRD technology. It is difficult to project images on eyes that are moving, the equipment is still not very mobile, and it is an expensive technology. Ultimately this could provide a virtual output technology that could be used anywhere and anytime, unlike a traditional computer monitor.

## Digital 3-D Imaging

Three dimensional movies have been shown in theaters for some time. Imagine the future where all digital output could be three dimensional. Rather than viewing a two dimension image of a product online, a customer could see a more realistic three dimensional representation. Electronic books could include more realistic pictures and online movies could provide a more realistic experience combining surround sound and 3D video. Games could also benefit from more realistic user experiences.

It is interesting how 3D digital image technology works. Essentially it is a projection intended to fool your eyes and brain into thinking that you are seeing a real-world image (Wilson, 2009b). Today's 3D imaging systems use polarization. Polarized lenses allow only light waves that are aligned in a particular direction to pass through the lenses. In a pair of 3D glasses, each lens is polarized differently. A screen is aligned to maintain the correct polarization when the projector light bounces off it. The polarized glasses combine the two image sources and you see what appears to be a three dimensional moving image. Unlike early 3D movie technology, this technique does not produce side effects like headaches.

This has become a popular addition to the movie experience and there are numerous applications in other digital product areas that may be available in the future. Consumer demand for 3D entertainment has created opportunities for sites that offer other 3D-related products and services. There are a number of sites that help people convert their 2D images into 3D. This technology will be particularly useful if 3D images can be viewed without special glasses or projection screens.

## Holographic Storage

Prior to the development of the digital computer, information was typically stored on paper and was organized using folders and file cabinets. This was an expensive way to store large amounts of data and it was often difficult to find the data needed to create a report or assist in making a decision. The earliest forms of computer-related data storage involved punch cards or magnetic tapes. The advent of the personal computer created an opportunity for a number of improvements in data storage from cassette tapes to diskettes to CDs and flash drives. Commercial and scientific data storage requirements have grown quickly from kilobytes to gigabytes to terabytes. Over time data was able to be stored more densely and cheaply which enabled a number of more complex software applications. The commercialization of the Internet and expansion of online search capabilities has led to the need for even more data storage and one new technology that may provide a solution is holographic storage.

Holographic storage uses an optical process where data is recorded in light patterns that are stored on a light sensitive material (Lohr, 2009). The holograms are like microscopic mirrors that reflect light patterns when a laser shines on them so that data can be read from the disc. General Electric has developed a holographic disc in a laboratory that can hold 500 gigabytes of data which is 100 times more than a standard DVD. The technology is not yet available for the commercial market, but it does show that there is potential to make great leaps in storage density and cost. Given the volumes of data being collected by scientific labs and online companies there is certainly a market for improved data storage products.

## Quantum and DNA Computers

Can a computer processor ever be too fast? Probably not. There will always be a complex software application that would benefit from producing results more quickly. In the earliest days of digital computers, one of the major issues that had to be addressed was how to improve the hardware used for processing. Vacuum tubes provided an early solution, but they were slow, expensive, unreliable, hot, and needed a lot of space. An early computer may have only included a few thousand tubes which greatly limited its capabilities. Transistors replaced vacuum tubes in the 1950s and 1960s and they enabled smaller, cheaper, and faster processors. Today's computers still use this same digital processing electromagnetic switch concept, but microprocessors now include billions of miniature transistors and they take up a very small amount of space. As software applications become more complex (for example, weather forecasting), and huge amounts of data must be stored and accessed, improving processing speed will continue to be an issue. Two new processor technologies are quantum computers and DNA computers.

Although they are years away from providing a practical computing platform, quantum computers are currently being developed that use atoms and molecules to perform storage and processing tasks (Bonsor & Strickland, 2009). In a typical electromagnetic computer, a switch described as a binary digit (bit) can hold one of two settings, on or off. In a quantum computer a quantum bit (qubit) can theoretically hold a value of one, zero, or both simultaneously. Because a quantum computer can have multiple states at the same time it has the potential to be millions of times more powerful than today's computers. There are a number of technological issues that must be addressed prior to manufacturing a commercially viable quantum computer, but there is great potential if it is available in the future. Processing power can increase while size can decrease. There is also a nearly unlimited amount of atomic material available which would reduce resource usage.

Another direction for future processor technology is to base storage and processing on biological material such as DNA. Scientists have made a computer from a circular piece of DNA, also known as a plasmid, inserted into a living cell and then they were able to show that it could solve a mathematical problem (Swaminathan, 2008). The unique ability of DNA computers would be that they are able to be inserted into living cells which would be very difficult for a traditional silicon-based electromagnetic computer. Just like quantum computers, DNA computers are years away from commercial availability but they potentially provide a smaller and faster processor and there is an unlimited amount of DNA material available because of their ability to replicate.

## Digital Smell Technology

Can a smell be digitized? The smell itself cannot, but if a representation of a smell can be digitized then it can be stored and distributed just like text, images, audio, and video. If this is possible then a virtual world of entertainment and shopping could expand its sensory environment to include smells. Instead of just seeing and hearing a movie, a person could also smell the environment shown on the screen. The same is true for shopping when a smell may increase a consumer's likelihood of buying a product such as perfume, candles, wine, or flowers. These days may not be far off.

Digital smell technology, scentography, is being developed and we should soon be able to access digitally encoded smell data using a device that produces smells (Platt, 2009). The device is about the size of an electric pencil sharpener and works as a peripheral device with a personal computer. A digital file is input into the device where the appropriate heated vials of oils interact to produce a smell that is blown out of by small fans. In current multimedia environments, a sound can be synchronized with a visual image. This digital smell technology would be able to add an additional sensory experience where smell would be synched to the sound and moving images.

A number of issues will need to be addressed as this technology becomes available. How much will the hardware and oil refills cost for a typical home user? How many unique smells can be digitized? And which product areas will benefit from the use of this technology? As with any new technology development, it will take some time to diffuse throughout all of the potential industries and users and each will be motivated to adopt the technology for different reasons. In the early twentieth century, movies did not include sound and many people did not see the need. A century later it is hard to imagine a movie that would be silent. Perhaps in the next decades the inclusion of smell with a movie experience will be just a normal as sound is today, and it will not require a moviegoer to get the smell sensation through a scratch off card. Home-based aromatherapy would also be made possible by this technology.

## Digital Touch Technology

Another sense that could be digitized is touch. In the real world people touch many different objects in a typical day and they incorporate the information gathered through these actions into their behaviors. In a shopping environment there are limitless possibilities for enabling customers to touch products online to enhance their shopping experience. Touch could also be incorporated into entertainment venues to create a more realistic experience.

One term used for the field of digital touch technology is computer haptics (Harris, 2009). When someone touches something, receptors on the body perceive several forms of stimulus: light touch, heavy touch, pressure, vibration and pain. A computer haptic system attempts to replicate the gathering, transmission, and analysis of these stimuli. A computer haptic system would include software to determine the forces involved in interacting with an object in a virtual environment, and hardware that applies these forces to the user. The process used for performing this action is called haptic rendering. Digital touch could be incorporated into a wide variety of entertainment, educational, social interaction, and commercial experiences with the goal being to more closely replicate a real-world experience. In the end, the full online sensory experience may require all senses – sight, sound, smell, touch, and taste – be provided to the user to round out the experience because they each individually complement each other (Madrigal, 2009).

## Deep Web Search

Digital technologies enable the collection and storage of vast amounts of data associated with human activities such as shopping, communications, financial transactions, and information search. The problem of the future will not be whether data is available, but how to use it to support problem solving. Creating a method for improved data collection and organization online is one example of an attempt to improve online information search in the future. Even with the tools available today, it is still frustrating to get thousands or millions of Web page results when using a search engine. Part of the problem is that even with the massive volume of information available to search engines, there is an even larger amount of information that is largely invisible to the crawler and spider technology that identifies and classifies information gathered from Web pages by following hyperlinks between sites (Wright, 2009).

Several organizations are attempting to mine online data to a deeper level using automated collection technologies. The challenge will be to incorporate deep Web content into search results without complicating the results shown to users. Turbo10 is available today as a deep Web search engine (Boswell, 2009). It is a metasearch engine that searches through many different niche-specific search engines plus government, business, and academic databases in an attempt to provide more useful results. The idea behind deep Web searching is promising, and utilization of higher speed processing technology should help it become a commercially viable future technology. There is a tremendous demand for improved online information search and many different perspectives on how to best solve the problem.

## Online Language Translation

Another digital technology related service that will bring global users closer together, and provide greater communications and commerce opportunities, is online language translation. Global communications requires digital systems that can connect people anywhere in the world, and the Internet provides this infrastructure. But there is still a major obstacle that must be overcome. There are hundreds of major languages and thousands of other regional dialects and obscure languages. When two people talk to each other face-to-face they are often speaking the same language because it has evolved that way in their locale over many centuries. When connecting people from one part of the world to another it is likely that they will not share the same first language. A question is whether technology can provide a solution to overcome this second obstacle to global communications. Can a digital service be created to automatically translate words and phrases from one language to another? And can it do it quickly enough to not only convert text and documents, but also real time voice and text communications.

Very basic online translation services are available on several Web sites and the service is typically free. More complex translations services are available for a fee. There are several existing Web sites that offer basic translation services for many common languages. One site, WorldLingo, (http://www.worldlingo.com/en/products_services/worldlingo_translator.html) offers free translation for up to 500 words. Type or paste in the text to convert, choose the language to translate from and to, and then click the translate button. To improve the quality of translations there is also a drop down list to choose the subject for the text. Their service includes about 140 languages and they work with companies in most major industries. Some of their services include translations for websites, text, e-mail, and chat.

A similar online translation service is available at Yahoo (http://babelfish.yahoo.com/). One free service will translate up to 150 words. Type or paste in the text, choose from about 40 combinations of from and to languages, and then click the translate button. It also offers a service to translate a Web page by providing the page's URL and then clicking translate. A third example is available on Google (http://translate.google.com). It provides translation for text, Web pages, or documents. Enter text, provide a URL, or upload a document, choose the form and into language, and then click the translate button. About 50 languages are included. They also offer a translator toolkit that supports other more complex translation services.

A number of other sites currently provide free or professional translation services online. They also provide dictionaries and language learning services. With continued technological developments in this area, there is the opportunity for automatic language translation in any situation where people are interacting online. Two people could have a phone conversation, with each person using their primary spoken language, and technology could automatically translate the voice in both directions. This would eliminate any language barriers. This could also automatically translate any online text or Web pages into an understandable language so that an online company would only ever have to develop their site in one language. This would save a lot of time and effort. These services also provide the potential for a Web site customization feature that would allow every user to choose the language they want to use when viewing the page's contents. With improvements in technology, there is also the potential for automatic translation for multimedia files such as audio clips and the audio portion of a video file.

A downside could be that ultimately some spoken and written languages will no longer be needed and will disappear. The cultural history embedded in language would be lost forever. To reduce the likelihood of this occurring, digital technology can capture and store examples for all existing written and spoken human languages which will mean that they will never ever truly be lost.

## Advanced Security Techniques

Many of the future digital technologies discussed have been associated with improving interaction between people and machines, or providing opportunities for more complex applications. As important are digital technologies that will improve security for users of these systems. The following discusses three advanced security techniques: quantum encryption, biometric access controls, and a system for simplifying the login process for multiple websites.

## Quantum Encryption

Businesses are increasingly connected by computer networks and the Internet. This is both a blessing and a curse because it enables all kinds of electronic communications and data transfer, but it is also much more difficult to ensure that business secrets are kept safe. For digital products this is particularly important because the product itself can be stolen through a network connection if security is not sufficient. One future technology solution that may improve security is to introduce more complex encryption techniques. Encryption scrambles data being stored or transmitted so that even if it is stolen it will not be in a usable form. The text can't be read and the multimedia file can't be played. The future may find that the next big thing in encryption is quantum encryption.

Encryption uses an algorithm or cipher to convert digital content into an unreadable form. A key is used to unscramble the encrypted file to bring it back to its original contents. The more bits in the key, the more difficult it is to break. This idea has been around for a long time, but a new method involves quantum encryption where photons are used to make up qubits. Qubits were described earlier in the section on quantum computers. Quantum encryption is based on the same particle behaviors that enable quantum computers, but the purpose is much different. Quantum encryption uses photons transmitted via fiber-optic cable (Clark, 2009; Mayne, 2009). A photon's polarization can be measured in three ways – the horizontal axis, diagonal axis, and the circular axis (Mayne, 2009), but measurement of a photon may alter its value which makes it a complex but highly useful method for encryption. Quantum encryption can be done in one of two ways: measurement protocols or entanglement protocols. Using the measurement protocol, the phenomenon that measuring an unknown quantum state changes the state makes it impossible to intercept a message without changing it. This makes it easier to identify messages that have been intercepted. In the entanglement protocol, the quantum states of two separate objects are linked so that they reflect each other's state. If one is measured then both are changed which again allows for the situation where a message has been intercepted to be recognized. Another advantage of quantum encryption is that once a photon-string key has been measured it ceases to exist which makes it nearly impossible to duplicate or reuse. Increased processing power makes this form of complex encryption a possibility, and it may be necessary in the future as hackers learn to break into systems that are using current security techniques.

## Biometric Access Controls

Another method for improving security is the use of biometric access controls. Many of these techniques have been available for years, but they are still not widely used. A biometric control is a method used to verify a person's identify based on their physical or behavioral characteristics (Turban & Volonino, 2010). Physical characteristic biometric systems include fingerprints, retinal scans, hand or finger geometry, vein geometry, and facial recognition. Behavioral examples include voice scans, handwriting, and keystroke dynamics.

Each system has some common characteristics (Wilson, 2009a). First, a system collects your basic identifying information (name, ID number, etc.) and then analyzes your trait and stores it in a digital format for later comparison. When using the system later, it compares the trait stored to the trait you present. It either accepts or rejects your attempt to enter the system. Fingerprint systems are fairly simple because they are matching a set of points on a finger pressed to a sensor to the fingerprint stored in their database. Retinal scans match blood vessel patterns in a person's retina to the pattern stored. Hand and finger geometry can be used to authenticate a person's identity in much the same way.

Some methods are more complex. Voice scans attempt to match words or phrases spoken by someone with a sample that was stored previously. The way a person says something will vary from time to time which complicates the matching process. The data used in a voiceprint is a sound spectrogram. A spectrogram is a graph that includes sound frequency and time. A benefit is that a person does not have to be at a particular location to authenticate their self. They could talk on a telephone to provide a means of authenticating their identity. Another complex biometric system is based on handwriting. As you know, each time you sign your name it will vary to some extent. These systems do not try to match a signature with a stored signature based on what is visible. Someone could learn to copy another person's signature. Handwriting biometric systems match based on the pressure a person uses and the speed and rhythm at which they write. This is harder to reproduce. There are enough variations in pressure, speed, and rhythm that handwriting systems can match a stored pattern for authentication.

Accuracy of stored traits and measurement systems is obviously an important issue, but in most situations the systems work for authenticating a user's identity. The benefit of using biometric access controls is that is does not require a user to remember an ID or password for each separate system. Most user traits do not change so they are able to be used for access to many systems for a long period of time. Future biometric access control systems may also utilize more than one trait to verify identity. This will improve overall security, but still not place much burden on the user. There are some concerns with these systems. The systems are not fool proof and overreliance on one system may be risky. Some systems are not accessible by all people which would limit their usefulness. And if different agencies use different systems there may be difficulties in sharing information about potential misuses. This is particularly important given the worldwide terrorism threats.

## OpenID

In the future, people will be doing more activities online and most of these systems will require some form of ID and password for access. Ideally each user would have a different ID and complex password for each system, but at some point people will not be able to remember them all. At the same time, the risks involved with someone stealing your personal or financial information is also growing. One method that could be used to simplify access is to have a common login and password for many, if not all, of the sites a user accesses. OpenID is one example of a system that may provide this now and in the future (openid.net). It enables users to enter multiple sites with a common login and basic profile information can be automatically entered into online forms.

The critical issue for OpenID is to reach a critical mass of participating sites. When starting out there was little value to having a common login that could only allow access to a few sites, but as the number of participants grows so does the value of the OpenID system. In May, 2009, Facebook began to allow users to log in using OpenID (Kirkpatrick, 2009). Other OpenID foundation corporate members include Google, IBM, Microsoft, Paypal and Yahoo. The benefits of this system are obvious, but so are the risks. If someone can steal your OpenID login information then they would be able to access multiple sites instead of just one. So far it appears that early adopters are willing to take these risks.

## Ubiquitous Payments

Imagine a world where payment is ubiquitous. There is no need to choose between cash, check, or credit card. Payments just happen automatically as they are required. Digital technology has been used in payment systems for decades because payment transaction information can be represented in a digital format and may only involve the need to add or subtract an amount from a person's account.

## Digital Technology in the 21st Century

The earliest forms of payment outside of barter included gemstones, animal hides, or salt. These were physical goods and were a cumbersome and inconvenient method for making a payment. Gold and silver coins were a better solution, and paper money allowed for possessing larger monetary values in a smaller space. The increased digitization of money has been occurring for awhile and has become a trusted form of payment, but many payment situations still require actions on the part of the payer. In a ubiquitous payment environment, no actions may ever be required.

Think of the example where you are shopping in a grocery store. Move the cart up and down the aisles to select the items to purchase, but then the shopper just walks through a device that identifies the products by their RFID tags, adds up the total amount owed, and automatically recognizes the shopper and chargers their account in the same way that a toll is paid by driving a car slowly through an automatic payment lane. No cash, checks, or credit cards are needed. It provides a highly convenient payment method, but a number of digital technologies must work with little room for error. Errors in product identification, pricing databases, individual recognition systems, and payment system account numbers could cause major problems. Grocery stores are currently reducing their number of employees by using self checkout lanes, but this would take it a step further. Employee reductions may be offset to some extent by the need for increased digital system security and maintenance. Grocery stores are just one example of a retail scenario where this could be used, but there are infinite other possibilities.

Another direction for future payment systems is to enable organizations and individuals to swap items of value through online barter systems. An accounting firm has extra time to work on someone's taxes, and an office supply store has some extra chairs in inventory. If both companies are in the same online bartering system then they can match their needs and trade without any cash trading hands. This is another example of a service that requires a critical mass of users to make it a viable and valuable service. An offline barter system would be nearly impossible, but it is a possibility given the massive number of current Internet users. Starting a barter site is very difficult, but once it reaches a certain size it will become more valuable to the next companies that join. A number of sites already provide these services either to businesses, or individuals, or both. Some sites also focus on niche industries which may make finding trading partners easier when needing a unique item.

**Interplanetary Internet**

Individual hardware and software technologies are useful by themselves, but they have even more impact when they are interconnected. The evolution of electronic communications technologies has seemingly reduced the size of the earth and this has occurred in several stages as new technologies were invented and then applied to real-world problems.

The means with which we communicate have changed dramatically in the past two centuries. Improved mail delivery systems in the mid-1800s allowed people to send and receive letters and other printed material to communicate with family and friends, and this also led to the development of new business models such as catalog retailing. At the same time, the railroads provided a method for efficiently delivering large volumes of correspondence and products. Later in the 1800s, telegraph technology provided a method for communicating simple information quickly and efficiently between interconnected points. News in one city could be shared with others cities in minutes instead of days or weeks. When first introduced it was not very useful because not enough people had the equipment, but over time new businesses were created that provided many people with the service at a reasonable cost. Phones, radio and television provided even richer interaction possibilities throughout the twentieth century. The Internet has jump-started an entire new age of electronic communications, but we typically only think of it in terms of connecting people on our planet. The 21st century will see the introduction of electronic communications systems that go beyond Earth. As with earlier communications systems, it will be expensive at the beginning and many people will not see a use for it, but over time more applications will emerge and it will ultimately be a cost-effective means of connecting Earth with space.

The next phase in the evolution of the Internet will take us outside our planet and lay the groundwork for a communications system that links our planet with space stations and manned missions to other parts of the solar system (Bonsor, 2010). At this early stage, the primary question is whether the system can work. What are some of the unique challenges? And what are some of the potential solutions?

A wired system is impossible for the interplanetary Internet, so the system must be wireless and this introduces some challenges (Bonsor, 2010). The first challenge is distance. Internet communications on Earth only covers a small distance relative to the millions of miles and light years that separate planets. This distance creates a significant time lag between sending and receiving information which can cause problems when combining digital data packets. This delay can be minutes or hours in some instances. Another challenge is line-of-sight obstructions. Wired systems do not have this problem, and Earth-based global wireless communications can be supported by a set of satellites. Each sending and receiving device can be connected to at least one satellite, and the satellites are high enough in the sky that they can directly communicate with the next satellite in the chain to move data and voice anywhere on Earth. Interplanetary systems are much more complex because there are more situations where line-of-sight is blocked at least for part of each day. A third challenge is that antenna equipment that is powerful enough to overcome some of these problems is too heavy to send into space. Thus, this system must rely on a larger number of less powerful transmission technologies.

One solution to these problems is to create a new Internet protocol that can transfer data effectively even in situations where time delays and obstructions are common. The parcel transfer packet (PTP) is one example where data could be stored and forwarded to a planet's gateway as connections become available. Current data transfer rates in space are very slow and this should help overcome this problem. Increased data transfer rates will support improved communications both in space and between space and Earth. This will provide opportunities for more space-related research and exploration and the collection and transmission of more data.

## NEW DIGITAL TECHNOLOGIES SUMMARY AND CONCLUSIONS

Internet applications can benefit from the capabilities enabled by new digital technologies. Information search can be faster and provide more useful results based on an analysis of vast amounts of stored information as well as an understanding of individual user requirements. Online products and services can incorporate more sensory-rich environments in an attempt to replicate the traditional shopping experience. How a product smells or feels may increase buying of certain products and it also creates new industries for the manufacture and sale of hardware and software used to provide these sensory experiences. New digital technologies will perform tasks that are unheard of today, and machines may be able to serve people better by performing a wider range of tasks that are hazardous or tedious for people. Machines will also use higher level reasoning to perform tasks rather than finding answers through simple deduction or blind search through all pos-

*Digital Technology in the 21st Century*

sible alternative solutions. It is hoped that future digital technologies may also be able to improve security and privacy for online environments using advanced encryption techniques.

The digital technologies discussed throughout this section provide a number of opportunities and threats for digital product companies. New information search methods, new digital products, and new digital services will certainly be developed to fully take advantage of digital technologies that are smaller, faster, cheaper, and more efficient that previous technologies. The cycle will continue where new technologies are created, companies incorporate these capabilities into new products and services to improve their competitive position, some existing firms fail while others survive, and then the cycle begins anew.

## managerial implications

Future digital technology capabilities provide digital product managers with external opportunities, and threats that must be incorporated into their strategic decision making. Each of the technologies discussed in this chapter has the potential to impact one or more of the digital product industries in the future. Based upon our previous discussion, examples of benefits and challenges for each digital product industry are listed in Table 1. There are many different impacts that the full range of new digital technologies will have on these industries in the future, but this small number of examples provides some idea of the extent of changes expected.

*Table 1. Future digital technology benefits and challenges for digital product companies*

| Digital Product Industry | Future Digital Technology Benefit | Future Digital Technology Challenge |
|---|---|---|
| Newspaper and Magazine | Electronic ink and paper will eliminate the need to physically distribute newspaper and magazine content which will save time and money | Improved online information search technology will enable users to find up-to-date news stories from any news provider making it difficult for many newspapers and magazines to maintain a competitive advantage and survive |
| Electronic book | Electronic ink and paper will allow readers to own one hardware device that can be reused for reading any content they want<br>Digital language translation technology could mean that books only need to be written in one language and they can be automatically translated as needed for each user | Electronic ink and paper technology will need to be standardized so that users do not have to own several proprietary reading devices |
| Music | Quantum and DNA computers will enable music playing devices to be very tiny while still being able to store many different songs | If one device can store and play millions of songs then some value chain participants may no longer be needed and overall profits may fall |
| Movies | Holographic storage enables many different movies to be stored on one disc<br>3-D imaging can make a movie appear more realistic | Some moviegoers may stop going to theaters if they do not want to experience smells and other sensory experiences that are part of future movies |
| Games | Brain-computer interfaces and speech recognition systems enable people to play games without the need for peripheral input devices like mice and joysticks<br>3-D imaging can create more realistic game environments | Users may demand that all games include smell and touch technology so game development may be much more complex in the future |
| Software | Quantum and DNA computers enable software to be much more complex without impacting response time | All future software development will be more complex if users expect a full sensory experience to be included |

To illustrate the impact of these new technology characteristics, think about how different the process will be for buying a book in the future. A person can initiate the buying process by thinking about buying a book in a particular subject area. A machine will quickly interpret the person's needs and offer several recommendations, or it may just go ahead and order the best book for the person. The book may include video or smell rather than just text and images. The book will be immediately delivered to the person's reusable e-book, and paid for automatically also. All aspects of the transaction will be stored by the book seller and the buyer, and all of this will be accomplished using technology that is very small and very cheap. Conceivably, this entire transaction could also be completed while a person is in space. This example illustrates the dramatic changes that will impact consumer's lives as well as business operations and strategy. Many current jobs will disappear, but new job categories will be created to work with the new technology.

The overall managerial implication arising from the development of new digital technologies is that product providers will be in a hypercompetitive environment where change occurs rapidly and media companies in one industry may cross over into new industries because the digital technology enables it. Consumers will be the winners as more and more unique and novel products and services become available and companies must compete for limited revenues.

## SOCIETAL INSTITUTION IMPLICATIONS

Beyond the business and economic impact, improvement in digital technology will have a profound influence on people's lives in the future because of its effect on a wide range of societal institutions. In this section six examples are discussed: healthcare, government services, higher education, political campaigns, cybercrime law enforcement, and life at home.

### Digital Technology and Healthcare

Digital technology will be an integral part of healthcare in the future. It will alter diagnosis, treatment, and interaction between patients and healthcare providers. Digital technologies can be used for automatically collecting relevant textual or image data at the source, combining data from different sources to support health care provider decisions, display of text and multimedia results in a usable format where needed, and transmission of data between locations and organizations. A variety of machines can create high density still and moving images for aid in diagnosis. Healthcare requires distributed systems and massive storage capabilities and future digital technology will be able provide it. The systems will also need to include all participants in healthcare including doctors and other healthcare personnel, patients, hospital organizations, insurance companies, and government agencies.

What can be expected for the future of information technology use in healthcare? Glaser and Foley (2008) identify several major initiatives. Electronic health records (EHRs) will be used to support activities such as care documentation, medication administration, and results management. Business intelligence systems can be used to reduce variations in care practices and improve overall performance by monitoring actual performance relative to care goals. Customer relationship management (CRM) systems will be used to know and anticipate individual customer needs to build loyalty and increase profits. Digital networks will create opportunities for connected care where medical care is delivered to patients who may not be at a hospital or doctor's office, and interorganizational systems will allow hospitals to exchange patient information with other hospitals, pharmacies, and organizations making payments for service. Future digital technology will also provide higher quality imaging systems to assist in diagnosis and treatment and improved networks will allow consultations with specialists who are located anywhere in the world.

In addition, the future of healthcare will be increasingly online according to Barras (2009). Health workers will be able to monitor home-based patients through online medical records that are updated every time the patient takes their own blood pressure, checks their weight, or takes a medication. Health records may merge with social network technologies to provide a space for health worker and patient interaction. There are obvious concerns about privacy, but social networking acceptance continues to grow and it may become the norm for future interactions.

## Digital Technology and Government Services

Local, state, and federal government agencies have adopted information systems to support a wide range of services. The earliest applications were used to automate and streamline processes. The Internet provided an opportunity for a new set of services related to collecting and sharing governmental information. According to Castro and Atkinson (2009), the digital technologies that will be available in the next decades will provide further opportunities for improving government services. Geographic information systems, global positioning systems, and distributed sensor networks make it possible for state and local governments to address problems for energy, public safety, and transportation. Information about all services, including any required forms, can be made available online and data that must be provided to government agencies can be collected online. Widespread inter-agency information sharing will also be possible, but jurisdictional and trust issues must be addressed to motivate these agencies to work together. It is possible that digital technology could lead to greater governmental transparency to improve relations with its citizens. The digital divide will be an increasingly important topic for the future as governments provide a greater portion of their services online and some people will not have equal access. Many of the benefits of these future systems will be lost if a sizable portion of the population does not have access.

Another impact of digital technology on government is the blurring of lines between boundaries and channels (Di Maio, 2009). Jurisdictions are blurring because there is so much economic and social activity across geographic boundaries and these activities happen so quickly and at any time. Government channels are blurring because there are so many potential points of contact between citizens and agencies. Agencies have offices that are open during work hours, but they also have Web sites that are expected to be available any hour of any day. Boundaries are also blurring between the roles of government as a holder of data versus being an owner of data. In the future the expectation may be that individuals and organizations outside of government own the data they provide and the government must ask permission to access these data. The data could be physically stored in many different places and tools would be used to integrate it. Finally, there is a blurring of technology boundaries between government and individuals. Everyone uses the same Internet and Web, but it is not clear where the boundary is between the government controlled infrastructure and the part it does not control.

## Digital Technology and Higher Education

A third institution that will be affected by digital technology is higher education and training. Technology will impact the nature of the classroom and all forms of interaction between teacher, student, and information sources. McCrea (2009) has identified several higher education technology trends that will be important in the next decade. Classrooms will be more interactive. Handheld devices for classroom interaction, multimedia, and streaming video will be increasingly in demand. Any technology that makes the classroom more interactive and engaging will be utilized to improve the learning environment. More information will be available in more ways. There will be no need to wait until class is over to look something up. Answers to questions can be found any time during class and this will change the dynamic for classroom interaction and the ways class material is covered. Technological equipment and software will be impacted by technological convergence which means that a smaller number of devices will be needed and they will each serve several purposes. Technology will be less isolated. Interactive technologies in the classroom can provide for a more social environment. And finally, tech savvy students will demand more from educational technology. They expect to use all of their devices as part of their education including wireless devices, online education, social networking, smart phones, and podcasting.

Higher education will be delivered in many ways to many people and many of the barriers from the past will be overcome. This will requires some cultural changes on the part of teachers and institutions, but the hope is that more people can be given more opportunities to continue their education. These changes will impact colleges and universities, but will also create increased opportunities for online training and continuing education.

## Digital Technology and Political Campaigns

In the 1800s, candidates for political offices typically traveled on the railroad and made speeches in town squares or at the depot. Prior to World War II, candidates were heard on the radio and, in the last half of the 1900s, television provided an opportunity to speak to a larger audience without having to spend time travelling. Radio and television had an impact on elections, but it was still a mass media approach where the same message went to everyone. In the 21st century the Internet and other digital technologies has again changed the ways that candidates and voters interact. There has been an explosion of online activity surrounding local, state, and especially federal elections, and this will continue to grow in the future.

Greengard (2009) and Elsasser (2008) identify several changes that have occurred because of digital technology and the Internet. In the late 1990s elections, candidates used the Web as little more than an e-brochure. Candidates posted news, information, and messages about issues, but this did little to improve their interaction with voters. Elections in 2004 and 2008 were a better indication of the things to come. E-mail addresses are collected to create large supporter databases. Donations can be solicited through several channels. Supporters can interact with candidates and other voters through social networking sites. These groups can be used to discuss key issues and policies and provide immediate feedback. Audio and video messages can be posted by candidates or by supporters either on their Web site or on sites like YouTube. Participation can be bottom-up as candidate's supporters individually create their own messages supporting a candidate. Voters are empowered in ways that were impossible before and their relationship with candidates will be much stronger. All of these activities can be continued after an election to provide greater interaction with constituents and potentially make government more transparent.

There are downsides to the digital political world. Everything a candidate has ever said or done will be available in a digital format in the future. There are many more opportunities where a candidate can be embarrassed. There will be less truly candid moments. Political organizations will need to adapt and be less hierarchical. Also, not every voter has the time or knowledge to participate in online political activities. These voters will continue to demand traditional communications before casting their vote. And ultimately, if every candidate uses all of these approaches, there will be an information overload as people get more information than they can process. It will be difficult to gain an advantage by using an online strategy.

Digital technology will also change the voting process. For the past several years there has been an increase in the use of electronic voting systems, but they also introduce some new concerns (Wallach, 2008). Will the right people be allowed to vote, and will the wrong people be able to vote more than once? Will the machines suffer glitches so that voter preferences are not properly recorded? And will hackers be able to break into the system to change votes? The goal is to have a system that accurately captures a person's votes. The digital technology used to capture the data, store, and transmit it works, but there are other issues that have caused problems. Some systems are hard to use. It is sometimes difficult to tell which candidate you are voting for when using a touch screen. Another concern is cost (Zetter, 2008). The old paper systems were relatively inexpensive and did not require much maintenance. Newer electronic voting system are expensive to purchase, or lease, and require training costs and maintenance costs for hardware that is typically only used once or twice a year.

## Digital Technology and Cybercrime Law Enforcement

Digital technology has created a whole new world of potential crimes, but it has also provided tools for fighting crime. Some of the most common cybercrimes involve electronically stealing or destroying various forms of digital information (personal, financial, business-related), electronically transferring money from bank accounts, stealing copies of software programs, attacking Web servers to bring them down (denial-of-service attacks), misusing network bandwidth and processing hardware, or stealing hardware devices. New methods of attack seem to appear daily. Cybercrimes can be conducted from anywhere and at any time, and the attacks can often inflict greater monetary damage that traditional crimes.

Some trends have been identified for recent cybercrime activity (Harwood, 2008; Voigt, 2009). Unsolicited message traffic, also known as spam, is increasing. In 2009, 87 percent of all e-mail traffic was spam. This amounts to 40 trillion messages. Malicious software, or malware, traffic is also rising as more attempts are made to hack into individual or business computers. Spam or malware are also being distributed using multimedia files such as Flash files. Denial-of-service attacks are used to target not only businesses, but increasing they are used against government agencies. In some cases it could even be considered terrorism as a country, or terrorist group, attacks another government. Attacks on non-PC operating systems such as the Mac operating system are also on the rise as their market share increases. Global financial crises and unemployment will mean that cybercrime activity will continue to rise.

The new digital technologies of the 21$^{st}$ century will create even more cybercrime threats. Imagine if brain wave activity can be intercepted from a brain-computer interface. And what will be the problems created by hackers who can misuse smell or touch technologies. Increased processing power and storage capabilities will also make it easier to commit more cybercrimes, more easily, and in more places around the world. Tremendous opportunities will be created for firms that can provide protections against future cybercrime activity. Law enforcement personnel will require a very different skill set in the future as they are increasingly called upon to assist in the detection and prosecution of cybercriminals. The legal system will also require constant updating to prosecute entirely new forms of cybercrime. The good news is that new digital technologies will also offer improved tools for fighting cybercrime which will hopefully help law enforcement stay ahead of the criminals.

## Digital Technology and Life at Home

As important as any of the issues discussed earlier in this chapter, the impact digital technologies have on our life at home is perhaps the most important and interesting because we spend so much time at home. Home is also the focal point for family life. Each decade new technologies are introduced for the home to make our lives easier. The 'home of the future' has been an exhibition at fairs and expositions ever since the late 1800s. In the early part of the 20th century much of these technological innovations involved mechanical devices – electric ovens, washing machines, etc. Many tedious tasks were automated and some new forms of entertainment (radio and television) made their way into the home. In the 21st century we can count on most innovations arising from new digital technology applications. These innovations can be expected to change our home life in many ways and several terms are used to describe this diffusion of technology into the home – smart homes, digital homes, digital living, home of the future, and so forth (Venkatesh, 2008).

What is the range of possible impacts that digital technology may have on our home lives? Venkatesh (2008) identifies three roles for technology in the home: (1) the enabling role of technology, (2) the mediating role of technology, and (3) the transformative role of technology.

Enabling technology makes things work faster, better, or in an improved way. Improvements in output technologies will provide increased opportunities for entertainment throughout the house. Family members will no longer have to be in a particular room to watch television or use a computer. The technology capabilities will not change, but use may be done more efficiently.

Mediating technology acts as a facilitator between the family members and their living environment. Many of the input and output technologies discussed earlier will have a role in the way the family interacts with home technologies. Heating, air conditioning, security systems, entertainment devices, kitchen devices, and many other technologies may now be controlled by voice commands or by thoughts. Instead of having to learn how to control each device separately, the control will be done through more natural human forms of the interaction.

Transformational technology alters family life in some fundamental way. Connecting home devices to the Internet will allow closer interconnectivity and control. For example, people will be able to check on their home's security from work which has never been possible before. They will also be able to be informed if there are water or fire problems that need immediate attention. Further, the home of the future may use interconnected technologies to monitor refrigerator inventory and make automatic replenishment purchases which could save time for other activities.

Each of these changes will impact home life to some extent, but there are some factors that will limit any dramatic changes in the near future (Venkatesh, 2008). The average family is unaware of all of the possible uses for new digital technologies in their homes. In tough economic times it is also less likely that they would buy these new technologies without seeing some cost savings. Technology is also very complicated so even if they know something exists they may not feel that could learn to use it. There is also limited incentive for companies to provide entirely new digital services at home because demand is not sufficient at this time. It will take some before all homes are even using high definition television and high bandwidth Internet access, let alone some of the advanced technologies that are just being developed.

## FUTURE DIGITAL TECHNOLOGY TRENDS

Previous sections have presented a wide range of future digital technology issues from several perspectives including the technology capabilities, the application of these technologies, and their impact on businesses, the economy, and society. In this section the question that is addressed is – what are the trends for future digital technology? Major trends that are expected for digital technology in the 21$^{st}$ century are identified below. Some of the issues have been discussed previously, while others have not. Understanding these trends will help digital product managers anticipate changes that will occur in the near term and in the next decades.

- Digital technologies based on electromagnetic microprocessors will continue to evolve, but at some point conversion to optical, DNA, or some other underlying technology will create dramatic improvements and revolutionary change.
- Technology change will continue to accelerate. More new technologies will be created in the 21$^{st}$ century than in all of the past centuries combined.
- Input devices will ultimately allow direct interaction between people and machines without the need for any digital hardware. Relying entirely on direct interaction will create some new risks and security issues.
- Digital processors will continue to get smaller, faster, and cheaper.
- Digital storage devices will continue to store more data in smaller areas. At some point the volume of data stored will overwhelm people leading to information overload. An important digital technology application for the future will be tools to make sense of vast amounts of data (knowledge management) so that humans can utilize it in decision making.
- Digital output devices in the future will be virtual. They will not require any hardware and they will be available anywhere a user is located.
- Future digital technologies will be wireless. There will no longer be any need to connect devices by wires and this will create opportunities for many new applications. Improvements in networking technology will enable everything to be connected including home, work, and everything in between. Technology will be ubiquitous.
- A greater proportion of news and information content will be created by users. The billions of individual users worldwide have the potential to create vast amounts of content, but the quality of this material will always be a concern.
- Many digital products will be free, while others will set prices through various hybrid mechanisms that allow digital product companies to price discriminate by charging users more if they are willing to pay more, and charging less for others. Ultimately, standard business models that are accepted by users will evolve for each digital product industry, but most new ventures will fail because of the hypercompetitive nature of the online marketplace.
- Many digital products will be sold through a service model. Information content, books, music, movies, games, and other software can all be rented or available by subscription. Data storage will become a service in the future.
- Financing digital product ventures will be a vital part of bringing a new idea to market and it will be increasingly difficult to obtain financing in the near future. Venture capital firms will not throw money at just any idea like they did in the early e-commerce days. Banks will also be hesitant to put money into highly risky ventures.
- The network effect (the value to a user grows as the number of other users grows, for example with an online auction site or a social networking site) will have a growing influence on digital product company success. Firms that reach a critical mass of users will be highly successful, while others will fail. The result will be oligopolies

in many industries because it will be more efficient to have a smaller number of large online sites each providing unique digital products and services to the vast number of users.
- Multichannel strategies will be most successful. Companies that provide choice for their customers will be chosen by more shoppers. People want to use online and traditional retail channels their own way. They can buy a product online or at a store, and they can get customer service online or at a store. They want all of the combinations to be available at times that fit their schedule.
- All forms of digital convergence will continue to provide opportunities in the digital product industries. Hybrid devices that serve several functions will be more common. Digital content and multimedia will become indistinguishable as text, images, audio, video and more advanced sensory-rich multimedia converges in all forms of interaction and communications. Industries will converge as partnerships across traditional industry boundaries become more common as a means of achieving competitive advantage and reaching a critical mass for users.
- Inventory may become a concept from the past. Digital products can be sold through value chains that include almost no product inventory. In the physical product industries, interorganizational information systems (including many of the new digital technologies discussed) will enable supply chain networks to use a demand-driven production and distribution process which will result in dramatic reductions in inventory. Reductions in inventory cost will allow firms to pass some of the savings on to their customers.
- The government will monitor and regulate more online activities. Digital technology will make this possible. The issue will still be to what extent the government should be monitoring people and businesses online. Ultimately, online commerce will be taxed because the government will need to tap any and all sources of tax revenue.
- Newly developed digital technologies will continue to make it possible to eliminate some human labor. But the same digital technologies will create new jobs that will require human workers. People will still be the best source of creative solutions to new and complex problems.
- Cultural differences will have an increasingly profound impact on worldwide digital technology use. New digital technologies will have the same potential benefits for everyone, but differences in culture, religion, customs, language, and other factors will produce situations where the technology is not used to its fullest potential. Views on government control, freedom of expression, privacy, and other issues will play a role in online communications, information sharing, and commerce. Legal and ethical differences in different parts of the world will also impact these activities. Compromises will be necessary for cooperation across borders and there may never be complete agreement on any of these issues related to the appropriate role technology should play in society.
- The environmental impact of future digital information technologies will be considered when determining whether to create or adopt a new technology. Future digital technologies will use less energy, but gains will be offset by the fact that there are many more digital hardware devices that require power. Digital hardware devices will need to be entirely recyclable.
- Finally, new digital technologies will not make the world better or worse, but they will make the world different.

## FUTURE RESEARCH DIRECTIONS

The technical research directions in this area are fairly clear. The goals for digital technology scientists and engineers are to develop hardware and applications that are smaller, faster, cheaper, more efficient, more user-friendly, and provide more realistic user experiences. In this section the focus is on looking beyond purely technical research directions to identify business and social science research directions from the perspective of business managers, consumers, and society as a whole. All methodologies will be useful including conceptual and theoretical research, empirical studies, experimental studies, and case studies.

From a strategic perspective, business managers need to know which technologies are commercially viable and which ones will complement their current products and services. Studies may also be needed to identify business models that produce competitive advantages in this high-technology environment. From the psychology perspective there are a number of research issues that arise from these new technologies. Which technologies will people value? And what factors will affect their decision to adopt a new technology? Also, will users trust new technologies in situations where they feel that they have lost control to machines to some extent? Understanding the psychology of individuals in highly complex technology environments will be critical. As the number of activities involving technology increases, we will need to understand issues like Internet addiction to help people cope with technological change and maintain their real-world relationships.

Digital technology ethics research also must confront a number of issues related to what actions are right, even if they are legal. Individual information rights must be protected while still enabling organizations to collect and use information for the benefit of their customers. Property rights are another important issue because digital technology enables information and products to be copied very easily and the law may not protect rights in all situations.

Governance over digital technologies and infrastructure must consider how best to allow an open exchange but also recognize that there are times where some regulations are necessary. Governance is complicated because digital technology is available globally which means that each of these issues must be addressed from multiple ethical perspectives realizing that actions in one place may be ethical, while in other areas they are not. Beginning around 2003, there has been some discussion of the idea that governance of the Internet could be placed under the control of the United Nations (Peake, 2009). Given the number of stakeholders, this is obviously a complex issue. One issue is whether governance involves only technical issues, or whether it should consider broader issues such as legal issues, e-commerce regulation, and privacy. A second issue is who should provide leadership. Should it be a large agency such as the United Nations, or should it continue to be led by a number of private sector organizations. The value of the Internet is that it offers one massive network that encompasses all online activities. It is important to keep all countries and organizations involved. These are very important issues because most nations view the Internet as a strategic issue and a potential threat to their sovereignty. In times of conflict, access to the Internet could be as important as energy, transportation, food, and water issues. Because of these risks, anyone currently controlling the Internet is probably unwilling to give it up. There are numerous technical, business, and social science research areas that can be addressed related to appropriate Internet governance.

From a societal standpoint several additional philosophical issues must be considered. From a human-computer interaction research perspective, to what extent do we want the virtual world to replicate the real world? Will this new virtual world positively or negatively impact our interactions with other humans and our quality of life?

How will it impact human employment levels? Will technology change accelerate to the point where humans cannot adapt to the change? What are the best ways to utilize the relative strengths of people and machines?

One study has addressed this question from the context of avatar use as online assistants in Web sites (McGoldrick et al., 2008). Avatars are not the same as humans, but they are images meant to represent a person in an online environment. Their study explored the potential uses for avatars in providing online shoppers with assistance and recommending products for purchase. They found that online consumers wanted avatars to tell them about special offers and let them know if they had made errors in selection of products. They did not want them to make recommendations based on other people's purchases. Overall, it was found that customers want to have a choice about whether to work with avatars or not, and care must be taken that avatars are not seen as inappropriate or offensive. The best avatars are ones that are adaptive and take into account user characteristics and desires when offering assistance.

Finally, given each of the issues identified above, and the inherent complexity involved in the interrelationship between technology, people, and society, this research needs to utilize an interdisciplinary perspective including all of the physical, biological, computer, and social sciences.

## CONCLUSION

All forms of digital technology will continue to evolve and improve. At some point the real and virtual worlds may become indistinguishable. The overall philosophical issue is not so much whether future digital technology will be able to do something, but whether we feel that it should do be used to do something. Are there tasks that we do not want technology to do in the future? These answers are complicated and may never be completely answered because every time we start to feel comfortable with our understanding of technology, something new will come along to change the status-quo.

## REFERENCES

Barras, C. (2009, July 24). Innovation: Is the future of healthcare online? *NewScientist*. Retrieved from http://www.newscientist.com/article/dn17513-innovation-is-the-future-of-healthcare-online.html

Bonsor, K. (2009). How electronic ink works. *HowStuffWorks*. Retrieved December 15, 2009, from http://electronics.howstuffworks.com/gadgets/high-tech-gadgets/e-ink.htm

Bonsor, K. (2010). How interplanetary Internet will work. *HowStuffWorks*. Retrieved January 10, 2010, from http://computer.howstuffworks.com/interplanetary-internet.htm

Bonsor, K., & Strickland, J. (2009). How quantum computers work. *HowStuffWorks*. Retrieved December 15, 2009, from http://computer.howstuffworks.com/quantum-computer.htm

Boswell, W. (2009). Turbo10, a deep Web search engine. Retrieved December 15, 2009, from http://websearch.about.com/od/invisibleweb/a/turbo10.htm

Castro, D., & Atkinson, R. (2009). The next wave of e-government. *Statetech*. Retrieved December 15, 2009, from http://statetechmag.com/events/updates/the-next-wave-of-e-government.html

Clark, J. (2009). How quantum cryptology works. *HowStuffWorks*. Retrieved December 15, 2009, from http://science.howstuffworks.com/quantum-cryptology.htm

Di Maio, A. (2009, July 16). The blurring of government. *Governing*. Retrieved from http://www.governing.com/column/blurring-government

Elsasser, J. (2009). Election 2008: The campaign of voter-generated content. *Public Relations Strategist*, *14*(1), 10–13.

Facebook. (2010). Statistics. Retrieved January 12, 2010, from http://www.facebook.com/press/info.php?statistics

Glaser, J., & Foley, T. (2008, November). The future of healthcare IT. What can we expect to see? *Healthcare Financial Management*, 82–88.

Grabianowski, E. (2009a). How brain-computer interfaces work. *HowStuffWorks*. Retrieved December 15, 2009, from http://computer.howstuffworks.com/brain-computer-interface.htm

Grabianowski, E. (2009b). How speech recognition works. *HowStuffWorks*. Retrieved December 15, 2009, from http://electronics.howstuffworks.com/gadgets/high-tech-gadgets/speech-recognition.htm

Greengard, S. (2009). The first Internet president. *Communications of the ACM*, *52*(2), 16–18. doi:10.1145/1461928.1461935

Harris, W. (2009). How haptic technology works. *HowStuffWorks*. Retrieved December 15, 2009, from http://electronics.howstuffworks.com/gadgets/other-gadgets/haptic-technology.htm

Harwood, M. (2008, December 10). Cybercrime trends will worsen in 2009, according to forecasts. *Security Management*. Retrieved from http://www.securitymanagement.com/

Kirkpatrick, M. (2009, May 18). The dam just broke: Facebook opens up to OpenID. *ReadWriteWeb*. Retrieved from http://www.readwriteweb.com/archives/the_dam_just_broke_facebook_opens_up_to_openid.php

Lohr, S. (2009, April 27). G.E.'s breakthrough can put 100 DVDs on a disc. *The New York Times*. Retrieved from http://www.nytimes.com/

Madrigal, A. (2009, March 4). Researchers want to add touch, taste and smell to virtual reality. *Wired.com*. Retrieved from http://www.wired.com/

Mayne, M. (2009, September 22). Encryption is becoming more elaborate to ensure confidential business data is kept secret. *SC Magazine*. Retrieved from http://www.scmagazineuk.com/encryption-is-becoming-more-elaborate-to-ensure-confidential-business-data-is-kept-secret/article/149413/

McCrea, B. (2009, December 9). 5 Higher Ed Tech Trends to Watch in 2010. *Campus Technology*. Retrieved from http://campustechnology.com/articles/2009/12/09/5-higher-ed-tech-trends-to-watch-in-2010.aspx

McGoldrick, P. J., Keeling, K. A., & Beatty, S. F. (2008). A typology for roles of avatars in online retailing. *Journal of Marketing Management*, *24*(3-4), 433–461. doi:10.1362/026725708X306176

Mearian, L. (2008, March 11). Study: Digital universe and its impact bigger than we thought. *Computerworld*. Retrieved from http://www.computerworld.com/

NeuroSky. (2009, March 26). NeuroSky to launch brainwave-based technology for stereo headsets for personal computers. *PR Newswire*. Retrieved from http://news.prnewswire.com/

Nusca, A. (2009, December 10). New 'electronic paper' technology allows for 1 million color choices for your walls, electronics. *Smartplanet*. Retrieved from http://www.smartplanet.com/

Peake, A. (2009). Internet governance. *Digital Review of Asia Pacific*. Retrieved December 15, 2009, from http://www.digital-review.org/themes/25-internet-governance.html

Swaminathan, N. (2008, May 30). DNA computer puts microbes to work as number crunchers. *Scientific American*. Retrieved from http://scientificamerican.com/

Turban, E., & Volonino, L. (2010). *Information Technology for Management: Improving Performance in the Digital Economy*. Hoboken, NJ: John Wiley & Sons.

Venkatesh, A. (2008). Digital home technologies and transformation of households. *Information Systems Frontiers*, *10*(4), 391–395. doi:10.1007/s10796-008-9097-0

Voigt, K. (2009, December 14). Cyber crime poses threat to e-commerce. *CNN.com*. Retrieved from http://www.cnn.com/

Wallach, D. (2008, November 3). My take: E-voting not user friendly. *Discovery Channel*. Retrieved from http://dsc.discovery.com/news/2008/11/03/tech-electronic-voting.html

Wilson, T. V. (2009). How biometrics works. *HowStuffWorks*. Retrieved December 15, 2009, from http://science.howstuffworks.com/biometrics.htm

Wilson, T. V. (2009). How is digital 3-D different from old 3-D movies? *HowStuffWorks*. Retrieved December 15, 2009, from http://www.howstuffworks.com/digital-3d.htm

Wolk, D. (2009, September 2). Future of the screen: Terminator-style augmented-reality glasses. *Wired.com*. Retrieved from http://www.wired.com/

Wright, A. (2009, February 23). Exploring a 'deep Web' that Google can't grasp. *The New York Times*. Retrieved from http://www.nytimes.com/

Zetter, K. (2008, April 4). The cost of e-voting. *Wired.com*. Retrieved from http://www.wired.com/

## ADDITIONAL READING

Ackerman, E. (2009). Interplanetary Internet tests. *IEEE Spectrum*, *46*(7), 13–14. doi:10.1109/MSPEC.2009.5109433

Bhasin, K. B. (2005). Interplanetary Internet. *Computer Networks*, *47*(5), 599–601. doi:10.1016/j.comnet.2004.08.007

Bhatt, G. (2004). Bringing virtual reality for commercial Web site. *International Journal of Human-Computer Studies*, *60*(1), 1–15. doi:10.1016/j.ijhcs.2003.07.002

Burleigh, S., Hooke, A., Torgerson, L., Fall, K., Cerf, V., & Durst, B. (2003). Delay tolerant networking: An approach to interplanetary Internet. *IEEE Communications Magazine*, *41*(6), 128–138. doi:10.1109/MCOM.2003.1204759

Chapman, N., & Chapman, J. (2009). *Digital Multimedia*. Hoboken, NJ: John Wiley & Sons.

Chen, A. J., Boudreau, M.-C., & Watson, R. T. (2008). Information systems and ecological sustainability. *Journal of Systems and Information Technology*, *10*(3), 186–201. doi:10.1108/13287260810916907

Cheng, J. (2007, October 15). Researchers help users control Second Life avatars via brain activity. *ARS Technica*. Retrieved from http://arstechnica.com/

Chia, E. (2010, January 18). Quantifying green IT. *Enterprise Innovation*, *4*(5), 20–21.

Deegan, M., & Tanner, S. (2006). *Digital Preservation*. London: Facet Publishing. Digital Scent Technology Blog, http://digiscents.com/blog/

Ess, C. (2009). *Digital Media Ethics*. Malden, MA: Polity Press.

Ghinea, G., & Chen, S. Y. (2006). *Digital Multimedia Perception and Design*. Hershey, PA: IGI Publishing.

Hur, W., & Kim, D. (2010). The future of digital imaging. *Communications of the ACM*, *53*(1), 131–135. doi:10.1145/1629175.1629207

Jackson, J. (2005). The interplanetary Internet. *IEEE Spectrum*, *42*(8), 30–35. doi:10.1109/MSPEC.2005.1491224

Kaye, P., Laflamme, R., & Mosca, M. (2007). *An Introduction to Quantum Computing*. New York: Oxford University Press.

Krepkiv, R. (2008). *Brain-Computer Interfaces*. Saarbrucken, Germany: VDM Verlag Dr. Mueller e.K.

Kressel, H. (2007). *Competing for the Future: How Digital Innovations are Changing the World*. New York: Cambridge University Press. doi:10.1017/CBO9780511611094

Krupp, J. (2008). Green pressures and Web 2.0 demands. *NetworkWorld Asia, 4*(8), 37–38.

Lazere, C., & Shasha, D. E. (2010). *Natural Computing: DNA, Quantum Bits, and the Future of Smart Machines*. New York: W. W. Norton & Co.

Lin, M. C., & Otaduv, M. (2008). *Haptic Rendering: Foundations, Algorithms and Applications*. Wellesley, MA: A K Peters, Ltd.

Mines, C. (2008). *The Dawn of Green IT Services*. Cambridge, MA: Forrester Research.

Molla, A., Pittayachawan, S., Corbitt, B., & Deng, H. (2009). An international comparison of green IT diffusion. *International Journal of e-Business Management, 3*(2), 3-23.

Mowshowitz, A. (2008). Technology as excuse for questionable ethics. *AI & Society, 22*(3), 271–282. doi:10.1007/s00146-007-0147-9

Pegrum, M. (2009). *From Blogs to Bombs: The Future of Digital Technologies in Education*. Perth, Australia: UWA Publishing.

Poole, D. (2007). A study of beliefs and behaviors regarding digital technology. *New Media & Society, 9*(5), 771–793. doi:10.1177/1461444807080341

Sarrel, M. D. (2010, January 18). The real deal on greening your data center. *eWeek, 27*(2), 16-18.

Sears, A., Lazar, J., Ozok, A., & Meiselwitz, G. (2008). Human-centered computing: Defining a research agendy. *International Journal of Human-Computer Interaction, 24*(1), 2–16.

Selker, T. (2008). Touching the future. *Communications of the ACM, 51*(12), 14–16. doi:10.1145/1409360.1409366

Sellen, A., Rogers, Y., Harper, R., & Rodden, T. (2009). Reflecting human values in the digital age. *Communications of the ACM, 52*(3), 58–66. doi:10.1145/1467247.1467265

Sharp, H., Rogers, Y., & Preece, J. (2007). *Interaction Design: Beyond Human Computer Interaction*. Hoboken, NJ: John Wiley & Sons.

Soars, B. (2009). Driving sales through shoppers' sense of sound, sight, smell and touch. *International Journal of Retail & Distribution Management, 37*(3), 286–298. doi:10.1108/09590550910941535

Song, Y., & Zhu. D. (2009). *High Density Data Storage: Principle, Technology, and Materials*. Singapore: World Scientific Publishing Company.

Stibel, J. M. (2009). *Wired for Thought: How the Brain is Shaping the Future of the Internet*. Boston: Harvard Business Press.

Stichler, R. N., & Hauptman, R. (2009). *Ethics, Information and Technology: Readings*. Jefferson, NC: McFarland & Company, Inc. Publishers.

Stoller, J. (2009). IT's growing role as a green enabler. *CMA Management, 83*(8), 37–38.

van den Hoven, J., & Weckert, J. (2008). *Information Technology and Moral Philosophy*. New York: Cambridge University Press. doi:10.1017/CBO9780511498725

Wands, B. (2007). *Art of the Digital Age*. London: Thames & Hudson.

Xie, I. (2008). *Interactive Information Retrieval in Digital Environments*. Hershey, PA: IGI Publishing.

## KEY TERMS AND DEFINITIONS

**Brain-Computer Interface (BCI):** Devices that enable people to interact with computers through thought.

**Computer Haptics:** The study of touch in a computer environment.

**Crawler Technology and Spider Technology:** Software that collects and classifies information gather from Web sites by moving through hyperlinks between pages and Web sites.

**Deep Web Search:** An improved online information search technology that provides search results using a wider range of information sources such as multiple niche-specific search engines and government, business, and academic databases.

**DNA Computer:** A computer where data is stored and processed using DNA molecules.

**Electronic Ink (e-ink):** A material designed to look like regular ink that uses microcapsules containing white and dark color chips that can be manipulated by an electric charge to display information on a surface.

**Electronic Paper (e-paper):** A material designed to look like regular paper that uses electric charges to manipulate embedded colored particles to display information.

**Holographic Storage:** A high-density data storage method that uses an optical process where data is recorded in light patterns stored on a light sensitive material.

**Interplanetary Internet:** A system using Internet protocols that is used to communicate beyond Earth.

**Quantum Computer:** A computer where data is stored and processed using quantum bits (qubits) that can hold a value of zero, one, or both values simultaneously.

**Quantum Encryption:** A complex encryption method that uses photon-string keys that are altered if they are intercepted. This encryption technique is based on the same qubit concept used in quantum computers.

**Speech Recognition:** A system that enables people to interact with machines through human speech. The speech is converted into a digital format to input text or to issue commands.

**Virtual Retinal Display (VRD):** A technology used to paint an image directly on a person's retina to display images using a combination of red, green, and blue ultra low power lasers.

# Compilation of References

"2nd jury rules against woman in music case." (2009, June 19) Associated Press. *Des Moines Register*, 11A.

Accenture (2008a). *Software as a Service (SaaS) Practice*. Retrieved February 26, 2009, from *Accenture Collaboration* database.

Accenture (2008b). Traditional licensing vs. SaaS (Software as a Service) vs. ASP (Application Service Provider). Retrieved January 28, 2009, from *Accenture Collaboration* database.

Accenture (2008c). SaaS Overview. Retrieved November 26, 2008, from *Accenture Collaboration* database.

Adams, R. (2010, January 14). AP, Yahoo near deal on content use. *The Wall Street Journal*. Retrieved from http://online.wsj.com/

Adams, R. (2010, January 21). New York Times to charge for Web. *The Wall Street Journal*. Retrieved from http://online.wsj.com/

Adobe. (2009). RIM joins Open Screen Project. Retrieved December 8, 2009, from http://www.adobe.com/aboutadobe/pressroom/pressreleases/pdfs/200910/100509RIMjoinsOSP.pdf

Advertising Association. (2009). UK media advertising expenditure 2008. Retrieved December 1, 2009, from http://www.adassoc.org.uk/aa/index.cfm/adstats/

Alexei, N. (2006). Information assurance seals: How they impact consumer purchasing behavior. *Journal of Information Systems*, *20*(1), 1. doi:10.2308/jis.2006.20.1.1

Aliaga, D., & Atallah, M. (2009). Genuinity signatures: Designing signatures for verifying 3D object genuinity. *Computer Graphics Forum*, *28*(2), 437. doi:10.1111/j.1467-8659.2009.01383.x

Alter, S. (2008). Service system fundamentals: Work system, value chain, and life cycle. *IBM Systems Journal*, *47*(1), 71–85. doi:10.1147/sj.471.0071

Alti, A. (2005). IPO market timing. *Review of Financial Studies*, *18*, 1105–1138. doi:10.1093/rfs/hhi022

Amazon. (2009). Profile for Amazon.com Inc. *Yahoo! Finance*. Retrieved December 8, 2009, from http://finance.yahoo.com/q/pr?s=amzn

Amit, R., & Zott, C. (2001). Value creation in e-business. *Strategic Management Journal*, *22*(6-7), 493–520. doi:10.1002/smj.187

Anonymous (2008a). Information forensics and security; New findings in information forensics and security described by M. Noorkami and co-researchers. *Computers, Networks & Communications*, 602.

Anonymous (2008b). Kodak; Kodak calls on businesses and industry to join the fight against counterfeiting. *Technology & Business Journal*, 59.

Anticybersquatting Consumer Protection Act of 1999, § 15 U.S.C. § 1125(d) (2000).

Apple Computer Inc. v. Franklin Computer Corp., 714 F.2d 1240 (3d Cir. 1983).

Apple. (2009). Profile for Apple Inc. *Yahoo! Finance*. Retrieved December 8, 2009, from http://finance.yahoo.com/q/pr?s=aapl

Aramand, M. (2008). Software products and services are high tech? New product development strategy for software products and services. *Technovation*, *28*(3), 154–160. doi:10.1016/j.technovation.2007.10.004

Armbrust, M., Fox, A., Griffith, R., Joseph, A. D., Katz, R. H., Konwinski, A., et al. (2009). *Above the Clouds: A Berkeley View of Cloud Computing*. Retrieved February 19, 2009, from http://www.eecs.berkeley.edu/Pubs/TechRpts/2009/EECS-2009-28.pdf

Arun, S. (2004). Managing digital piracy: Pricing and protection. *Information Systems Research*, *15*(3), 287. doi:10.1287/isre.1040.0030

Ascensão, J. (2008). Sociedade da informação e liberdade de expressão. In *Direito da Sociedade da Informação* (pp. 51–73). VII.

Aston, G. (2009). *What Is A Sales Strategy?* Retrieved June 28, 2009, from http://www.freshbusinessthinking.com/business_advice.php?AID=2668&Title=What+Is+A+Sales+Strategy

Atari Games Corp. v. Oman, 979 F.2d 242 (D.C. Cir. 1992).

Atton, C. (2008). Alternative media theory and journalism practice. In Boler, M. (Ed.), *Digital Media and Democracy: Tactics in Hard Times* (pp. 213–226). Cambridge, MA: MIT Press.

Avdic, A. (2002). User and developer-user systems development using a spreadsheet program. In Khosrow-Pour, M. (Ed.), *Issues and Trends of Information Technology: Management in Contemporary Organizations* (Vol. 1). Hershey, PA: Idea Group Publishing.

Baca, F. (2009). Considering HR outsourcing? Consider SaaS. *Financial Executive*, *25*(8), 59–60.

Bakos, Y., & Brynjolfsson, E. (1999). Bundling information goods: Pricing, profits, and efficiency. *Management Science*, *45*(12), 1613–1630. doi:10.1287/mnsc.45.12.1613

Bakos, J. Y. (1991). A strategic analysis of electronic marketplaces. *Management Information Systems Quarterly*, *15*, 294–310. doi:10.2307/249641

Bakos, J. Y. (1991). A strategic analysis of electronic marketplaces. *Management Information Systems Quarterly*, *15*(3), 295–310. doi:10.2307/249641

Bakos, Y., & Brynjolfsson, E. (1999). Bundling information goods: Pricing, profits, and efficiency. *Management Science*, *45*(12), 1613–1630. doi:10.1287/mnsc.45.12.1613

Barclay, M. J., Morellec, E., & Smith, C. W. (2001). On the debt capacity of growth options. Unpublished manuscript, University of Rochester, New York.

Barnatt, C. (2008). Explaining Web 2.0. *ExplainingComputers.com*. Retrieved October 9, 2008, from http://www.youtube.com/watch?v=7BAXvFdMBWw

Barras, C. (2009, July 24). Innovation: Is the future of healthcare online? *NewScientist*. Retrieved from http://www.newscientist.com/article/dn17513-innovation-is-the-future-of-healthcare-online.html

Barua, A., Konana, P., Whinston, A., & Yin, F. (2000). Making e-business pay: Eight key drivers for operational success. *IT Professional*, *2*(6), 22–30. doi:10.1109/6294.888013

Basch, C. E. (1987). Focus group interview: An underutilized research technique for improving theory and practice in health education. *Health Education Quarterly*, *14*(4), 411–448.

Basole, R. C., & Rouse, W. B. (2008). Complexity of service value networks: Conceptualization and empirical investigation. *IBM Systems Journal*, *47*(1), 53–70. doi:10.1147/sj.471.0053

Bates, B. (2008). Commentary: Value and digital rights management - A social economics approach. *Journal of Media Economics*, *21*(1), 53. doi:10.1080/08997760701806850

Beam, R. A., Brownlee, B. J., Weaver, D. H., & DiCicco, D. T. (2009). Journalism and public service in troubled times. *Journalism Studies*, *10*(6), 734–753. doi:10.1080/14616700903274084

## Compilation of References

Begun, S. J. (1949). *Magnetic Recording*. New York: Rinehart & Company.

Benkler, Y. (2004). 'Sharing nicely': On shareable goods and the emergence of sharing as a modality of economic production. *The Yale Law Journal, 114*, 273–358. doi:10.2307/4135731

Biddick, M. (2010, January 18). Time for a SaaS strategy. *InformationWeek, 1254*, 27–32.

Blog note update. (2009, September 14). Retrieved September 16, 2009, from

Blokdijk, G. (2008). *SaaS 100 Success Secrets - How Companies Successfully Buy, Manage, Host and Deliver Software as a Service (SaaS)*. Brisbane, Australia: Emereo Pty Ltd.

Blokdijk, G. (2008). *SaaS 100 Success Secrets - How Companies Successfully Buy, Manage, Host and Deliver Software as a Service (SaaS)*. Brisbane, Australia: Emereo Pty Ltd.

Bluedog (2009). *Workbench "Always on the Job!"*. Retrieved March 17, 2009, from http://www.bluedog.net/wb/93/wo/90kVL48YV9Ww5MvGIx7Naw/0.5

Blythe, P. A., Sr. (2005). *Biometric authentication system for secure digital cameras*. Unpublished Ph.D., State University of New York at Binghamton, United States -- New York.

Boehm, J. (2009). Copyright reform for the digital era: Protecting the future of recorded music through compulsory licensing and proper judicial analysis. *Texas Review of Entertainment & Sports Law, 10*(2), 169–211.

Bonsor, K. (2009). How electronic ink works. *HowStuffWorks*. Retrieved December 15, 2009, from http://electronics.howstuffworks.com/gadgets/high-tech-gadgets/e-ink.htm

Bonsor, K. (2010). How interplanetary Internet will work. *HowStuffWorks*. Retrieved January 10, 2010, from http://computer.howstuffworks.com/interplanetary-internet.htm

Bonsor, K., & Strickland, J. (2009). How quantum computers work. *HowStuffWorks*. Retrieved December 15, 2009, from http://computer.howstuffworks.com/quantum-computer.htm

Borchers, A. (2003). Intrinsic and contextual data quality: The effect of media and personal involvement. In Grant, G. (Ed.), *ERP & Data Warehousing in Organizations: Issues and Challenges*. Hershey, PA: Idea Group Publishing.

Boswell, W. (2009). Turbo10, a deep Web search engine. Retrieved December 15, 2009, from http://websearch.about.com/od/invisibleweb/a/turbo10.htm

Boyd, D. M., & Ellison, N. B. (2007). Social network sites: Definition, history, and scholarship. *Journal of Computer-Mediated Communication, 13*(1), article 11. Retrieved on December 20, 2009, from http://jcmc.indiana.edu/vol13/issue1/boyd.ellison.html

Braude, E. (2008). Software-as-a-service and offshoring. *International Journal of Business Insights & Transformation, 2*(1), 93–95.

Brogan, C., & Smith, J. (2009). *Trust Agents: Using the Web to Build Influence, Improve Reputation, and Earn Trust*. New York: John Wiley & Sons, Inc.

Buyya, R., Yeo, C. S., Venugopal, S., Broberg, J., & Brandic, I. (2009, June). Cloud computing and emerging IT platforms: Vision, hype, and reality for delivering computing as the 5th utility. *Future Generation Computer Systems, 25*(6), 599–616. doi:10.1016/j.future.2008.12.001

Byrne, C. (2009). Funding the regional news of tomorrow. *Press Gazette*, 13-15.

Cable/Home Communications Corp. v. Network Prods. Inc., 902 F.2d 829 (11th Cir. 1990).

Caldwell, F., & Eid, T. (2007). Is SaaS safe for financial governance, risk, and compliance solutions. *Gartner*, Article G00150913. Retrieved November 13, 2008, from Gartner database.

Campbell v. Acuff-Rose Music, 510 U.S. 569 (1994).

Campbell-Kelly, M. (2009). Historical reflections: The rise, fall, and resurrection of software as a service. *Communications of the ACM*, *52*(5), 28–30. doi:10.1145/1506409.1506419

Carr, N. (2006). *Here comes HaaS*. Retrieved October 18, 2008, from http://www.roughtype.com/archives/2006/03/here_comes_haas.php

Carraro, G., & Chong, F. (2006). *Architecture strategies for catching the long tail*. Microsoft. Retrieved October 4, 2008, from http://msdn.microsoft.com/en-us/library/aa479069.aspx

Carter, R., & Manaster, S. (1990). Initial public offerings and underwriter reputation. *The Journal of Finance*, *45*, 1045–1068. doi:10.2307/2328714

Carter, R., & Van Auken, H. (2005). Bootstrap financing and the characteristics of small business owners and their business environment. *Entrepreneurship and Regional Development*, *17*, 129–144. doi:10.1080/08985620500067548

Carter, R. B., Strader, T. J., & Dark, F. H. (2009). *Digital product & service firms: fixed-to-variable costs and the timing of IPOs*. Unpublished manuscript, Iowa State University.

Cashmore, P. (2006, July 11). Myspace: America's number one. Retrieved on October 11, 2009, from http://mashable.com/2006/07/11/myspace-americas-number-one/

Castro, D., & Atkinson, R. (2009). The next wave of e-government. *Statetech*. Retrieved December 15, 2009, from http://statetechmag.com/events/updates/the-next-wave-of-e-government.html

Catone, J. (2009, November 2). How to use Twitter lists. Retrieved January 13, 2010, from http://mashable.com/2009/11/02/twitter-lists-guide/

Chaney, P. (2009, October 13). Social media works for small business. I have proof. Retrieved on December 19, 2009, from http://www.mpdailyfix.com/2009/10/social_media_works_for_small_b.html

Chang, C., & Lin, P. (2008). Adaptive watermark mechanism for rightful ownership protection. *Journal of Systems and Software*, *81*(7), 1118. doi:10.1016/j.jss.2007.07.036

Chapman, M. R. (2008). *The 2008 Softletter SaaS Report*. Retrieved March 25, 2009, from *Softletter.com* database.

Chau, P. Y. K. (1995). Factors used in the selection of packaged software in small businesses: Views of owners and managers. *Information & Management*, *29*(2), 71–78. doi:10.1016/0378-7206(95)00016-P

Chaudhuri, S. (2008). SaaS pricing and metering. Retrieved May 20, 2009, from http://sumanchaudhuri.wordpress.com/2008/02/28/saas-pricing-and-metering/

Chen, H., Guo, F., Chen, C., Chen, J., & Kuo, T. (2001). Review of telemedicine projects in Taiwan. *International Journal of Medical Informatics*, *61*(2-3), 117–129. doi:10.1016/S1386-5056(01)00134-4

Chen, P. (1997). Pricing Strategies for digital information goods and online services on the Internet. Retrieved September 4, 2009, from http://www.mba.ntu.edu.tw/~jtchiang/StrategyEC/eec/report1/report1.htm

Cheng, H. K., Sims, R. R., & Teegen, H. (1997). To purchase or to pirate software: An empirical study. *Journal of Management Information Systems*, *13*(4), 49–60.

Cheung, S., Chiu, D., & Ho, C. (2008). The use of digital watermarking for intelligence multimedia document distribution. *Journal of Theoretical and Applied Electronic Commerce Research*, *3*(3), 103. doi:10.4067/S0718-18762008000200008

Chia, E. (2008). Marketing firm adopts software-as-a-service path. *Enterprise Innovation*, *4*(1), 23.

Chong, B. (2008). Stop data leaks through SaaS. *NetworkWorld Asia*, *4*(8), 4.

Chou, D. C., & Chou, A. Y. (2008). Software as a service (SaaS) as an outsourcing model: An economic analysis. *Proceeding of the 39th Southwest Decision Science Institute Conference*, Houston, Texas. Retrieved October 3, 2008, from http://www.swdsi.org/swdsi08/paper/SWDSI%20Proceedings%20Paper%20S469.pdf

Choudary, V. (2007). Software as a service: Implications for investment in software development. *Proceedings of the 40th Hawaii International Conference on System Sciences*, Waikoloa, Hawaii.

Choudhary, V. (2007). Comparison of software quality under perpetual licensing and software as a service. *Journal of Management Information Systems*, *24*(2), 141–165. doi:10.2753/MIS0742-1222240206

Clark, C. (1996). The answer to the machine is in the machine. In Hugenholtz, P. B. (Ed.), *The Future of Copyright in a Digital Environment*. The Hague: Kluwer Law International.

Clark, D. (2004, April 12). Microsoft strikes new patent accord. *Wall Street Journal*, p. A3.

Clark, J. (2009). How quantum cryptology works. *HowStuffWorks*. Retrieved December 15, 2009, from http://science.howstuffworks.com/quantum-cryptology.htm

Clayton, S. (2009). *Google talks software plus services*. Retrieved August 16, 2009, from http://blogs.msdn.com/stevecla01/archive/2009/01/28/google-talks-software-plus-services.aspx

Clemons, E. K., Reddi, S. P., & Row, M. C. (1993). The impact of information technology on the organization of economic activity: The 'move to the middle' hypothesis. *Journal of Management Information Systems*, *10*(2), 9–35.

Cohen, J. E. (1996). A right to read anonymously: A closer look at 'copyright management' in cyberspace. *28 Conn. L. Rev 981*. Retrieved October 15, 2009, from http://ssrn.com/abstract=17990

Collberg, C., Huntwork, A., Carter, E., Townsend, G., & Stepp, M. (2009). More on graph theoretic software watermarks: Implementation, analysis, and attacks. *Information and Software Technology*, *51*(1), 56. doi:10.1016/j.infsof.2008.09.016

Concha, D., Espadas, J., Romero, D., & Molina, A. (2010). The e-HUB evolution: From a custom software architecture to a Software-as-a-Service implementation. *Computers in Industry*, *61*(2), 145–151. doi:10.1016/j.compind.2009.10.010

Constance, E. B., & Dauch, C. E. (2008). *The Entrepreneur's Guide to Business Law*. Mason Ohio: South-Western.

Cook, J. (2001, January 12). Venture capital: Where Mercata led, consumers were unwilling to follow. *Seattle Post-Intelligencer*. Retrieved from http://www.seattlepi.com/business/vc122.shtml.

Copyright Act of 1976, Pub. L. No. 95-553, 90 Stat. 2541 (1976).

Cortese, A., & Stepanek, M. (1998, May 4). Good-bye to fixed pricing? *Business Week*, 70-84.

Crane, D. (2009). Intellectual Liability. *Texas Law Review*, *88*(2), 253.

Creeger, M. (2009). Cloud computing: An overview. *ACM Queue; Tomorrow's Computing Today*, *7*(5), 1–5.

Cronan, T., & Al-Rafee, S. (2008). Factors that influence the intention to pirate software and media. *Journal of Business Ethics*, *78*(4), 527. doi:10.1007/s10551-007-9366-8

Crosman, P. (2009). No slow down for SaaS. *Wall Street & Technology*, *27*(8), 22.

Crovitz, L. G. (2009, October 11). Media moguls and creative destruction. *The Wall Street Journal*. Retrieved from http://online.wsj.com/

Cusumano, M. A. (2007). The changing labyrinth of software pricing. *Communications of the ACM*, *50*(7), 19–22. doi:10.1145/1272516.1272531

Cusumano, M. A. (2008). The changing software business: Moving from products to services. *Computer*, *41*(1), 20–27. doi:10.1109/MC.2008.29

Da Lio, E., Fraboni, L., & Leo, T. (2005, October). *TWiki-based facilitation in a newly formed academic community of practice*. Proceedings of the 2005 International Symposium on Wikis. San Diego: ACM.

Dainow, B. (2009). Comparing offline with online advertising. Retrieved October 8, 2009 from http://www.visibilitymagazine.com/think-metrics/brandt-dainow/comparing-offline-with-online-advertising.

Danielson, K. (2007). *Confusing SaaS with Web 2.0*. Retrieved August 28, 2009, from http://www.ebizq.net/blogs/saasweek/2008/05/confusing_saas_with_web_20/

D'Aubeterre, F., Singh, R., & Iyer, L. (2008). Secure activity resource coordination: Empirical evidence of enhanced security awareness in designing secure business processes. *European Journal of Information Systems*, *17*(5), 528. doi:10.1057/ejis.2008.42

David, P. A. (2002). Understanding the digital economy's evolution and the path of measured productivity growth: present and future in the mirror of the past. In Brynolfsson, E., & Kahin, B. (Eds.), *Understanding the Digital Economy* (pp. 49–95). Boston, MA: MIT Press.

Davila, T. (2003). Salesforce.com: The evolution of marketing systems. *Harvard Case #E-145*, March 18. Wine Snob. Retrieved November 28, 2009, from http://www.iwinesnob.com/

Davis, F. D. (1989). Perceived usefulness, perceived ease of use, and user acceptance of information technology. *Management Information Systems Quarterly*, *13*(3), 319–339. doi:10.2307/249008

Dear, J. (2006, January 2). Put people before profits. *The Guardian*, 8.

Dejean, S. (2009). What can we learn from empirical studies about piracy? *CESifo Economic Studies*, *55*(2), 326. doi:10.1093/cesifo/ifp006

Delta, G., & Matsuura, J. (2009). *Law of the Internet* (3rd ed.). Frederick, Maryland: Aspen.

DeMark, E. F. (2004). Revenue recognition issues in a digital economy. *CPA Journal*, *74*(5). Retrieved August 14, 2009, from http://www.nysscpa.org/cpajournal/2004/504/perspectives/nv3.htm

Demirkan, H., Kauffman, R. J., Vayghan, J. A., Fill, H.-G., Karagiannis, D., & Maglio, P. P. (2008). Service-oriented technology and management: Perspectives on research and practice for the coming decade. *Electronic Commerce Research and Applications*, *7*(4), 356–376. doi:10.1016/j.elerap.2008.07.002

Demopoulos, T. (2007). *What No One Ever Tells You About Blogging and Podcasting: Real-Life Advice from 101 People Who Successfully Leverage the Power of the Blogosphere*. Kaplan Publishing.

Department of Innovation and Skills (and Department of Culture). (2009). *Digital Britain*. Retrieved November 27, 2009, from http://www.culture.gov.uk/images/publications/digitalbritain-finalreport-jun09.pdf

Depoorter, B. (2009). Technology and uncertainty: The shaping effect on copyright law. *University of Pennsylvania Law Review*, *175*, 1831–1868.

Dewan, R. M., & Freimer, M. L. (2003). Consumers prefer bundled add-ins. *Journal of Management Information Systems*, *20*(2), 99–111.

Di Maio, A. (2009, July 16). The blurring of government. *Governing*. Retrieved from http://www.governing.com/column/blurring-government

Diamond v. Diehr, 450 U.S. 175 (1981).

Digital Millennium Copyright Act of 1998, Pub. L. No. 105-304, 112 Stat. 2860 (1998).

Doctorow, C. (2003). *Down and Out in the Magic Kingdom*. New York: Tor Books.

Doherty, N. F., Anastasakis, L., & Fulford, H. (2009). The information security policy unpacked: A critical study of the content of university policies. *International Journal of Information Management*, *29*(6), 449–457. doi:10.1016/j.ijinfomgt.2009.05.003

Domingo, D., Quandt, T., Heinonen, A., Paulussen, S., Singer, J. B., & Vujnovic, M. (2008). Participatory journalism practices in the media and beyond: An international comparative study of initiatives in online newspapers. *Journalism Practice*, *2*(3), 326–341. doi:10.1080/17512780802281065

Dörte, B. (2008). Digital rights description as part of digital rights management: A challenge for libraries. *Library Hi Tech*, *26*(4), 598. doi:10.1108/07378830810920923

Dunham, M. (2009). SaaS metrics – Saasonomics-101. Retrieved April 1, 2009, from http://blog.sciodev.com/2009/02/10/saas-metrics-saasonomics-101/

Dym, R. (2009). Why software as a service? Helping our customers reduce costs and increase revenue. *OpSource - The Business of Web Operations* database. Retrieved March 22, 2009, from http://www.opsource.net/saas/wp_why_saas.pdf

E. & J. Gallo Winery v. Spider Webs Ltd., 286 F.3d 270 (5th Cir. 2002).

Economic Espionage Act, Pub. L. No. 104-294, 110 Stat. 3488 (1996).

Economides, N., & Katsamakas, E. (2005). Linux vs. Windows: a comparison of application and platform innovation incentives for open source and proprietary software platforms. New York University, Law and Economics Research Paper No. 05-21. Retrieved August, 12, 2009, from http://papers.ssrn.com/sol3/papers.cfm?abstract_id=822894

El-Affendi, M. A. (2008). Completing the circuit in e-government process automation. *Business Process Management Journal*, *14*(1), 96. doi:10.1108/14637150810849436

Electronic Arts. (2009). Profile for Electronic Arts Inc. *Yahoo! Finance*. Retrieved December 8, 2009, from http://finance.yahoo.com/q/pr?s=erts

Elfatatry, A. (2007). Dealing with change: Components versus services. *Communications of the ACM*, *50*(8), 35–39. doi:10.1145/1278201.1278203

Elsasser, J. (2009). Election 2008: The campaign of voter-generated content. *Public Relations Strategist*, *14*(1), 10–13.

Elvis Presley Enterprises, Inc. v. Passport Video, 249 F.3d 622 (2003).

Emarketer, Twitter Tally. (2009, April 28). Retrieved on October 10, 2009, from

Ewalt, D. M. (2005, November 30). Judge sours BlackBerry settlement. *Forbes*. Retrieved January 11, 2010, from http://www.forbes.com/2005/11/30/rim-blackberry-lawsuit-cx_de_1130rimm.html.

E-zest. (2009). E-zest: Company Overview. Retrieved August 1, 2009, from http://www.e-zest.net/technical_consulting.html

Facebook visits increased 194 percent in past year. (2009). *Experian Hitwise*. Retrieved on October 13, 2009, from http://www.hitwise.com/us/press-center/press-releases/social-networking-sept-09/

Facebook. (2009, May 27). Retrieved October 2, 2009, from

Facebook. (2010). Statistics. Retrieved January 12, 2010, from http://www.facebook.com/press/info.php?statistics

Facebook: Largest, fastest growing social network. (2008, August 13). Retrieved September 30, 2009, from http://www.techtree.com/India/News/Facebook_Largest_Fastest_Growing_Social_Network/551-92134-643.html

Family Entertainment and Copyright Act of 2005, § 17 U.S.C. § 110(11).

Fan, M., Kumar, S., & Whinston, A. B. (2009). Short-term and long-term competition between providers of shrink-wrap software and software as a service. *European Journal of Operational Research*, *196*(2), 661–671. doi:10.1016/j.ejor.2008.04.023

Fast Guide to CD/DVD. (2009). Retrieved October 15, 2009, from http://whatis.techtarget.com/definition/0,sid9_gci514667,00.html#

Federal Trademark Dilution Act of 1995 (FTDA), Pub. L. No. 104-98, 109 Stat. 985 (1995).

Feller, J., Finnegan, P., Fitzgerald, B., & Hayes, J. (2008). From peer production to productization: A study of socially enabled business exchanges in open source service networks. *Information Systems Research*, *19*(4), 475–493. doi:10.1287/isre.1080.0207

Fernando, C., Gatchev, V., & Spindt, P. (2005). Wanna dance? How firms and underwriters choose each other. *The Journal of Finance*, *60*, 2437–2469. doi:10.1111/j.1540-6261.2005.00804.x

Ferrante, D. (2006). Software licensing models: What's out there? *IT Professional*, *8*(6), 24–29. doi:10.1109/MITP.2006.147

Fetscherin, M. (2002). Present state and emerging scenarios of digital rights management systems. *International Journal on Media Management*, *4*(3), 164–171.

Financial Accounting Standards Board (FASB). (2006). Joint Conceptual Framework Project. Retrieved September 5, 2009, from http://www.fasb.org/fasac/Conceptual_Framework_12-06.pdf

Financial Accounting Standards Board (FASB). (2009). Position paper on IASB / FASB revenue recognition project. Retrieved September 5, 2009, from http://www.fasb.org/cs/BlobServer?blobcol=urldata&blobtable=MungoBlobs&blobkey=id&blobwhere=1175819078334&blobheader=application%2Fpdf

Fisher, W. W. (2004). *Promises to Keep. Technology, Law, and the Future of Entertainment*. Palo Alto, CA: Stanford University Press.

Fonovisa, Inc. v. Cherry Auction, Inc., 76 F.3d. 259 (9th Cir. 1996).

Ford, S., & Ratoza, M. (2008). Landmark trademark infringement awarded to Adidas. Bullivant Houser Bailey PC. Retrieved January 12, 2010, from http://www.bullivant.com/Landmark-trademark-infringement-awarded-to-Adidas.

Fowler, G. A. (2010, January 7). More makers jump into the e-reader market. *The Wall Street Journal*. Retrieved from http://online.wsj.com/

Fowler, G. A. (2010, January 7). Texas Instruments to enter e-reader market. *The Wall Street Journal*. Retrieved from http://online.wsj.com/

Fowler, G. A. (2010, January 31). Amazon backs down from e-book publisher fight. *The Wall Street Journal*. Retrieved from http://online.wsj.com/

Fox, L. (2009). Integrating SaaS and legacy apps: 5 steps for success. *NetworkWorld Asia*, *5*(1), 30.

Franklin, B., & Murphy, D. (1991). *What News? The Market, Politics and the Local Press*. London: Routledge.

Freer, J. (2007). UK regional and local newspapers. In Anderson, P., & Wood, G. (Eds.), *The Future of Journalism in the Advanced Democracies* (pp. 89–103). London: Ashgate.

Gallaway, T., & Kinnear, D. (2001). Unchained melody: A price discrimination-based policy proposal for addressing the MP3 revolution. *Journal of Economic Issues*, *35*(2), 279–287.

Galluzzo, V. (2009). When "now known or later developed" fails its purpose: How P2P litigation has turned the distribution right upside-down. *Florida Law Review*, *61*, 1165–1200.

Gayer, A., & Shy, O. (2003). Internet and peer-to-peer distributions in markets for digital products. *Economics Letters*, *81*(2), 197–203. doi:10.1016/S0165-1765(03)00170-8

Geiger, C. (2008b). The answer to the machine should not be the machine: Safeguarding the private copy exception in the digital environment. *European Intellectual Property Review*, *30*(4), 121–129.

Geiger, C., Hilty, R. M., Griffiths, J., & Suthersanen, U. (2008). A balanced interpretation of the "three-step test" in copyright law. *Max Planck Institute for Intellectual Property, Competition and Tax Law*. Retrieved October 15, 2009, from http://www.ip.mpg.de/shared/data/pdf/declaration_three_step_test_final_english.pdf

Geisman, J. (2008). *SaaS pricing for prosperity* [Webinar]. *MarketShare, Inc*. Retrieved April 16, 2009, from http://www.softwarepricing.com/readingroom/Content/OpSource%20Pricing%20for%20Prosperity%20Webinar.pdf

Geisman, J., & Nelson, B. (2008). *Pricing a SaaS product – what's the big deal?* [Webinar]. *PragmaticMarketing.com*. Retrieved June 01, 2009, from http://www.pragmaticmarketing.com/resources/archived-webinars/pricing-a-saas-product-2013-what2019s-the-big-deal

Gelatt, R. (1955). *The Fabulous Phonograph: From Tin Foil to High Fidelity*. Philadelphia: J. B. Lippincott Company.

Gezmer v. Public Health Trust of Miami-Cty, 219 F. Supp. 2d 1275, 1280 (S.D. Fla. 2002).

Ghose, A. (2009). Internet exchanges for used goods: An empirical analysis of trade patterns and adverse selection. *Management Information Systems Quarterly*, *33*, 263–291.

Gill, T., & Lei, J. (2009). Convergence in the high-technology consumer markets: Not all brands gain equally from adding new functionalities to products. *Marketing Letters*, *20*(1), 91–103. doi:10.1007/s11002-008-9050-5

Giurata, P. (2008). SaaS, PaaS, cloud computing, on-demand - what do they all mean? Retrieved October 15, 2008, from http://www.catalystresources.com/saas-blog/saas_paas_cloud_computing_on_demand_what_do_they_all_mean/

Glaser, J., & Foley, T. (2008, November). The future of healthcare IT. What can we expect to see? *Healthcare Financial Management*, 82–88.

Glen, D. (2008). Safer digital information. *Broadcasting & Cable*, *138*(27), 16.

Gold, N., Knight, C., Mohan, A., & Munro, M. (2004). Understanding service-oriented software. *IEEE Software*, *21*(2), 71–77. doi:10.1109/MS.2004.1270766

Goldstein, P. (2003). *Copyright's Highway: From Gutenberg to the Celestial Jukebox*. Palo Alto, CA: Stanford University Press.

Google. (2009). Run your Web apps on Google's infrastructure. Retrieved December 5, 2008, from http://code.google.com/appengine/

Gospill, T. (2009). Look after the pennies. *Free Press*, *1*, 8.

Grabianowski, E. (2009a). How brain-computer interfaces work. *HowStuffWorks*. Retrieved December 15, 2009, from http://computer.howstuffworks.com/brain-computer-interface.htm

Grabianowski, E. (2009b). How speech recognition works. *HowStuffWorks*. Retrieved December 15, 2009, from http://electronics.howstuffworks.com/gadgets/high-tech-gadgets/speech-recognition.htm

Green, K., Klemenhagen, B., & Hoch, F. (2004). Software as a service: Changing the paradigm in the software industry. *Software & Information Industry Association (SIIA) and TripleTree*. Retrieved January 23, 2009, from http://www.pangeafoundation.org/pdf/saas_aug04.pdf

Greengard, S. (2009). The first Internet president. *Communications of the ACM*, *52*(2), 16–18. doi:10.1145/1461928.1461935

Greer, M. B. (2009). *Software as a Service Inflection Point: Using Cloud Computing to Achieve Business Agility*. Bloomington, IN: iUniverse.

Greschler, D., & Mangan, T. (2002). Networking lessons in delivering 'software as a service' – Part II. *International Journal of Network Management*, *12*(5), 317–321. doi:10.1002/nem.446

Guadamuz, A. (2009, May). *If you build it, they won't come: Placing user-generated content in context of commercial copyright policy*. Presented at the Mashing-Up Culture: The Rise of User-generated content Workshop, Uppsala University, Sweden.

Guibault, L., & Helberger, N. (2005). Copyright Law and Consumer Protection. *European Consumer Law Group*. Retrieved October 15, 2009, from http://www.ivir.nl/publications/other/copyrightlawconsumerprotection.pdf

Guptill, B., & McNee, W. S. (2008). SaaS sets the stage for 'cloud computing'. *Financial Executive*, *24*(5), 37–44.

Gurbaxani, V., & Whang, S. (1991). The impact of information systems on organizations and markets. *Communications of the ACM*, *34*(1), 59–73. doi:10.1145/99977.99990

Han, J. K., Chung, S. W., & Sohn, Y. S. (2009). Technology convergence: When do consumers prefer converged products to dedicated products? *Journal of Marketing*, *73*(4), 97–108. doi:10.1509/jmkg.73.4.97

Haralambous, Y. (2007). *Fonts & Encodings*. O'Reilly Media.

Harper & Row Publishers v. Nation Enterprises, 471 U.S. 539 (1985).

Harris, W. (2009). How haptic technology works. *HowStuffWorks*. Retrieved December 15, 2009, from http://electronics.howstuffworks.com/gadgets/other-gadgets/haptic-technology.htm

Harwood, M. (2008, December 10). Cybercrime trends will worsen in 2009, according to forecasts. *Security Management*. Retrieved from http://www.securitymanagement.com/

Hatch, R. (2008). *SaaS Architecture, Adoption and Monetization of SaaS Projects using Best Practice Service Strategy, Service Design, Service Transition, Service Operation and Continual Service Improvement Processes*. Brisbane, Australia: Emereo Pty Ltd.

Hayes, B. (2008). Cloud computing. *Communications of the ACM*, *51*(7), 9–11. doi:10.1145/1364782.1364786

Heinzel, M. (2006, March 6). BlackBerry case could spur patent-revision efforts. *Wall Street Journal*, B4.

Helberger, N., & Hugenholtz, P. B. (2007). No place like home for making a copy: Private copying in European copyright law and consumer law. *Berkeley Technology Law Journal*, *22*, 1061–1098.

Helberger, N. (2005, August 26). Not so silly after all – new hope for private copying. *INDICARE Monitor*. Retrieved October 15, 2009, from http://www.indicare.org/tiki-read_article.php?articleId=132

Hendrickson, B. (2009, November 20). Evolution in the media revolution. Retrieved on December 20, 2009, from http://theshortestdistance.biz/?p=476

Herbert, L., Ross, C.F., Thresher, A., & Bartolomey, F. (2007). The components of SaaS pricing and negotiations. *Forrester Research*, Document # 43581. Retrieved February 20, 2009, from Forrester database.

Herman, B. (2008). Breaking and entering my own computer: The contest of copyright metaphors. *Communication Law and Policy*, *13*(2), 231. doi:10.1080/10811680801941276

Herman Hollerith. (2009). Retrieved October 15, 2009, from http://www.columbia.edu/acis/history/hollerith.html

Hermida, A., & Thurman, N. (2007, March). *Comments please: How the British news media is struggling with user-generated content*. Presented at the 8th International Symposium on Online Journalism, University of Texas, Austin.

Hewlett-Packard. (2010). Profile for Hewlett-Packard Company. *Yahoo! Finance*. Retrieved January 26, 2010, from http://finance.yahoo.com/q/pr?s=hpq

Hill, S. Jr. (2008). SaaS seems to favor users more than vendors. *Manufacturing Business Technology*, *26*(1), 48.

Hoch, F., Kerr, M., & Griffith, A. (2001). *Software as a service: strategic backgrounder*. Retrieved October 3, 2008, from http://www.siia.net/estore/ssb-01.pdf

Hoffman, L. (2009). Content control. *Communications of the ACM*, *52*(6), 16. doi:10.1145/1516046.1516052

Hogan, J., & Nagle, T. (2006). *The Strategy and Tactics of Pricing: A Guide to Growing More Profitably* (4th ed.). Upper Saddle River, NJ: Prentice Hall.

Hogan, J., & Nagle, T. (2005). What is strategic pricing? *Strategic Pricing Group Insights*. Retrieved March 15, 2009, from http://www.monitor.com/Portals/0/MonitorContent/documents/Monitor_What_Is_Strategic_Pricing.pdf

Hoogvliet, M. T. (2008). SaaS interface design. Retrieved April 28, 2009, from http://one3rd.nl/whitepaper_maartenhoogvliet_saasinterfacedesign.pdf

Hosting your party. (2009, August 29). Video retrieved September 30, 2009, from http://www.youtube.com/watch?v=1cX4t5-YpHQ.

Howe, J. (2008). *Crowdsourcing: Why the Power of the Crowd is Driving the Future of Business*. New York: Crown.

Howe, J. (2006). The rise of crowdsourcing. *Wired*. Retrieved May 21, 2009, from http://www.wired.com/archive

Hoxmeier, J. A. (2000). Software preannouncements and their impact on customers' perceptions and vendor reputation. *Journal of Management Information Systems*, *17*(1), 115–139.

http://theonlinephotographer.typepad.com/the_online_photographer/2009/09/blog-note update.html

http://topics.nytimes.com/topics/news/business/companies/facebook_inc/index.html

http://www.collegian.psu.edu/archive/2009/09/30/tweeting_provides_benefits_for.aspx

http://www.emarketer.com/Article.aspx?R=1007059

Hui, K. L., & Chau, P. Y. K. (2002). Classifying digital products. *Communications of the ACM, 45*(6), 73–79. doi:10.1145/508448.508451

Hunger, J. D., & Wheelen, T. L. (2007). *Essentials of Strategic Management*. Upper Saddle River, NJ: Pearson Prentice Hall.

Hunt, T. (2009). *The Whuffie Factor: Using the Power of Social Networks to Build Your Business*. New York: Crown Publishing Group.

IBM (2008). Making sense of SOA and today's IT innovations. *IBM Smart SOA solutions*. Retrieved November 20, 2008, from IBM database.

In re Nuijten, 500 F.3d 1346 (Fed. Cir. 2007).

Intacct (2009). Intacct Support. Retrieved on August 10, 2009, from http://us.intacct.com/services/support.php

International Accounting Standard Board (IASB). (2009). *IAS 2- Inventories*. Retrieved December 5, 2009, from http://www.iasplus.com/standard/ias02.htm

International Accounting Standard Board (IASB). (2009). *IAS 18- Revenue*. Retrieved December 5, 2009, from http://www.iasplus.com/standard/ias18.htm

International Accounting Standard Board (IASB). (2009). *Framework for the Preparation and Presentation of Financial Statements*. Retrieved December 5, 2009, from http://www.iasplus.com/standard/framewk.htm

Jaisingh, J., See-To, E. W. K., & Tam, K. Y. (2008). The impact of open source software on the strategic choices of firms developing proprietary software. *Journal of Management Information Systems, 25*(3), 241–275. doi:10.2753/MIS0742-1222250307

Jansen, S., Brinkkemper, S., & Finkelstein, A. (2009). Business network management as a survival strategy: A tale of two software ecosystem. *Proceedings the International Workshop on Software Ecosystem (IWESECO 2009)*, Virginia, USA.

Jaszi, P. (1998). Intellectual property legislative update: Copyright, paracopyright, and pseudo-copyright. Paper presented at the Association of Research Libraries conference: The Future Network: Transforming Learning and Scholarship. Eugene, Oregon, May 13-15, Retrieved from http://www.arl.org/resources/pubs/mmproceedings/132mmjaszi~print.shtml

Jean-Noël, E., Elspeth, M., & David, B. (2007). Mastering the art of corroboration. *Journal of Enterprise Information Management, 20*(1), 96.

Jerri, A. (1977). The Shannon sampling theorem - its various extensions and applications: A tutorial review. *Proceedings of the IEEE*, 1567-1596.

Jiang, J. J., Klein, G., & Carr, C. L. (2002). Measuring information system service quality: SERVQUAL from the other side. *Management Information Systems Quarterly, 26*(2), 145–166. doi:10.2307/4132324

Jin, L., & Robey, D. (1999). Explaining cybermediation: An organizational analysis of electronic retailing. *International Journal of Electronic Commerce, 3*(4), 47–65.

Jones, D. (2008). The five qualities of good software pricing. *Forrester Research*, Document # 46218. Retrieved February 20, 2009, from Forrester database.

Joystiq. Retrieved October 1, 2009, from http://www.joystiq.com/

Kahlow, A. (2009, July 23). Marketing to people. Retrieved October 5, 2009, from http://www.clickz.com/3634482

Kanaracus, C. (2009). NetSuite, ex-reseller locked in ugly legal battle. *Techworld*. Retrieved August 1, 2009, from http://www.techworld.com.au/article/304143/netsuite_ex-reseller_locked_ugly_legal_battle

Kane, Y. I. (2009, November 25). EA chief wagers on digital future for games. *The Wall Street Journal*. Retrieved from http://online.wsj.com/

Kannan, P. K., & Kopalle, P. K. (2001). Dynamic pricing on the Internet: Importance and implications for consumer behavior. *International Journal of Electronic Commerce*, *5*(3), 63–83.

Kaplan, J. M. (2009). SaaS movement accelerating. *Business Technology Trends & Impacts Advisory Service Executive Update*, *8*(22).

Kaplan, T. (2008, July 20). Music industry zealous in tracking tune thieves. *St. Petersburg Times*, 1B.

Katyal, S. (2004a). The new surveillance. *Case Western Law Review*, *54*(297). Retrieved September 25, 2009, from http://ssrn.com/abstract=527003

Katyal, S. (2004b). Privacy vs. piracy. *Yale Journal of Law & Technology*, *7*. Retrieved September 25, 2009, from http://ssrn.com/abstract=722441

Kee, T. (2009, June 17). EA COO John Pleasants: 'going digital' is key to returning to profitability. *Paidcontent.org*. Retrieved December 4, 2009, from http://paidcontent.org/article/419-can-eas-new-digital-strategy-bring-back-its-glory-days/

Keller, E. (2007). How software application pricing models are likely to change. *Manufacturing Business Technology*, *25*(1), 42–43.

Kettinger, W. J., & Lee, C. C. (2005). Zones of tolerance: Alternative scales from measuring information systems service quality. *Management Information Systems Quarterly*, *29*(4), 607–623.

Khouja, M., & Park, S. (2007-8). Optimal pricing of digital experience goods under piracy. *Journal of Management Information Systems*, *24*(3), 109–141. doi:10.2753/MIS0742-1222240304

Kindle, A. (2008)... *Technology Review*, *111*(2), 94–95.

Kirkpatrick, M. (2009, May 18). The dam just broke: Facebook opens up to OpenID. *ReadWriteWeb*. Retrieved from http://www.readwriteweb.com/archives/the_dam_just_broke_facebook_opens_up_to_openid.php

Kittlaus, H., & Clough, P. N. (2009). *Software Product Management and Pricing: Key Success Factors for Software Organizations*. Berlin: Springer.

Klaassen, A. (2007, November 26). Facebook's bid ad plan: if users like you, they'll be your campaign. *Advertising Age*, retrieved on September 30, 2009, from http://adage.com/digital/article?article_id=121806&search_phrase=%22social+ads%22

Knol, P., Spruit, M., & Scheper, W. (2008). Web 2.0 revealed - business model innovation through social computing. *Proceedings of the Seventh AIS SIGeBIZ Workshop on e-business (WeB 2008)*, Paris, France.

Koelman, K. J. (2006). Fixing the three-step test. *European Intellectual Property Review*, *8*, 407–412.

Kotha, S. (1995). Mass customization: implementing the emerging paradigm for competitive advantage. *Strategic Management Journal*, *16*(7), 21–42. doi:10.1002/smj.4250160916

KPN. (2009a). *Exact Online*. Retrieved on February 26, 2009, from http://zakelijk.kpn.com/web/file?uuid=d386cdd0-b010-4bd3-b66d-2e183510674a&owner=3c096f1f-64ae-471a-a72b-161373c8b70b

KPN. (2009b). *KPN Online Boekhouden*. Retrieved on February 26, 2009, from http://zakelijk.kpn.com/business/meer-diensten/softwareonline/alle-software-online/online-boekhouden.htm

Krueger, R. A., & Casey, M. A. (2000). *Focus Groups. A Practical Guide for Applied Research*. Thousand Oaks, CA: Sage Publications.

Kruff, J. (2008). Ocimum benefits from SaaS model. *Enterprise Innovation*, *4*(5), 22–23.

Kuan, A. H. Y. (2006). *Planning Intellectual Property for Marketing Strategies in the Digital Content Industry*. Unpublished masters dissertation, National Chengchi University, Taiwan.

Lal, R., & Sarvary, M. (1999). When and how is the Internet likely to decrease price competition? *Marketing Science*, *18*(4), 485–503. doi:10.1287/mksc.18.4.485

Lamont, J. (2010). SaaS: Integration in the cloud. *KM World*, *19*(1), 12–22.

Landy, G. K. (2008). *The IT/Digital Legal Companion-A Comprehensive Business Guide to Software, IT, Internet, Media and IP Law*. Burlington, MA: Syngress Publishing.

Lang, K. R., & Vragov, R. (2005). A pricing mechanism for digital content distribution over computer networks. *Journal of Management Information Systems*, *22*(2), 121–139.

Langholz, G., Kandel, A., & Mott, J. L. (1998). *Foundations of digital logic design*. Singapore: Word Scientific Publishing Co. Pte. Ltd.

Lassila, A. (2006).Taking a service-oriented perspective on software business: How to move from product business to online service business. *IADIS International Journal on WWW/Internet*, *4*(1), 70-82.

Laudon, K. A. (2009). *Essentials of Management Information Systems*. Upper Saddle River, NJ: Prentice Hall.

Laudon, K., & Traver, C. (2004). *E-Commerce: Business, Technology, Society* (2nd ed.). Boston, MA: Addison-Wesley.

Laudon, K. C., & Traver, C. G. (2010). *E-Commerce: Business. Technology. Society*. Upper Saddle River, NJ: Pearson Prentice Hall.

Lee, C. H., Barua, A., & Whinston, A. (2000). The complementarity of mass customization and electronic commerce. *Economics of Innovation and New Technology*, *9*(2), 81–110. doi:10.1080/10438590000000005

Lenk, A., Klems, M., Nimis, J., Tai, S., & Sandholm, T. What's inside the cloud? An architectural map of the cloud landscape. *ICSE Workshop on Software Engineering Challenges of Cloud Computing 2009*, Vancouver, Canada.

Lessig, L. (2001). *The Future of Ideas: The Fate of the Commons in a Connected World*. New York: Random House.

Lessig, L. (2006). *Code 2.0*. New York: Basic Books.

Lessig, L. (2008). *Remix: Making art and commerce thrive in the hybrid economy*. New York: Penguin Press.

Lev, B. (1974). On the association between operating leverage and risk. *Journal of Financial and Quantitative Analysis*, *9*, 627–641. doi:10.2307/2329764

Li, C., & Bernoff, J. (2008). *Groundswell: Winning in a World Transformed by Social Technologies*. Boston, MA: Harvard Business Press.

Liebowitz, S. J., & Margolis, S. E. (1994). Network externality: An uncommon tragedy. *The Journal of Economic Perspectives*, *8*(2), 133–150.

Lin, G., Fu, D., Zhu, J., & Dasmalchi, G. (2009). Cloud computing: IT as a service. *IT Professional*, *11*(2), 10–13. doi:10.1109/MITP.2009.22

Lin, C., & Chen, H. (2001). Chen, C., & Hou, S. Implementation and evaluation of a multifunctional telemedicine system in NTUH. *International Journal of Medical Informatics*, *61*(2-3), 175–187. doi:10.1016/S1386-5056(01)00140-X

Linde, F. (2009). Pricing information goods. *Journal of Product and Brand Management*, *18*(5), 379–384. doi:10.1108/10610420910981864

Lohr, S. (2009, April 27). G.E.'s breakthrough can put 100 DVDs on a disc. *The New York Times*. Retrieved from http://www.nytimes.com/

Loughran, T., & Ritter, J. (1995). The new issues puzzle. *The Journal of Finance*, *50*, 23–51. doi:10.2307/2329238

Lovelock, C., Wirtz, J., & Chew, P. (2008). *Essentials of Service Marketing*. Pearson Edition.

Luit Infotech. (2008). *Difference between the ASP model and the SaaS model*. Retrieved November 12, 2008, from http://www.luitinfotech.com/downloads/saas-asp-difference.pdf

Luxem, R. (2000). The impact of trading digital products on retail information systems. Paper presented at the meeting of the *33rd Hawaii International Conference on System Sciences (HICSS)*, Hawaii.

Madrigal, A. (2009, March 4). Researchers want to add touch, taste and smell to virtual reality. *Wired.com*. Retrieved from http://www.wired.com/

Mahowald, R. P. (2003). Do service providers deliver value and reduce enterprise costs? *IDC*. Retrieved November 5, 2008, from IDC database.

Mahowald, R. P. (2009). SaaS, PaaS, and cloud: choices for success. *Proceedings of the IDC SaaS Summit Spring 2009*, Document #217935. Retrieved April 5, 2009, from IDC database.

Majchrzak, A. (2009). Comment: Where is the theory in wikis? *Management Information Systems Quarterly*, *33*(1), 18–20.

Malone, T. W., Yates, J., & Benjamin, R. I. (1987). Electronic markets and electronic hierarchies. *Communications of the ACM*, *30*(6), 484–497. doi:10.1145/214762.214766

Maximilien, E. M., Ranabahu, A., & Gomadam, K. (2008). An online platform for Web APIs and service mashups. *IEEE Internet Computing*, *12*(5), 32–43. doi:10.1109/MIC.2008.92

Mayne, M. (2009, September 22). Encryption is becoming more elaborate to ensure confidential business data is kept secret. *SC Magazine*. Retrieved from http://www.scmagazineuk.com/encryption-is-becoming-more-elaborate-to-ensure-confidential-business-data-is-kept-secret/article/149413/

Mazziotti, G. (2008). *EU Digital Copyright Law and the End-User*. New York: Springer.

McCrea, B. (2009, December 9). 5 Higher Ed Tech Trends to Watch in 2010. *Campus Technology*. Retrieved from http://campustechnology.com/articles/2009/12/09/5-higher-ed-tech-trends-to-watch-in-2010.aspx

McDermott, C. (2009, September 30). Tweeting provides benefits for businesses. *The Daily Collegian Online*. Retrieved on October 6, 2009, from

McGoldrick, P. J., Keeling, K. A., & Beatty, S. F. (2008). A typology for roles of avatars in online retailing. *Journal of Marketing Management*, *24*(3-4), 433–461. doi:10.1362/026725708X306176

McManis, C. R. (1999). The privatization (or shrink-wrapping) of American copyright law. *California Law Review*, *87*, 173–190. doi:10.2307/3481006

Mearian, L. (2008, March 11). Study: Digital universe and its impact bigger than we thought. *Computerworld*. Retrieved from http://www.computerworld.com/

Menell, P. (2002). Can our current conception of copyright law survive the Internet age?: Envisioning copyright law's digital future. *New York Law School Law Review. New York Law School*, *46*, 63–199.

Menken, I. (2008). *SaaS - The Complete Cornerstone Guide to Software as a Service Best Practices Concepts, Terms, and Techniques for Successfully Planning, Implementing and Managing SaaS Solutions*. Brisbane, Australia: Emereo Pty Ltd.

Merchant, H., & Schendell, D. (2000). How do international joint ventures create shareholder value? *Strategic Management Journal*, *21*(7), 723–737. doi:10.1002/1097-0266(200007)21:7<723::AID-SMJ114>3.0.CO;2-H

Merchant, N., & Geisman, J. (2006). Solving the puzzle: pricing, licensing and business models. *Rubicon Consulting, Inc & Market Share, Inc*. Retrieved March 5, 2009, from http://rubiconconsulting.com/downloads/whitepapers/Rubicon_Solving-Puzzle.pdf

Merriam-Webster Online Dictionary. (2008). Retrieved June 25, 2008, from http://www.merriam-webster.com/dictionary/blog

Mersey, R. (2009). Online news users' sense of community: Is geography dead? *Journalism Practice*, *3*(3), 347–360. doi:10.1080/17512780902798687

Metro-Goldwyn-Mayer Studios Inc. v. Grokster, Ltd., 259 F. Supp. 2d 1029 (C.D. Cal. 2003), aff'd, 380 F.3d 1154 (9th Cir. 2004).

Metro-Goldwyn-Mayer Studios Inc. v. Grokster, Ltd., 545 U.S. 913 (2005).

Meyer, P. (2004). *The Vanishing Newspaper: Saving Journalism in the Information Age*. Columbia, MO: University of Missouri Press.

Meyer, P. (2008). The elite newspaper of the future. *American Journalism Review*, 32-35.

Microsoft. (2009a). *Cloud computing infrastructure*. Retrieved August 12, 2009, from http://www.microsoft.com/virtualization/en/us/cloud-computing.aspx

Microsoft. (2009b). Software + services. *The Architecture Journal, 13*. Retrieved August 15, 2009, from *MSDN Architecture Center* database via http://msdn.microsoft.com/en-us/architecture/bb906058.aspx

Miller, C., & Wells, F. (2007). Balancing security and privacy in the digital workplace. *Journal of Change Management, 7*(3/4), 315. doi:10.1080/14697010701779181

Miller, C. H. (2008). *Digital Storytelling: A Creator's Guide to Interactive Entertainment* (2nd ed.). Focal Press.

Ming-Chiang, H., Der-Chyuan, L., & Ming-Chang, C. (2007). Dual-wrapped digital watermarking scheme for image copyright protection. *Computers & Security, 26*(4), 319. doi:10.1016/j.cose.2006.11.007

Mintel. (2007). Regional newspapers. *Mintel Report*, London: Mintel.

Monty, S. (2010, January 2). Social media predictions for 2010. Retrieved on January 18, 2010, from http://www.scottmonty.com/2010/01/social-media-predictions-for-2010.html#ixzz0czJJdsg1

Moon, J. Y., & Sproull, L. S. (2008). The role of feedback in managing the Internet-based volunteer work force. *Information Systems Research, 19*, 494–515. doi:10.1287/isre.1080.0208

Moore, J. F. (1993). Predators and prey: A new ecology of competition. *Harvard Business Review, 71*(3), 75–86.

Moore, G. (1965). Cramming more components onto integrated circuits. *Electronics, 38*(8), 114–117.

Moore, B., & Mahmoud, Q. H. (2009). A service broker and business model for SaaS applications. *Proceedings of the IEEE/ACS International Conference on Computer Systems and Application*, Rabat, Morocco.

Moores, T., & Dhillon, G. (2000). Software piracy: A view from Hong Kong. *Communications of the ACM, 43*(12), 88–93. doi:10.1145/355112.355129

Mullins, L. J. (2005). Managing intellectual property in the digital product market. *Journal of Digital Asset Management, 1*(1), 59–66. doi:10.1057/palgrave.dam.3640010

Murfin, J. (2005). Business case for entering SaaS. Retrieved August 3, 2009, from *Microsoft Solution for Hosted Messaging and Collaboration* database.

Murphy, S., & Samir, W. (2009). 'In the cloud' IT creates new opportunities for network service providers. *Journal of Telecommunications Management, 2*(2), 107–120.

Murphy, D. (2008). Earthquake undermines structure of local press ownership: Many hurt. In Franklin, B. (Ed.), *Pulling Newspapers Apart: Analysing Print Journalism*. London: Routledge.

Myers, S., & Majluf, S. (1984). Corporate financing and investment decisions when firms have information that investors do not have. *Journal of Financial Economics, 13*, 187–221. doi:10.1016/0304-405X(84)90023-0

Myers, S. C. (1977). Determinants of corporate borrowing. *Journal of Financial Economics, 5*, 147–175. doi:10.1016/0304-405X(77)90015-0

Myers, S. C. (1984). The capital structure puzzle. *The Journal of Finance, 39*, 575–592. doi:10.2307/2327916

National Conference of Commissioners on Uniform State Laws. (1985). Uniform Trade Secrets Act (1985).

National Readership Survey. (2007). Readership of newspapers and online news. Retrieved December 1, 2009, from www.nrs.co.uk

National Readership Survey. (2009). Readership of newspapers and online news. Retrieved December 15, 2009, from www.nrs.co.uk

Neil, M. (2009). McCain says sorry for campaign use of signature Jackson Browne song. ABA Journal. Retrieved January 10, 2010, from http://www.abajournal.com/news/article/mccain_says_sorry_for_campaign_use_of_signature_jackson_browne_song/

Neoseeker. Retrieved October 6, 2009, from http://www.neoseeker.com/.

Netanel, N. W. (2003).Impose a Noncommercial Use Levy to Allow Free Peer-to-Peer File Sharing. *Harvard Journal of Law & Technology, 17*. Retrieved October 15, 2009, from http://ssrn.com/abstract=468180

NeuroSky. (2009, March 26). NeuroSky to launch brainwave-based technology for stereo headsets for personal computers. *PR Newswire*. Retrieved from http://news.prnewswire.com/

Newspaper Society. (2009). Local and regional newspaper historical data. Retrieved December 5, 2009, from http://jiab.jicreg.co.uk/JIAB.cfm?NoHeader=1

Nichols, J. (2009). It's crunch-time for journalism. *Free Press*, 1-2.

Nihon Keizai Shimbun, Inc. v. Comline Bus. Data, Inc., 166 F.3d 65, 72 (2d Cir. 1999).

No Electronic Theft (NET) Act of 1997. Pub. L. No. 105-147, 111 Stat. 2678 (1997).

Nusca, A. (2009, December 10). New 'electronic paper' technology allows for 1 million color choices for your walls, electronics. *Smartplanet*. Retrieved from http://www.smartplanet.com/

Obannon, I. (2009). What Web 2.0, SaaS and cloud computing mean for tax & accounting professionals. Retrieved August 28, 2009, from http://getanewbrowser.com/2006/05/web-20-soa-saas/

Office for National Statistics. (2009). Social trends, No. 39. Basingstoke, Hampshire: Palgrave Macmillan.

Orr, B. (2006). SaaS just may be the end of software as we know it. *ABA Banking Journal, 98*(8), 51–52.

Orr, B. (2007). Microsoft begins its radical shift to software as a service. *ABA Banking Journal, 99*(12), 46–47.

Owen, B., & Wildman, S. (1992). *Video Economics*. Cambridge, MA: Harvard University Press.

Owyang, J. (2009, April 27). The future of the social Web: in five eras. Retrieved January 16, 2010, from http://www.web-strategist.com/blog/2009/04/27/future-of-the-social-web/

Panko, R. (2009). *Business Data Networks and Telecommunications*. Upper Saddle River, NJ: Prentice Hall.

Papazoglou, M. (2003). Service-oriented computing: Concepts, characteristics and directions. *Proceedings of the Fourth International Conference on Web Information System Engineering (WISE'03)*, Rome, Italy.

Park, Y., & Scotchmer, S. (2005). Digital rights management and the pricing of digital products. National Bureau of Economic Research, NBER Working Paper No. 11532. Retrieved September 4, 2009, from http://www.nber.org/papers/w11532.pdf?new_window=1

Peake, A. (2009). Internet governance. *Digital Review of Asia Pacific*. Retrieved December 15, 2009, from http://www.digital-review.org/themes/25-internet-governance.html

Peitz, M., & Waelbroeck, P. (2006). Piracy of digital products: A critical review of the theoretical literature. *Information Economics and Policy, 18*(4), 449–476. doi:10.1016/j.infoecopol.2006.06.005

Perfect 10, Inc. v. Amazon.com, Inc., 508 F.3d 1146 (9th Cir. 2007).

Picard, R. (2004). Commercialism and newspaper quality. *Newspaper Research Journal, 25*(1), 54–64.

Pincus, W. (2009). Newspaper narcissism: Our pursuit of glory led us away from readers. *Columbia Journalism Review*. Retrieved June 1, 2009, from http://www.cjr.org/archive

Pine, B. J. II, Victor, B., & Boyton, A. C. (1993). Making mass customization work. *Harvard Business Review, 71*(5), 108–122.

Poon, S., & Joseph, M. (2000). Product characteristics and Internet commerce benefit among small businesses. *Journal of Product and Brand Management, 9*(1), 21–34. doi:10.1108/10610420010316311

Porter, M. E. (1985). *Competitive Advantage: Creating and Sustaining Superior Performance*. New York: Free Press.

Porter, M. E. (2001). Strategy and the Internet. *Harvard Business Review*. Retrieved March 26, 2009, from http://www.soum.com.br/sonaomuda/ecommerce/Arquivos/Artigos/Strategy_and_the_internet.pdf

Postmus, D., Wijngaard, J., & Wortmann, H. (2009). An economic model to compare the profitability of pay-per-use and fixed-fee licensing. *Information and Software Technology*, *51*(3), 581–588. doi:10.1016/j.infsof.2008.08.004

Postmus, D., Wijngaard, J., & Wortmann, H. (2009). An economic model to compare the profitability of pay-per-use and fixed-fee licensing. *Information and Software Technology*, *51*(3), 581–588. doi:10.1016/j.infsof.2008.08.004

Poston, R. S., Kettinger, W. J., & Simon, J. C. (2009). Managing the vendor set: Achieving best pricing and quality service in IT outsourcing. *MIS Quarterly Executive*, *8*(2), 45–58.

predictions round-up. (2009, December 31). Retrieved on January 18, 2010, from http://www.emarketer.com/Articles/Print.aspx?1007446

Press Release. (Nov. 4, 2003). E-Poll, E-poll study looks at consumer's attitudes before and after RIAA lawsuits. Retrieved from http://www.prnewswire.com/cgi-bin/stories.pl?ACCT=104&STORY=/www/story/11-04-2003/0002050963&EDATE.

Pring, B., & Stahlman, M. (2009). Sizing the cloud: The world's first comprehensive cloud computing services forecast [Webinar]. *Gartner*. Retrieved March 19, 2009, from Gartner database.

Pring, B., Desisto, R.P., & Bona, A. (2007). The cost and benefits of SaaS vs. on-premise deployment. *Gartner*, Article G00151171. Retrieved November 13, 2008, from Gartner database.

ProgressSoftware. (2008). SaaS billing & metering. *Progress Software Corporation*. Retrieved May 20, 2009, from http://communities.progress.com/pcom/servlet/JiveServlet/download/12057-2-11235/SaaS_BillingMetrics.pdf

Project, D. V. B. (2009). DVB Fact Sheet – April 2009.

Qualman, E. (2009, August 11). Statistics show social media is bigger than you think. Retrieved from http://socialnomics.net/2009/08/11/statistics-show-social-media-is-bigger-than-you-think/

Quibble, Z. K. (2005). Blogs and written business communication courses: A perfect union. *Journal of Education for Business*, *80*, 327–331. doi:10.3200/JOEB.80.6.327-332

Raghu, T. S., Sinha, R., Vinze, A., & Burton, O. (2009). Willingness to pay in an open source software environment. *Information Systems Research*, *20*(2), 218–236. doi:10.1287/isre.1080.0176

Raghu, T. S., Sinha, R., Vinze, A., & Burton, O. (2009). Willingness to pay in an open source software environment. *Information Systems Research*, *20*, 218–236. doi:10.1287/isre.1080.0176

Ram, D. G., & Sanders, G. L. (2000). Global software piracy: You can't get blood out of a turnip. *Communications of the ACM*, *43*(9), 82. doi:10.1145/348941.349002

Rayner, N. (2008). The impact of SOA and SaaS on financial systems. *Gartner*, Article G00157191. Retrieved November 13, 2008, from Gartner database.

Rayport, J. F., & Sviokla, J. J. (1995). Exploiting the virtual value chain. *Harvard Business Review*, *75*, 75–85.

Rayport, J. F., & Sviokla, J. J. (1995). Exploiting the virtual value chain. *Harvard Business Review*, *75*(6), 75–85.

Redman, T. C. (1998). Impact of poor data quality on the typical enterprise. *Communications of the ACM*, *41*(2), 79–82. doi:10.1145/269012.269025

Restatement (Second) of Agency § 228 (1958).

Retrieved May 21, 2009, from http://www.trinitymirror.com/documents/2009%20Interim%20Announcement%20final.pdf

Ritter, J. (1984). The 'Hot' issue market of 1980. *The Journal of Business, 57*, 23–51. doi:10.1086/296260

Ritter, J., & Welch, J. (2002). A review of IPO activity, pricing, and allocations. *The Journal of Finance, 57*, 1795–1827. doi:10.1111/1540-6261.00478

Rowell, J. (2009). A step-by-step guide to starting up SaaS operations. *OpSource – The SaaS Delivery Experts* database. Retrieved March 22, 2009, from http://www.opsource.net/saas/starting_up_saas_operations.pdf

Rust, R. T., & Kannan, P. K. (2003). E-service: A new paradigm for business in the electronic environment. *Communications of the ACM, 46*(6), 36–42. doi:10.1145/777313.777336

Sääksjärvi, M., Lassila, A., & Nordstrom, H. (2005). Evaluating the software as a service business model: From CPU time-sharing to online innovation sharing. *Proceedings of the IADIS International Conference e-Society 2005*, Qawra, Malta.

Safko, L., & Brake, D. K. (2009). *The Social Media Bible: Tactics, Tools, and Strategies for Business Success*. New York: John Wiley & Sons, Inc.

Sainio, L.-M., & Marjakoski, E. (2009). The logic of revenue logic? Strategic and operational levels of pricing in the context of software business. *Technovation, 29*(5), 368–378. doi:10.1016/j.technovation.2008.10.009

Salazar, C. (2007, May 23). Can Facebook win the battle over MySpace? Retrieved October 9, 2009, from http://ebizz.wordpress.com/2007/05/23/can-facebook-win-the-battle-over-myspace/

Salesforce.com. (2008). Developer Force: Apex Code. Retrieved December 3, 2008, from http://wiki.developerforce.com/index.php/Apex

Salesforce.com. (2009a). Contact Partner. Retrieved on August 5, 2009, from http://www.salesforce.com/partners/opportunities/consulting-partners/profiles/a0x30000000CfKu.jsp

Salesforce.com. (2009b). Job detail of Customer Success Manager. Retrieved on August 20, 2009, from http://www.salesforce.com/company/careers/locations/a0800000000Ab42AAC/a017000000AIvVJ.jsp

Salter, C. (2009, February 11). The Fast Company 50 – 2009, #9 Amazon. *Fast Company*. Retrieved December 8, 2009 from http://www.fastcompany.com/list/amazon

Samtani, R. (2009, March). Ongoing innovation in digital watermarking. *Computer*, 111–113.

Schill, M., & Zhjou, C. (2001). Pricing an emerging industry: Evidence from Internet subsidiary carve-outs. *Financial Management, 29*, 5–33. doi:10.2307/3666374

Schilling, M. (1998). Technological lockout: An integrative model of the economic and strategic factors driving technology success and failure. *Academy of Management Review, 98*, 267–284. doi:10.2307/259374

Schilling, M. (2002). Technology success and failure in winner-take-all markets: The impact of learning orientation, timing and network externalities. *Academy of Management Journal, 45*, 387–398. doi:10.2307/3069353

Schmurr, A., & Crawley. (2003). Cybercrime in the United States criminal justice system: Cryptography and steganography as tools of terrorism. *Journal of Security Administration, 26*(2), 51–76.

Schreiber, J. (2001, March 7). Five reasons to co-brand your Web site. *MarketingProfs*. Retrieved December 4, 2009, from http://www.marketingprofs.com/2/cobrand5reasons.asp

Schultz, M. F. (2009, January). Live performance, copyright, and the future of the music business. *University of Richmond Law Review. University of Richmond, 43*, 685–764.

Schultz, P. (2003). Pseudo market timing and the long-run underperformance of IPOs. *The Journal of Finance, 58*, 483–518. doi:10.1111/1540-6261.00535

Schultze, A. (2007). *Channel Excellence*. USA: Axel Schultze.

Schulz, G. (2009). *The Green and Virtual Data Center*. Boca Raton, FL: Auerbach Publications. doi:10.1201/9781420086676

Scott, M. D. (2007). *Scott on Information Technology Law* (3rd ed.). United States: Aspen Publishers.

Scupola, A. (2005). Strategies of e-commerce business value optimization. In Khosrow-Pour, M. (Ed.), *Encyclopedia of Information Science and Technology*. Hershey, PA: Idea Group Reference.

Seddon, P. (1998). Digital products and processes: A critique of Whinston, Stahl, and Choi's Chapter 2. Retrieved August 14, 2009, from http://disweb.dis.unimelb.edu.au/staff/peterbs/research/DigitalProductsAndProcesses.doc

Sega Enter. Ltd. v. MAPHIA, 857 F. Supp. 679 (N.D. Cal. 1994).

Senftleben, M. (2004). *Copyright, Limitations and the Three-Step Test. An Analysis of the Three-Step Test in International and EC Copyright Law*. The Hague: Kluwer Law International.

Sessions, R. (2006). Software as a service: Another perspective. *The ObjectWatch Newsletter, 52*. Retrieved October 3, 2008, from http://www.objectwatch.com/newsletters/ObjectWatchNewsletter052.pdf

Shaw, M. J., Gardner, D. M., & Thomas, H. (1997). Research opportunities in electronic commerce. *Decision Support Systems, 21*(3), 149–156. doi:10.1016/S0167-9236(97)00025-0

Sifry, D. (2007). The state of the live Web. Retrieved June 25, 2008, from www.sifry.com/alerts/archives/000493.html

Smith, C. W., & Watts, R. L. (1992). The investment opportunity set and corporate financing, dividend and compensation policies. *Journal of Financial Economics, 32*, 263–292. doi:10.1016/0304-405X(92)90029-W

Smith, M. D., & Telang, R. (2009). Competing with free: The impact of movie broadcasts on DVD sales and internet piracy. *Management Information Systems Quarterly, 33*, 321–338.

Smith, E. (2010, January 19). MTV games seeks buzz for 'Rock Band'. *The Wall Street Journal*. Retrieved from http://online.wsj.com/

Smith, E., & Kane, Y. I. (2009, December 4). Apple acquires Lala Media. *The Wall Street Journal*. Retrieved from http://online.wsj.com/

*Software Net Corporation Form 424B1*. (1998, June 18). Retrieved September 14, 2009 from http://www.sec.gov/Archives/edgar/data/1060531/0000891618-98-002957.txt

Sohl, J. (1999). The early stage equity market in the USA. *Venture Capital: An International Journal of Entrepreneurial Finance, 1*, 101–1120. doi:10.1080/136910699295929

Sony Corp. v. Universal City Studios, Inc., 464 U.S. 417 (1984).

Spohrer, J., Maglio, P. P. Bailey, J., & Gruhl, D. (2007). Steps toward a science of service systems. *IEEE Computer Society*, January, 71-77.

Sreenivasan, S. (2009). Web retailers finding allies at sites with nothing to sell. Retrieved October 8, 2009 from http://www.nytimes.com/1997/04/14/business/web-retailers-finding-allies-at-sites-with-nothing-to-sell.

Srinivasan, R., Lilien, G., & Rangaswamy, A. (2004). First in, first out? The effects of network externalities on pioneer survival. *Journal of Marketing, 68*, 41–58. doi:10.1509/jmkg.68.1.41.24026

State of Washington, Department of Revenue (DOR). (2009). *Digital Products Bill (ESHB 2075)*. Retrieved September 5, 2009, from http://dor.wa.gov/Content/GetAFormOrPublication/PublicationBySubject/TaxTopics/DigitalProductsQA.aspx

Statement No, F. A. S. B. 151, Inventory Costs. Retrieved December 5, 2009, from http://asc.fasb.org/

Stats, F. 2010. Retrieved September 26, 2010, from http://www.facebook.com/press/info.php?statistics

Stats, F. Retrieved October 10, 2009, from http://blog.facebook.com/blog.php?post=136782277130

Stats, T. Retrieved October 13, 2009, from http://www.crunchbase.com/company/twitter

Stiffler, D., & Bois, R. (2008). Consulting in the cloud: The emerging SaaS consulting, product development, and outsourcing ecosystem (Excerpt). *AMR Research, Inc.* Retrieved December 2, 2008, from http://www.accenture.com/NR/rdonlyres/D2BC2C93-38BB-41AB-9F3E-48B26C91D26C/0/ConsultingintheCloudEmerginSaaSConsulting.pdf

Stone, B., & Barnes, B. (2008, November 10). MGM to post full films on YouTube. *The New York Times*. Retrieved from http://www.nytimes.com/

Strader, T. J., & Shaw, M. J. (1997). Characteristics of electronic markets. *Decision Support Systems*, *21*(3), 185–198. doi:10.1016/S0167-9236(97)00028-6

Strauss, J., & Frost, R. (1999). *Marketing on the Internet—Principles of Online Marketing*. Upper Saddle River, NJ: Prentice-Hall.

Subramanian, R., & Yen, M. Y. (2002). Digital asset management: Concepts and issues. In Gangopadhyay, A. (Ed.), *Managing Business with Electronic Commerce: Issues and Trends*. Hershey, PA: Idea Group.

Sundararajan, A. (2004). Managing digital piracy: Pricing and protection. *Information Systems Research*, *15*(3), 287–308. doi:10.1287/isre.1040.0030

Sundararajan, A. (2004). Nonlinear pricing of information goods. *Management Science*, *50*(12), 1660–1673. doi:10.1287/mnsc.1040.0291

Susarla, A., Barua, A., & Whinston, A. B. (2009). A transaction cost perspective of the 'software-as-a-service' business model. *Journal of Management Information Systems*, *26*(2), 205–240. doi:10.2753/MIS0742-1222260209

Swaminathan, N. (2008, May 30). DNA computer puts microbes to work as number crunchers. *Scientific American*. Retrieved from http://scientificamerican.com/

Synercom Tech., Inc. v. University Computing, 462 F. Supp. 1003 (N.D. 1979).

Sysmans, J. (2006). Software as a service: A comprehensive look at the total cost of ownership of software applications. *Software & Information Industry Association (SIIA)* database. Retrieved April 3, 2009, from via http://www.siia.net/estore/ssb-01.pdf

Takeda, H., Veerkamp, P. J., Tomiyama, T., & Yoshikawa, H. (1990). Modeling design processes. *AI Magazine*, *11*(4), 37–48.

Tao, G., Yi-jun, L., Jing, G., & Long, G. (2006). Research on the economic features and pricing of digital products. In Y. Ji-rong (Ed.) Proceedings of the 2006 International Conference on Management Science & Engineering (13th) (pp.152-156). Harbin: Institute of Technology Press.

Tapscott, D. (2008). *Grown Up Digital*. New York: McGraw Hill.

Tarzey, B., Longbottom, C., & Stimson, T. (2007). On-premise to on-demand: The software as a service opportunity for independent software vendors. *Quocirca Insight Report*. Retrieved March 22, 2009, from Quaocirca database.

Teboul, J. (2006). *Service is Front Stage: Positioning Services for Value Advantage*. Insead Business Press.

Technorati Website Statistics. Retrieved July 30, 2008, from http://www.technorati.com/about/

Te'eni, D. (2009). Comment: The wiki way in a hurry--The icis anecdote. *Management Information Systems Quarterly*, *33*, 20–22.

Tekla, S. P. (2007). Imagine there's no DRM... I wonder if you can. *IEEE Spectrum*, *44*(7), 14. doi:10.1109/MSPEC.2007.4286549

Texas Instruments. (2010). Profile for Texas Instruments Inc. *Yahoo! Finance*. Retrieved January 8, 2010, from http://finance.yahoo.com/q/pr?s=txn

The ethics of reviewing. (2009, September 14). Retrieved September 16, 2009, from http://theonlinephotographer.typepad.com/the_online_photographer/2009/09/the-ethics-of-reviewing.html

The Lanham Trade-Mark Act, 15 U.S.C. § 1125 (2000).

Theodorou, P. (2006). A DSS model that aligns business strategy and business structure with advanced information technology: A case study. In Khosrow-Pour, M. (Ed.), *Cases on Information Technology and Organizational Politics & Culture*. Hershey, PA: Idea Reference Group.

Thompson, J. K. (2009). Business intelligence in a SaaS environment. *Business Intelligence Journal*, *14*(4), 50–55.

Thurman, N., & Myllylahti, M. (2009). Taking the paper out of news: A case study of Taloussanomat, Europe's first online-only newspaper. *Journalism Studies*, *10*(5), 691–708. doi:10.1080/14616700902812959

Tocci, R., Windmer, N., & Moss, G. (2006). *Digital Systems: Principles and Applications* (10th ed.). Upper Saddle River, NJ: Prentice Hall.

Toys "R" Us, Inc. v. Akkaoui, LEXIS 17090 (N.D. Cal. 1996).

Trammell, K. D., & Ferdig, R. E. (2004). Pedagogical implications of classroom blogging. *Academic Exchange Quarterly*, *8*(4), 60–64.

Trinity Mirror. (2009). Half yearly financial report for the 26 weeks ended 28 June 2009.

Tukey, J. W. (1958). The teaching of concrete mathematics. *The American Mathematical Monthly*, *65*(1), 1–9. doi:10.2307/2310294

Turban, E., & Volonino, L. (2010). *Information Technology for Management: Improving Performance in the Digital Economy*. Hoboken, NJ: John Wiley & Sons.

Turbotax. Retrieved January 26, 2010, from http://turbotax.intuit.com/

Turner, M., Budgen, D., & Brereton, P. (2003). Turning software into a service. *Computer*, *36*(10), 38–44. doi:10.1109/MC.2003.1236470

Two Peso, Inc. v. Taca Cabana, Inc., 505 U.S. 763 (1992). E.F. Johnson Co. v. Uniden Corp. of America, 623 F. Supp. 1485 (D. Minn. 1985).

U.S. Constitution, Art. I, § 8.

U.S.C. § 1832 (2000).

U.S.C. §§ 102, 103, 112 (2000).

Unicode Consortium. (2009). Retrieved October 15, 2009, from http://www.unicode.org/

Uniform Trade Secrets Act (1985). 12 U.L.A. 433.

United States v. Whitehead. (C.D. Cal. 2003). Retrieved January 12, 2010, from http://www.usdoj.gov/criminal/cybercrime/whiteheadConviction.htm).

United States v. Williams, 526 F.3d 1312 (11th Cir. 2008).

Universal City Studios, Inc. v. Corley, 273 F.3d 429, 435 (2d Cir. 2001).

Universal City Studios v. Reimerdes, 111 F. Supp. 294 (S.D.N.Y. 2000).

Vaquero, L. M., Rodero-Merino, L., Caceres, J., & Linder, M. (2009). A break in the clouds: Towards a cloud definition. *ACM SIGCOMM Computer Communication Review*, *39*(1), 50–55. doi:10.1145/1496091.1496100

Vargo, S. L., & Lusch, R. F. (2004). Evolving to a new dominant logic for marketing. *Journal of Marketing*, *68*, 1–17. doi:10.1509/jmkg.68.1.1.24036

Vargo, S. L., & Lusch, L. S. (2006). Service-dominant logic: What it is, what it is not, what it might be. In Lusch, L. S., & Vargo, S. L. (Eds.), *The Service-Dominant Logic of Marketing: Dialog, Debate, and Directions* (pp. 43–56). New York: ME Sharpe.

Venkatesh, R., & Chatterjee, R. (2006). Bundling, unbundling, and pricing of multiform products: The case of magazine content. *Journal of Direct and Interactive Marketing*, *20*(2), 21–40. doi:10.1002/dir.20059

Venkatesh, V., & Davis, F. D. (2000). A theoretical extension of the technology acceptance model: Four longitudinal field studies. *Management Science*, *46*(2), 186–204. doi:10.1287/mnsc.46.2.186.11926

Venkatesh, A. (2008). Digital home technologies and transformation of households. *Information Systems Frontiers*, *10*(4), 391–395. doi:10.1007/s10796-008-9097-0

Vgchartz. Retrieved October 6, 2009, from http://www.vgchartz.com/

Vicente, D. M. (2004). Cópia privada e sociedade da informação. Retrieved October 15, 2009, from http://www.apdi.pt/APDI/DOUTRINA/c%C3%B3pia%20privada%20e%20sociedade%20da%20informa%C3%A7%C3%A3o.pdf

Vincent, S. W. (2004). WTO, e-commerce, and information technologies: from the Uruguay Round through the Doha Development Agenda. In J. McIntosh (Ed.) *A Report for the UN ICT Task Force*. Retrieved September 14, 2009, from http://www.iie.com/publications/papers/wunsch1104.pdf

Viswanathan, S. (2005). Competing across technology-differentiated channels: The impact of network externalities and switching costs. *Management Science, 51*, 483–496. doi:10.1287/mnsc.1040.0338

Vitorovich, L. (2010, January 25). H-P enters Europe music venture. *The Wall Street Journal*. Retrieved from http://online.wsj.com/

Voigt, K. (2009, December 14). Cyber crime poses threat to e-commerce. *CNN.com*. Retrieved from http://www.cnn.com/

Vorisek, J., & Feuerlicht, G. (2004). Is it the right time for the enterprise to adopt software-as-a-service model? *Information & Management, 17*(3-4), 18–21.

Vouk, M. A. (2008). Cloud computing – issues, research and implementations. *Proceedings of the 30th International Conference on Information Technology Interfaces*, Dubrovnik, Croatia.

Wagner, D. (2007). A comprehensive approach to security. *Sloan Management Review, 48*(4), 8.

Wallach, D. (2008, November 3). My take: E-voting not user friendly. *Discovery Channel*. Retrieved from http://dsc.discovery.com/news/2008/11/03/tech-electronic-voting.html

Walsh, K. R. (2003). Analyzing the application ASP concept: Technologies, economies, and strategies. *Communications of the ACM, 46*(8), 103–107. doi:10.1145/859670.859677

Walsh, M. (2009, March 26). Facebook users growing up fast. Retrieved September 29, 2009, from http://www.mediapost.com/publications/?fa=Articles.showArticle&art_aid=102973

Wang, Y., Tang, T., & Tang, J. E. (2001). An instrument for measuring customer satisfaction toward Web sites that market digital products and services. *Journal of Electronic Commerce Research, 2*(3), 89–102.

Wang, R. Y., Strong, D. M., & Guarascio, L. (1996). Beyond accuracy: What data quality means to data consumers. *Journal of Management Information Systems, 12*(4), 5–34.

Wang, R. (2009). Shape your apps strategy to reflect new SaaS licensing and pricing trends. *Forrester Research*, Document # 46602. Retrieved February 28, 2009, from Forrester database.

Waters, B. (2005). Software as a service: A look at the customer benefits. *Journal of Digital Asset Management, 1*(1), 32–39. doi:10.1057/palgrave.dam.3640007

Waugh, E. (1938). *Scoop*. Boston: Little, Brown and Co.

Webster, B. (2009). In Re Bilski appealed to the Supreme Court. Retrieved January 13, 2010, from http://bfwa.com/2009/01/29/in-re-bilski-appealed-to-the-supreme-court/.

Werra, J. (2001). Le régime juridique des mesures techniques de protection des oeuvres selon les Traités de l'OMPI, le Digital Millennium Copyright Act, les Directives Européennes et d'autres legislations (Japon, Australie). *Revue Internationale du Droit d'Auteur, 189*, 66–213.

Whinston, A. B., Stahl, D. O., & Choi, S. Y. (1997). *The Economics of Electronic Commerce*. Indianapolis, IN: McMillan Technical Publishing.

White, C. (2007). *Data Communications and Computer Networks*. Boston, MA: Thomson Course Technology.

Wilbon, A. (2002). Predicting survival of high-technology initial public offering firms. *The Journal of High Technology Management Research, 13*, 127–141. doi:10.1016/S1047-8310(01)00052-9

Wilson, D., & Basiliere, P. (2008). The flavors of e-procurement extend beyond software as a service and on-premise. *Gartner*, Article G00146768. Retrieved November 13, 2008, from Gartner database.

Wilson, T. V. (2009). How biometrics works. *HowStuffWorks*. Retrieved December 15, 2009, from http://science.howstuffworks.com/biometrics.htm

Wilson, T. V. (2009). How is digital 3-D different from old 3-D movies? *HowStuffWorks*. Retrieved December 15, 2009, from http://www.howstuffworks.com/digital-3d.htm

Wolk, D. (2009, September 2). Future of the screen: Terminator-style augmented-reality glasses. *Wired.com*. Retrieved from http://www.wired.com/

Wong, K. (2008). SaaS vendors buying innovation rather than developing it themselves. *NetworkWorld Asia*, *4*(8), 35.

Wordreference.com. (2009). *WordNet 2.0*. Princeton, NJ: Princeton University. Retrieved July 5, 2009, from http://www.wordreference.com/definition/software

Wright, A. (2009, February 23). Exploring a 'deep Web' that Google can't grasp. *The New York Times*. Retrieved from http://www.nytimes.com/

Wu, S.-Y., & Chen, P.-Y. (2008). Versioning and piracy control for digital information goods. *Operations Research*, *56*(1), 157–172. doi:10.1287/opre.1070.0414

Yahoo. (2006). Yahoo! forms strategic partnership with consortium of more than 150 newspapers across the U.S. Retrieved December 8, 2009, from http://yhoo.client.shareholder.com/press/releasedetail.cfm?ReleaseID=219204

Yamamoto, H., Okada, I., Kobayashi, N., & Ohta, T. (2002). The information channel effect in the winner-take-all: a multi-agent simulation. *Proceedings of the 6th World Multi-Conference on Systemics, Cybernetics and Informatics, 2*, 510-513.

Yarrow, J. (2010, April 14). Twitter finally reveals all its secret stats. Retrieved on September 26, 2010, from http://www.businessinsider.com/twitter-stats-2010-4#ixzz10fR5TZAu

York, J. (2008). Contrasting software-as-a-service and enterprise software business models. Retrieved January 27, 2009, from http://chaotic-flow.com/2008/09/02/contrasting-software-as-a-service-and-enterprise-software-business-models-2/

Youseff, L., Butrico, M., & Da Silva, D. (2008). Toward a unified ontology of cloud computing. *Proceedings on Grid Computing Environments Workshop (GCE) 2008*, Austin, Texas.

YouTube Fact Sheet. Retrieved October 10, 2009, from http://www.youtube.com/t/fact_sheet

YouTube Facts and Figures. (May 2010). Retrieved September 26, 2010, from http://www.website-monitoring.com/blog/2010/05/17/youtube-facts-and-figures-history-statistics/

Zetter, K. (2008, April 4). The cost of e-voting. *Wired.com*. Retrieved from http://www.wired.com/

Zhen, J. (2008). Defining SaaS, PaaS, IaaS, etc. Retrieved November 15, 2008, from http://cloudfeed.net/2008/06/03/defining-saas-paas-iaas-etc/

Zimmerman, D. (2007). Interdisciplinary living without copyright in a digital world. *Albany Law Review*, *70*, 1375–1397.

# About the Contributors

**Troy J. Strader** is Professor of Information Systems in the Drake University College of Business and Public Administration. Dr. Strader received his PhD in Business Administration (Information Systems) from the University of Illinois at Urbana-Champaign in 1997. He has taught computer programming and technology strategy courses and his research interests include digital product management, online consumer behavior, information technology adoption, mobile commerce, and the impact of the Internet and e-business on initial public offerings. Dr. Strader has published in the *International Journal of E-Commerce, Communications of the ACM*, the *European Journal of Information Systems*, the *Journal of the Association of Information Systems, Decision Support Systems,* and other academic and practitioner journals and books. He has co-edited two books, the *Handbook on Electronic Commerce*, and *Mobile Commerce: Technology, Theory and Applications.* Prior to beginning his faculty career he worked as a computer programmer and information systems analyst

\* \* \*

**Nizar Abdat** is a master business informatics (MBI) student at Utrecht University, the Netherlands. Currently he is working on his master thesis research about 'Software as a Service and its Pricing Strategy'. From his previous study at Inholland University, he had acquired broad IT knowledge including web programming, database management, and networking. In the end of 2008, he had also finished writing a chapter on Testing Management for the book 'Distributed Software Product Development' by Prof. Dr. Sjaak Brinkkemper, which will expectedly be published in the beginning of 2010. http://www.linkedin.com/in/nizarabdat

**Menne Bos**, Msc is currently a manager of the Network Group at Accenture, the Netherlands. He has six years experience in working as a Project Manager and IP network Planner at Versatel Netherlands. He has also gained broad knowledge about infrastructure technology such as Routing, Load balancing, Servers, Monitoring, ITIL, virtualization, IP services, and product development of Hosting and Internet. He acquired experience as a Sale Engineer in ICT product on media archiving as well. http://www.linkedin.com/pub/menne-bos/0/480/914

**Richard B. Carter** is the Dean's Professor of Finance in the College of Business at Iowa State University. Dr. Carter currently teaches real estate finance courses at both the undergraduate and graduate levels. He received his MBA in 1985 and his PhD in Finance in 1987, both from the University of Utah. Prior to receiving his graduate degrees, Dr. Carter was Supervisor of Data Processing and of Policy-owner

### About the Contributors

Service at Mutual of New York and Terminal Manager at Stott and Davis Motor Express. His research interests include investment banking, security analysis, the effects of e-commerce on financial services and capital acquisition. He has published in such journals as the *Journal of Financial and Quantitative Analysis*, the *Journal of Finance*, the *Journal of Financial Research*, the *Financial Review*, the *Journal of Portfolio Management*, the *Journal of Accounting and Public Policy* and *Financial Management*.

**Hsin-Lu Chang** is an associate professor in the Department of Management Information Systems, National Chengchi University. She received a PhD in information systems at the School of Commerce, University of Illinois at Urbana-Champaign. Her research areas are on B2B e-commerce, IT value, and service strategy and innovation. She has numerous publications in past issues of *Information Systems Journal, Industrial Management and Data Systems, International Journal of Electronic Commerce, Journal of Organizational Computing and Electronic Commerce, Mathematical and Computer Modelling etc.*

**Frederick H. Dark** is the interim Chair of the Accounting and Finance Departments in the College of Business at Iowa State University and holds a Dean's Professorship in Finance. Dr. Dark currently teaches corporate finance courses at both the undergraduate and graduate levels. He received his PhD in Finance in 1988 from the University of Utah. Prior to receiving his graduate degree, Dr. Dark served in the US Air Force and was in various locations around the world. His research interests include franchising, investment banking and financial markets. He has published in such journals as the *Journal of Financial and Quantitative Analysis*, the *Journal of Finance*, the *Journal of Financial Economics*, the *Journal of Law and Economics*, *Financial Management* and the *Financial Review*.

**J. Royce Fichtner** is an Assistant Professor of Business Law in the College of Business and Public Administration at Drake University where he teaches in the areas of business law, accounting law, and company law. He received his Bachelor of Arts in Business Administration from the University of Northern Iowa and his Juris Doctorate from Drake University. He is a former law clerk for Justice Michael Streit of the Iowa Supreme Court and a former staff attorney for the Honorable Terry Huitink of the Iowa Court of Appeals and the Honorable Robert Mahan of the Iowa Court of Appeals. He has published his research in the *International Journal of Disclosure and Governance* and *Drake Law Review*. His major research interests are in the areas of company law, corporate governance, and information technology law.

**Gary Graham** is a lecturer in Operations Management at Manchester Business School. Most of his work is to do with the Internet and its impact on supply chain management. He has written two books entitled "Wired Marketing: Energising Business for eCommerce" (published by Wiley) and "Creative Destruction: Transformation of the Music Industry" (published by Lightning Source). His main areas of expertise are withing supply chain related activity in the newspaper industry.

**Gary Hackbarth** is an Assistant Professor of Management Information Systems in the College of Informatics, Department of Business Informatics, at Northern Kentucky University. He holds a PhD in Management Information Systems from the Moore School of Business at the University of South Carolina. His teaching, research, and consulting areas of interest include Project Management, Health Informatics, Information System Strategy, and Organizational Memory. He has published in the European Journal of Information Systems, Information & Management, Database, Information Systems

Management, Journal of Web-Based Learning and Teaching Technologies, Journal of Electronic Commerce Research, the Financial Times, and in both national and international conference proceedings.

**Anthony R. Hendrickson** is Dean of the College of Business at Creighton University. Previously, Dr. Hendrickson was in the College of Business at Iowa State University where he served as Associate Dean for Academic Programs, Chair of the Department of Logistics, Operations, and Management Information Systems. He received his PhD in Computer Information Systems and Quantitative Analysis from the University of Arkansas, in 1994. Prior to entering academia, Dr. Hendrickson spent 12 years in a variety of executive positions. He has published leading research in a variety of journals and national and international conferences including *MIS Quarterly, Decision Sciences, Information Systems Research, Academy of Management Executive, The Journal of International Business Studies, Communications of the ACM, The Journal of Labor Research, Information & Management, Journal of International Information Management, The Journal of Computer Information Systems, DATA BASE, Electronic Markets, Computer Personnel,* and *HR Magazine.*

**Philip A. Houle** is professor of information systems in the College of Business & Public Administration at Drake University. He received his PhD in computer and information control sciences from the University of Minnesota. Professor Houle teaches database management, data communications and networking, and introductory information systems. He also served as the top information technology administrator at Drake University from 1999-2001, a time during which the university upgraded its campus network to improve campus security and network performance. Currently he is working on issues involving email address identity/mobility and e-commerce, including the use of spy-ware.

**Yasemin Zengin Karaibrahimoğlu** graduated with a bachelor's degree in economics from Ege University, Izmir, Turkey. Karaibrahimoğlu spent her sophomore yeat at De Rijksuniversiteit Groningen (University of Groningen) in Netherlands as an exchange student. After graduation, she has worked at an enginnering company in manufacturing industry as an administrator director. Since 2006, she has been working as research assistant and continues her PhD studies in accounting at Izmir University of Economics, İzmir, Turkey. Karaibrahimoğlu spent the 2008 summer term at University of Texas at Dallas, Dallas, USA as visitor researcher. During her PhD, she published different articles in the different fields of accounting. Her research area of interest is financial reporting and managerial and cost accounting.

**Delaney J. Kirk**, PhD, has taught management courses for the past 28 years at both public and private universities and is currently teaching at the University of South Florida in Sarasota. She has over 50 publications on various management topics and serves as an expert witness on court cases dealing with discrimination. In addition, Dr. Kirk conducts workshops to help other college professors improve their teaching and is the author of *Taking Back the Classroom: Tips for the College Professor on Becoming a More Effective Teacher.* Dr. Kirk blogs at www.delaneykirk.com, tweets at twitter.com/DelaneyKirk, networks at www.linkedin.com/in/delaneykirk, and has friends at www.facebook/DelaneyKirk.

**Feipei Lai** is a professor in the Graduate Institute of Biomedical Electronics and Bioinformatics, the Department of Computer Science & Information Engineering and the Department of Electrical Engineering at National Taiwan University. He received a B.S.E.E. degree from National Taiwan University in 1980, and MS and PhD degrees in computer science from the University of Illinois at Urbana-Champaign

*About the Contributors*

in 1984 and 1987, respectively. He was a vice superintendent of National Taiwan University Hospital and the chairman of Taiwan Network Information Center. Lai holds 7 Taiwan patents and 4 USA patents currently. He is one of the founders of the Institute of Information & Computing Machinery and a member of Phi Kappa Phi, Phi Tau Phi, Chinese Institute of Engineers. He was also the chairman of Taiwan Internet Content Rating Foundation, and received the Taiwan Fuji Xerox Research award in 1991 and K-T Li's Breaking-through award in 2008. Lai is a senior member of IEEE and included in "Who's Who in Science and Engineering" and "Who's Who in the World." His current research interests are SOC low power computing and medical information system.

**Brook Matthews** is an MBA student and director of marketing for the College of Business at Creighton University. She received her undergraduate degree in journalism from the University of Arkansas. Matthews has a background in non-profit and for-profit marketing. Prior to coming to Creighton, she was communications director for the American Heart Association's Midwest Affiliate.

**Chip Miller** has a PhD in marketing from the University of Washington. He has held positions at Drake University, the University of Montana and the University of Puget Sound. Recently he was a Fulbright scholar in the Philippines at the University of San Carlos. He has consulted for several businesses regarding marketing strategy and is an active researcher. His articles have appeared in the *Journal of Marketing, Journal of International Marketing, Journal of Marketing Theory and Practice, European Journal of Marketing* and *International Journal of Cases on Electronic Commerce* among others. Research interests include global communication in marketing, consumer attitude change and promotion effectiveness.

**Pedro Pina** is a lawyer and a law teacher in the Oliveira do Hospital School of Technology and Management at the Polytechnic Institute of Coimbra. He holds a law degree from the University of Coimbra Law School and a post-graduation in Territorial Development, Urbanism and Environmental Law from the Territorial Development, Urbanism and Environmental Law Studies Center (CEDOUA) at the University of Coimbra Law School. He holds a master degree in Procedural Law Studies from the University of Coimbra Law School and is currently a PhD student in the Doctoral Programme "Law, Justice, and Citizenship in the Twenty First Century" from the University of Coimbra Law School and Economics School.

**Michael J. Shaw** is a Professor and Hoeft Endowed Chair of Information Systems in the Department of Business Administration at the University of Illinois at Urbana-Champaign. He is also the Director of Center for IT and e-Business there. He has published extensively with over 140 papers in the information systems, e-business, and decision support areas. He was an Associate Editor of Information Systems Research, 1991-2003, and has been the Co-Editor-in-Chief of the journal Information Systems and e-Business Management, published by Springer, since its inauguration in 2003. Shaw's research has focused on studying how to make information technology beneficial to large enterprises. He has been working with companies such as Motorola, GE, John Deere, and Caterpillar on research projects. Shaw was a Fulbright Scholar and continues to keep a global perspective in his research and teaching. Over the years Shaw has been involved in multidisciplinary research. He is affiliated with, in addition to his home department, National Center for Supercomputing Applications (NCSA) and Information Trust

Institute (ITI). His most recent research has focused on the functions of Chief Information Officers, service management, and IT governance.

**LouAnn Simpson** is a Professor of Business Law in the College of Business and Public Administration at Drake University where she teaches the introductory course in business law along with advanced courses in property and employment law. She received her Bachelor of Science in Business Administration as well as her Juris Doctorate from Drake University. After practicing law for a few years in Des Moines (private practice in Des Moines and City Prosecutor for the city of West Des Moines), she joined the faculty at Drake. Her major research interests are in the areas of employment law and information technology. She frequently reviews Business Law textbooks.

**Dave Sly**, PE, MBA is President of Proplanner, a web-based manufacturing process engineering software firm, located in Ames, Iowa. Dave is also a Lecturer within the Industrial Engineering and Logistics departments at Iowa State University. Dave received his BS, MS, and PhD degrees in Industrial Engineering and his MBA from Iowa State as well. Dave is currently a member of Rotary, and a Senior member of the Society of Industrial Engineers and a registered Professional Engineer in the State of Iowa.

**Marco Spruit** is an Assistant Professor in the Organization & Information research group at the Institute of Information and Computing Sciences of Utrecht University where he lectures in "Information systems" and "Method Engineering", among others. His information systems research currently focuses on business aspects regarding Natural Language Technologies, Data Mining & Business Intelligence, Web technologies, IT Security, and Service Computing. He was awarded an Association for Linguistic and Literary Computing (ALLC) Bursary Award for his PhD research in the field of Quantitative Linguistics which he completed at the University of Amsterdam. http://www.cs.uu.nl/staff/spruit.html

**Trent Wachner** is an Assistant Professor of Marketing at Creighton University in Omaha, Nebraska. He received his PhD in Marketing from Washington State University in 2008. His research focuses on knowledge development, marketing strategy, sales force effectiveness and the governance of boundary-spanning interfirm relationships, with an emphasis on the interaction of emergent and disruptive technologies on all of these areas. Prior to academia, Trent spent over 10 years in industry focusing on industrial technology. Trent held positions in outside sales and strategic marketing, consulting with Lucent Technologies' largest service provider customers including Verizon Wireless, AT&T, Sprint, British Telecom and China Telecom amongst others. His most recent role at Lucent Technologies prior to academia was managing international sales training teams dedicated to the Asia Pacific/China regions. Trent has been published in the *Journal of Personal Selling and Sales Management, Industrial Marketing Management and the Journal of the Academy of Marketing Sciences (JAMS)* as well as conference proceedings.

# Index

## A

access control technologies  20
after-market returns  77
after-market return variance  77
airline tickets  54, 59, 67
Amazon  55, 56, 59, 60, 63, 64, 206, 208, 213
Amazon Elastic Compute Cloud (EC2)  160, 161
Amazon Simple Storage Service (S3)  160, 161
American National Standard Code for Information Interchange (ASCII)  4, 5, 18
AOL  55
Apex  161, 189, 192
Apple  206
Application Service Provider (ASP)  154, 155, 157, 159, 162, 163, 185, 188, 190
artistic works  197
Associated Press (AP)  118, 137, 145

## B

binary digits  2, 4, 18
biometric access controls  244, 245, 246
BlackBerrys  34, 48, 49
blogs  97, 98, 99, 100, 101, 103, 105, 106, 108, 110, 111
book readers  213
bootstrap financing  75
Brain-Computer Interfaces (BCI)  238, 262
break-even point (BEP)  169
bundling  53, 60, 64, 68, 69
business entities  20
business models  225, 229, 230, 232
business to business (B2B)  55

## C

capital markets  77
case studies  73, 74, 81
CD-quality recordings  7
Coca-Cola  35
CODECs  7, 14
community-building  100
Competitive Advantage  140, 141
Computer Haptics  262
computer-literate public  214
computer security  19
computer systems  19
concert tickets  54
consumer organizations  195
content based products  85
Content Convergence  141
control plan quality management  219
copyrighted works  195, 196, 200, 205
copyrights  32, 33, 44, 47, 193, 194, 195, 196, 198, 199, 200, 201, 202, 203, 204, 205
Crawler Technology  262
Creative Commons  227
creative works  32, 33, 44, 47, 48
cross-media publishing  229
Crowdsourcing  150, 152
Customer Relationship Management (CRM)  160, 173, 176, 208, 250
CyberCash  80
cybermarkets  193
cybermediaries  126, 129
cybermediary  113, 127, 128, 129, 136

## D

Daily Mail and General Trust (DMGT)  143

data quality 229, 231, 232, 234
data security 19
decimal systems 3, 4
Deep Web Search 243, 262
Digital Age, the 53
digital attributes 54, 69, 72
digital books 193, 213
digital computer systems 1
digital content 19, 20, 25, 30
digital convergence 113, 114, 117, 123, 124
digital copyrights 193, 204
digital files 129
digital formats 32, 33, 36, 38, 47, 51, 206
digital goods 53, 54, 56, 58, 59, 61, 63, 65, 67, 72
digital information 19, 21, 24, 25, 26, 28
Digital Millennium Copyright Act (DMCA) 30
Digital Millennium Copyright Act, the 194, 203, 204
digital music 73
digital music services 206
digital piracy 20, 27
digital platforms 213
digital processing systems 4, 10
digital products (DP) 32, 34, 39, 45, 47, 73, 74, 75, 76, 77, 78, 79, 80, 81, 85, 86, 87, 88, 89, 90, 91, 92, 93, 96, 97, 98, 103, 117, 118, 129, 132, 134, 135, 141, 206, 207, 213, 214, 226, 227, 229, 230
digital products (DP) firms 73, 74, 75, 77, 78, 79, 81
digital representations 2, 8, 9, 10, 15, 16
digital rights management (DRM) 19, 20, 23, 26, 28, 29, 30, 194, 195, 205
digital security 19, 20
digital systems 2, 3, 5, 10, 11, 15, 16
digital technologies 1, 2, 3, 9, 13, 15, 16, 18, 33, 35, 46, 86, 142, 225
digital technology companies 35
digital trunk networks 7
digital video broadcasting project (DVB) 22, 23, 28, 30
digitization 208, 209, 214, 227
digitized services 208, 209
DNA Computers 242, 249, 262

## E

e-book devices 206
e-books 32, 34, 47, 51, 73, 113, 119, 120, 124, 125, 126, 128, 129, 130, 131, 133, 134, 135, 136, 137, 206
Economic Espionage Act (EEA) 35, 36
Edison, Thomas 6
educational institutions 20
electrical signals 6, 11, 13
Electronic Arts (EA) 118, 122, 137
Electronic commerce (e-commerce) 85, 86, 87, 93, 193
Electronic Health Records (EHRs) 250
Electronic Ink (e-ink) 239, 258, 262
Electronic Paper (e-paper) 240, 262
encryption 19, 23, 24
e-newspapers 73
engineering change management 219

## F

Facebook 96, 97, 98, 99, 100, 101, 103, 104, 105, 106, 107, 108, 110, 111
face-to-face 218, 220
financial institutions 20
Financial Times (FT), the 149
fixed-to-variable cost ratio (FCVC) 75
flow analysis 220
FMEA 219
forums 97, 108
Free Culture movement 227
freelance photographers 147

## G

game industries 129
Gannett crowdsourcing strategy (GCS) 145
General Motors' (GM) 208
global customers 53
globalization 208
Google 220, 221
graphics interchange format (GIF) 8, 18

## H

hardware manufacturers 20
health services 208, 209, 210, 213
Hewlett-Packard (HP) 121

## Index

Holographic Storage  241, 262
Horizontal Integration Strategy  126, 141
Hyper-Local News  152

### I

IBM  5
identity theft  20
immaterial goods  193
indirect sales  218
industrial engineering  218
Industry Convergence  141
Influencers  108
information assurance (IA)  20
information goods  53, 54, 56, 58, 62, 64, 67, 68, 69, 70, 71
information security  19, 27
information systems (IS)  19
Information Technology (IT)  154, 159, 163, 164, 166, 168, 170, 172, 173, 186, 187, 188, 191, 192, 207, 208, 212, 214, 215, 219, 220, 221, 223, 224, 226, 228, 230, 231
initial public offering (IPO)  73, 74, 75, 76, 77, 78, 79, 80, 81, 82, 83, 84
Intangible Asset  94
intensive care unit (ICU)  209
interactive transactions  53
Interplanetary Internet  247, 260, 262
intranets  220, 222
iPods  20, 206

### J

joint photographic experts group (JPEG)  8, 18
J. P. Morgan  79, 80
judicial proceedings  197

### K

Key Performance Indicator (KPI)  192
Kindle  206, 208, 213, 214

### L

large-enterprise markets  80
legal challenges  32, 47
license agreements  194
licensing materials  32, 47
line balancing  219, 220, 221
Linux  74
Liverpool Daily Post  143, 144
long-term solutions  75

### M

Mass Customization  130, 131, 132, 140, 141
Microsoft (MS)  221, 223
monthly product subscription fees  220
monthly subscriptions  222
moving pictures experts group (MPEG)  9, 18
MP3s  7, 131
MS.NET  221
MySpace  97, 98, 99, 106

### N

Nasdaq  79, 80
National Readership Survey (NRS)  144, 153
National Taiwan University Hospital (NTUH)  209, 210, 211, 212, 213, 214
natural systems  2
net realisable value (NRV)  90
Network Externalities  234
newspaper readership  142, 144, 147
newspapers  96, 99, 118, 119, 125, 127, 129, 130, 131, 132, 133, 140, 142, 143, 144, 145, 146, 147, 148, 149, 150, 151, 152, 153, 225, 228, 229, 230
Newspaper Society (NS)  144
news supplys  142, 143

### O

one-to-one marketing  229
online games  124, 125, 126, 129
online newspapers  99
OpenID  246, 259
Open Mic nights  103
Operant Resources  234

### P

parcel transfer packet (PTP)  248
paywalls  147
Paywalls  152
peer-to-peer (P2P)  194, 198, 199, 201, 202, 204

PepsiCo 35
physical goods 54, 57, 58, 64
pirating 53
Platform as a Service (PaaS) 160, 161, 162, 167, 170, 172, 185, 187, 188, 189, 190, 192
podcasts 108, 109
pricing strategies 53, 54, 55, 69
Pricing Strategy Guideline Framework (PSGF) 154, 155, 178, 184, 185, 192
Private Copy 193, 205
product engineering 220
professional journalism 142
Proplanner 218, 219, 220, 221, 222
public service journalism 142, 152
publishers 226, 227

# Q

Quantum Computer 242, 262
Quantum Encryption 245, 262

# R

record industry 6
red-green-blue (RGB) 240
Regional Newspapers 142, 146, 152
remote engineering services 222
Research in Motion (RIM) 34, 122, 137
Reseller 192,
Return on Investment (ROI) 166, 179, 184, 185
revenues 220, 221, 222
rightholders 193, 194, 195, 196, 197, 198, 199, 200, 201, 202

# S

sales methods 53
self-publishing technologies 227
sensitive information 19
service elements 207, 216
Service Level Agreement (SLA) 170, 172, 179, 180, 182, 184, 185
service metrics 206, 207, 213
service models 206, 207, 213
service systems 206, 207, 208, 209, 210, 212, 213, 214, 215, 216

short-term debt 75
social capital 98, 109
social media 96, 100, 102, 103, 105, 106, 107, 143, 148
Social Media 96, 97, 98, 99, 100, 101, 102, 103, 104, 105, 106, 107, 108, 109, 111, 112
social media networks 96
social media tools 96, 98, 103
Social Netiquette 109
social networking 96, 97, 98, 100, 101, 102, 103, 104, 107, 108, 111, 112
software as a product (SaaP) 218
software as a service (SaaS) 154, 155, 156, 157, 158, 159, 160, 161, 162, 163, 164, 165, 166, 167, 168, 169, 170, 171, 172, 173, 174, 175, 176, 177, 178, 179, 180, 181, 182, 183, 184, 185, 186, 187, 188, 189, 190, 191, 192, , 218, 219, 220, 221, 222, 223, 224
software developers 73
software ecosystem (SECO) 171
software programmers 98
software sales 219, 221
Software + Service (S+S) 154, 163, 185
software subscriptions 221
speaker technologies 6
Speech Recognition 239, 262
speedometers 3
Spider Technology 262
Sports Illustrated (SI) 119
stockholders 54
Strategic Management 115, 138, 139, 141

# T

Taxation 91, 94
Technological Convergence 141
technological protection measures (TPM) 194, 195, 196, 198, 199, 201, 202, 205
technology acceptance model 212, 215
telephone systems 6, 7, 13, 14
telephone traffic 7
Texas Instruments (TI) 120
text messages 131
Three-Step Test, the 196, 198, 199, 200, 202, 203

## Index

Ticketmaster 54
time estimations 219, 220, 221
trademarks 32, 33, 36, 37, 38, 45, 48
trade-related aspects of intellectual property rights (TRIPS) 196
trade secrets 32, 33, 35, 36, 41, 50
traditional firms 225
traditional networks 221
Traditional Systems Development (TSD) 228
Twitter 97, 99, 100, 101, 103, 105, 106, 107, 108, 110, 111

## U

U-Health service system 206, 209, 210, 212, 213, 216
unbundling 53, 64, 70
Unicode standard 4, 5
uniform resource locator (URL) 15, 18
Uniform Trade Secrets Act 34, 49
U.S. Congress 32
U.S. Constitution 32, 50
User-generated Content (UGC) 225, 226, 227, 228, 229, 231
User Systems Development (USD) 228, 234
U.S. Patent and Trademark Office 33
U.S. Supreme Court 34, 43, 44, 46

## V

value networks 207, 216
versioning 53, 55, 60, 62, 66, 71, 72
viral marketing 220
Virtual Retinal Displays (VRD) 240, 262

## W

Wall Street Journal, the 149
Web 2.0 142, 145, 148, 149, 153, 154, 159, 162, 185, 186, 188, 189
web-deployed models 221
web sitcoms 99
web technologies 222
Whuffie 98, 105, 109
wikis 227, 231
windowing 53, 59, 60, 62
Windows 7 100
Wine Snobs 208, 214
winning services 212
work systems 207, 216
World Intellectual Property Organization (WIPO) 195, 196

## Y

YouTube 97, 98, 99, 100, 106